Hazardous Pollutants in Biological Treatment Systems

Hazardous Pollutants in Biological Treatment Systems

Fundamentals and a Guide to Experimental Research

Edited by

Ferhan Çeçen and Ulaş Tezel

IWA PUBLISHING

Published by **IWA Publishing**
Alliance House
12 Caxton Street
London SW1H 0QS, UK
Telephone: +44 (0)20 7654 5500
Fax: +44 (0)20 7654 5555
Email: publications@iwap.co.uk
Web: www.iwapublishing.com

First published 2018

British Library Cataloguing in Publication Data
A CIP catalogue record for this book is available from the British Library

ISBN: 9781780407708 (Paperback)
ISBN: 9781780407715 (eBook)
ISBN: 9781780408903 (ePUB)

Contents

Chapter 4
Removal and behavior of hazardous pollutants in biological treatment systems *123*
Ferhan Çeçen

About the Editors

Ferhan Çeçen is a professor in the Institute of Environmental Sciences at Boğaziçi University, Istanbul. She received her B.Sc. degree in Chemical Engineering from the Boğaziçi University, M.Sc. and Ph.D. degrees in Environmental Engineering from the Technical University in Istanbul. She has authored many publications on environmental biotechnology with a special emphasis on biodegradation of hazardous pollutants in activated sludge and biofilm systems, impacts of heavy metals and organochlorine compounds, and integration of adsorption and biodegradation for removal of pollutants from water and wastewater. She has authored the Wiley-VCH book titled *Activated Carbon for Water and Wastewater Treatment: Integration of Adsorption and Biological Treatment* which was published in 2011.

Ulaş Tezel is an assistant professor in the Institute of Environmental Sciences at Boğaziçi University, Istanbul. He received his B.Sc. and M.Sc. degrees in Environmental Engineering from the Middle East Technical University, and the Ph.D. degree from the School of Civil and Environmental Engineering at Georgia Institute of Technology where he worked as a postdoctoral research scientist for 1.5 years before joining Boğaziçi University in 2012. His research is mainly on the occurrence and fate of emerging organic pollutants in the environment, pollutant biotransformation pathways and mechanisms, evalution and dissemination of antimicrobial resistance, microbial ecology and genetics. He has co-authored over 35 articles and 2 book chapters on environmental microbial processes and discovered a novel enzyme that detoxifies biocides in the environment.

Chapter 1

A guide book to studying the fate and effect of hazardous pollutants in biological treatment systems

Ferhan Çeçen

Boğaziçi University, Institute of Environmental Sciences, 34342 Bebek, Istanbul, TURKEY

Corresponding author: cecenf@boun.edu.tr

1.1 HAZARDOUS POLLUTANTS AND BIOLOGICAL REMOVAL

The present book deals specifically with the occurrence, effects, removal and monitoring of various types of hazardous pollutants during biological treatment. In that regard, the main focus is on the biological units of drinking water and wastewater treatment plants.

This introductory chapter provides an overview of all of the topics covered within the scope of the book. This chapter also presents the rationale behind the selection of topics and delineates in what aspects the book differs from others. While providing a brief look at the content of each chapter, also suggestions are made for the reader on how to best use the book.

As will be seen throughout this book, hazardous pollutants can be defined in many ways. Commonly, the term "hazardous pollutant" refers to a substance that is likely to threaten the health of humans and the environment upon its release to the environment. Hazardous pollutants may stem from both point- and non-point sources. However, point sources are the primary pathways by which hazardous compounds enter the water environment. Among point sources, the most important one is the discharges made from wastewater treatment plants (WWTPs) (Sathyamoorthy

 Hazardous Pollutants in Biological Treatment Systems: Fundamentals and a Guide to Experimental Research
Ferhan Çeçen, Ulaş Tezel
doi: 10.2166/9781780407708_1

et al. 2013). To avoid the entrance of hazardous pollutants into the water environment, consisting of seas, oceans, lakes, and streams, wastewaters should be adequately treated in WWTPs. Another option to control the release of hazardous substances into the water environment is prevention of pollutant formation, an even more difficult task to accomplish. Since hazardous pollutants reach water bodies, many of them are inevitably found in waters that are extracted for drinking purposes. Thus, the presence of hazardous pollutants is also of concern in drinking water treatment.

1.1.1 Emerging need to control hazardous pollutants

At the beginning of water pollution control, the key issue in wastewater treatment was merely the removal of bulk organic matter and disinfection of treated effluent. Later, nutrient removal too became an important component of biological wastewater treatment. Beginning from the 1970s, hazardous pollutants began to receive attention both in the natural water environment as well as in water and wastewater treatment works. The underlying reason for this was the production of thousands of synthetic organic chemicals (SOCs). A large number of those pollutants were recognized, or at least believed, to have hazardous effects in animals and in humans, such as carcinogenicity and/or mutagenicity. In addition, hazardous inorganic pollutants too have begun to receive attention in recent decades.

Hazardous pollutants can be eliminated or transformed in many ways. However, among all processes the biological ones deserve special attention. Biological processes come into play both in the natural as well the engineered water environment, namely in drinking water and wastewater treatment systems. In both cases biotransformation or biodegradation of hazardous pollutants is possible. If full mineralization (ultimate biodegradation) does not take place, the stable products of transformation can even be more persistent and/or toxic than the parent compounds (Rücker & Kümmerer, 2012). Therefore, today not only the hazardous pollutants in their original form but also their transformation products are studied in biological treatment systems. They are believed to pose risks for ecological systems and human health. While the effects of parent compounds are relatively better understood, there is still a lack of information on generation and effects of transformation products. Particularly, in biological treatment systems it is often difficult to predict and measure transformation products.

Among hazardous pollutants, many organic pollutants such as hormones, pharmaceuticals and personal care products (PPCPs), pesticides, UV filters, fragrances, antimicrobial agents, illicit drugs, brominated flame retardants, industrial products, household products, pesticides, disinfectants and antiseptics, have received special attention in the last few decades (Margot *et al.* 2015). A recent look at the conferences organized and papers written on the subject, indicates that a significant part of the research on hazardous pollutants centers around the removal of PPCPs from water and wastewater, due to their increased use in recent years. A substantial proportion of PPCPs originate from hospitals, while the rest are discharged domestically (Forrez *et al.* 2011). PPCPs, as well as many other compounds, cause concern because of their persistent and bioaccumulative properties, and health and ecological impacts. In order to determine such impacts, risks have to be assessed. On the other hand, risk assessment is a comprehensive task involving fields such as chemistry, process technology, toxicology and biology (van Leeuwen & Vermeire, 2007).

Nowadays, a large number of organic hazardous substances enter domestic wastewater treatment plants. These pollutants are often found at extremely low concentrations, namely at the level of a few μg/L or even ng/L. Therefore, they are referred to as *micropollutants*. If found at those levels a hazardous pollutant may not have an adverse effect on the biological processes going on in a bioreactor. However, if these pollutants are inadequately removed, their effects in the receiving water environment are quite important due to their toxicity, bioaccumulation, long-range transport properties etc. Therefore, in any case, they have to be removed by biological or other means. Also, inorganic hazardous pollutants may affect the treatment systems and then the water environment.

Despite the effects of hazardous pollutants on receiving waters, there is still no consensus about which hazardous substances and discharges need to be regulated and reduced. This arises partly due to the still existing uncertainties about the fate and effects of hazardous pollutants, both in water as well as in wastewater treatment systems, and thereafter in the natural water environment. In some places, bans have been introduced for the use of some specific organics such as alkylphenol ethoxylates with the aim to protect flora and fauna in receiving waters. Yet, regulations have not been developed for the discharge of micropollutants from WWTPs. In few cases limits have been set for specific organics such as benzene which serves as a surrogate parameter for other organic micropollutants. It is believed that upcoming regulations will be increasingly stricter regarding the release of such pollutants (Clouzot *et al.* 2012).

Similarly, in drinking water treatment systems, hazardous pollutants are encountered at very reduced concentrations, at even lower levels than in wastewater treatment. The hazardous pollutants found in the raw water flowing into a drinking water plant can stem from point- or non-point sources. In addition, beginning from the 1970s it was also recognized that the common practices applied in the treatment of drinking water, such as chlorination and ozonation, could lead to generation of a number of hazardous organic and inorganic pollutants. While wastewater discharges indirectly affect human health, the presence of hazardous pollutants in a finished drinking water may have direct effects.

1.1.2 Challenges in removal of hazardous pollutants

Different treatment options are available for the removal of hazardous pollutants. In engineered systems biological treatment is hardly ever applied alone, it often finds use in combination with physicochemical treatment.

Compared to biological processes, physicochemical processes often transfer the pollutant from one phase to another. Biological processes are relatively cheaper. In addition, they lead to complete destruction of hazardous pollutants whenever they can be effectively applied. Thus, the complete disappearance of many hazardous pollutants would largely depend on the effectiveness of biological treatment units. However, in contrast to physicochemical processes that are more straightforward to apply and whose outcomes are more predictable, in biological treatment there are still many unresolved issues regarding the effects, removal and fate of hazardous pollutants.

Determining the appropriate regime for the removal of hazardous pollutants from drinking water and wastewater continues to be a challenge. Although in both cases hazardous pollutants often exist as micropollutants, their removal and effects are quite problematic. In fact, most of the problems encountered in biological removal of such pollutants arise due to their extremely low concentrations and not the opposite.

With the increased use of hazardous substances and their discharge into WWTPs, there is a great need today for optimum removal in WWTPs. Yet, biological units of conventional treatment plants were initially designed to remove organic carbon; then they were expanded to include nutrient removal. But, they did not particularly aim to remove hazardous pollutants. Consequently, hazardous pollutants are mostly poorly removed in existing WWTPs and a significant number of them are released into the water environment with the effluent. In addition, most of them are simply transferred from liquid to solid phase by sorption onto sludge.

Many laboratory-scale studies are conducted to assess the fate and effect of hazardous pollutants in biological treatment. However, in most research studies hazardous pollutants are examined as individual compounds. In reality, in full-scale treatment, hazardous pollutants exist in a matrix composed of hazardous as well as non-hazardous organic and inorganic substances. Therefore, laboratory-scale experimenting on single pollutants may not yield adequate results if extrapolated to full scale. Also in the natural water environment, hazardous pollutants do not exist alone. Due to their high numbers, their combined effects are usually measured rather than their concentrations (Clouzot *et al.* 2012).

1.2 PURPOSE OF THE BOOK

The basic features of this book that distinguish it from others may be summarized as follows:

(a) *Targeting engineered systems only*: Many studies are available on the fate and transport of hazardous pollutants once released from point- and non-point sources into the water environment. However, this book narrows down the subject and addresses mainly the biological processes taking place in the engineered water environment, and everything that centers around biological treatment. The natural water environment is only mentioned when needed, for example, when origins, occurrence and effects of hazardous pollutants are discussed.

(b) *Focus on hazardous pollutants in biological treatment*: Another aim of the book is to provide a comprehensive document that entirely focuses on biological water and wastewater treatment alone. Although many books have been written on hazardous pollutants, such a unified document on biological treatment alone did not exist to date. To fill this gap, this book delineates the fundamental characteristics of hazardous pollutants and concentrates on their behavior and effects in biological treatment systems. To achieve this objective, it is divided into successive chapters that follow each other in a logical order, but also can be read alone.

(c) *Focus on fundamental principles and processes rather than specific groups*: While most books written on hazardous pollutants examine a specific hazardous pollutant, a hazardous pollutant group or a specific biological process alone, the aim of this book is different. In the present book, all subjects related to hazardous pollutants are put into a context where the main focus is on principles that would apply to all kinds of hazardous pollutants. In that respect, pollutants are also not differentiated as macro- or micropollutants, whenever appropriate.

(d) *Focus on experimental methods and application*: The book aims to cover all aspects of hazardous pollutants as far as biological treatment is of interest. First, it contains basic information on the characteristics, occurrence and origin of hazardous pollutants. In addition, it addresses the fate and effects of hazardous pollutants in biological wastewater and drinking water treatment units (surface and groundwater), both at laboratory- and pilot/full-scale. As also reflected in the title, one of the major goals of this book is to serve as a guide in experimental research. Therefore, details are provided about analysis of various types of hazardous pollutants, experimental methodologies and evaluation of experimental data. Computational methods are also mentioned since for experimenters they provide valuable information on biodegradability and inhibitory properties of hazardous pollutants before conduction of experiments. The book also tries to delineate the link between the biological processes and the microbial ecology of systems containing hazardous pollutants. In that respect, it includes a discussion on the use of molecular microbiology and genetic tools.

1.3 CONTENT OF THE BOOK

Figure 1.1 illustrates the major fields on which the book focuses. Complementary to this figure, the following discussion clarifies the content of each chapter. To provide a quick idea of the content of each chapter, Table 1.1 lists some of the most common concepts in each chapter.

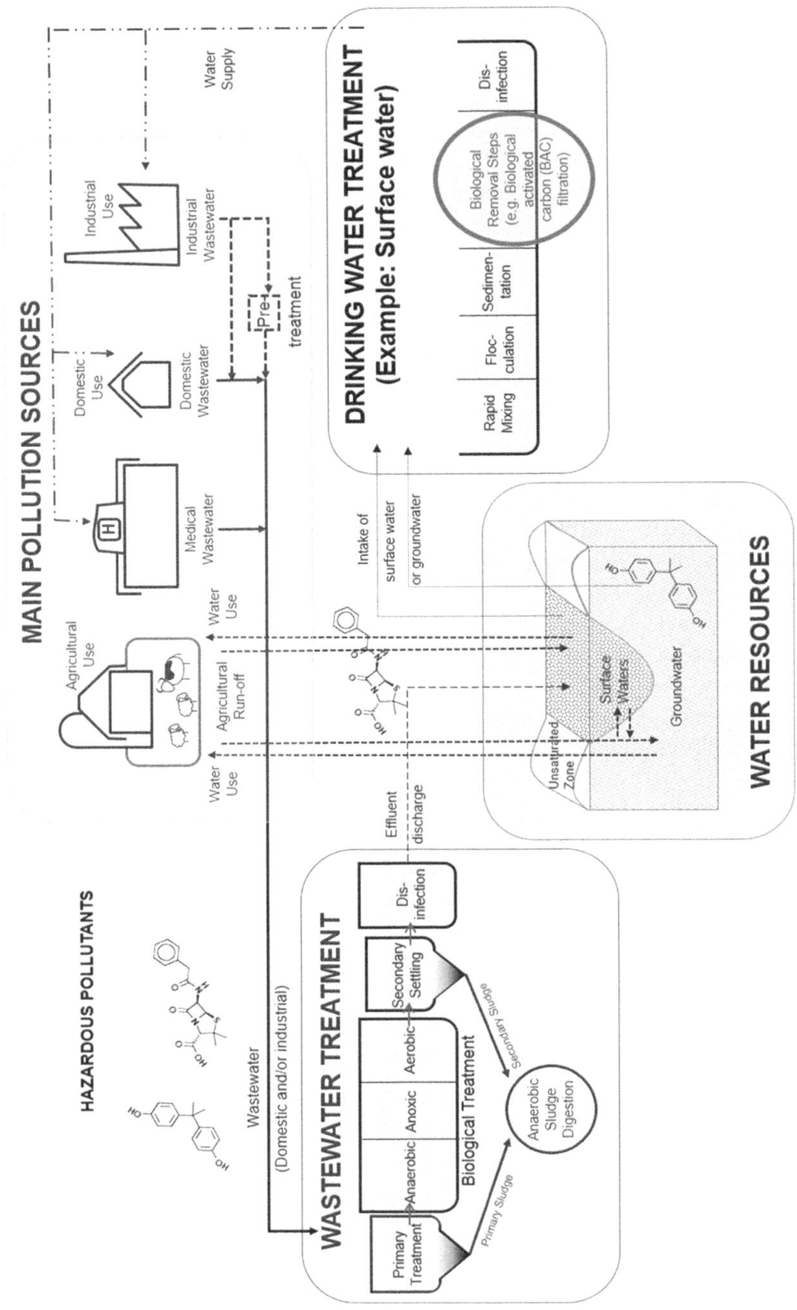

Figure 1.1 Hazardous pollutants in the water environment.

Table 1.1 A brief look at each chapter.

	Main Terms in the Chapter
Chapter 1	• Emergence of hazardous pollutants • Importance of hazardous pollutants for biological treatment • Purpose of the book • Targeted readership • Overview of chapters' content • Tips for reading the book
Chapter 2	• Definition of hazardous pollutants • Origins and occurrence • Fate and transport • Toxicity and exposure hazards • Physicochemical properties • Computational tools • Pollutant groups • Regulatory frameworks
Chapter 3	• Environmental quantitative analysis • Environmental regulations • Maximum contaminant levels (MCLs) • Good laboratory practices • Quality assurance (QA) • Quality control (QC) • Standard methods • Sampling from biological systems • Sample preparation • Instrumental determination • Quantification of hazardous pollutants
Chapter 4	• Biodegradation • Biotic removal • Abiotic removal • Direct metabolism • Cometabolism • Biosorption • Inhibition • Biological treatment units • Suspended and attached-growth (biofilm) reactors • Biodegradation rates of pollutants • Organic hazardous pollutants • Inorganic hazardous pollutants
Chapter 5	• Need for experimenting • Characteristics of experimental systems • Inhibition tests • Inhibition kinetics

(*Continued*)

Table 1.1 A brief look at each chapter (*Continued*).

	Main Terms in the Chapter
	• Standard tests • Biodegradability tests (ready biodegradability, inherent biodegradability, simulation tests) • OECD tests • Shortcomings of standard tests • Computational tools assisting experimenting • Databases about hazardous pollutants
Chapter 6	• Experiences gained from full-scale biological treatment • Pharmaceuticals and personal care products (PPCPs) • Endocrine disrupting compounds (EDCs) • Antibiotics and antibiotic resistance • Antimicrobial agents and sunscreens • Volatile organic compounds (VOCs) • Surfactants • Agricultural pollutants • Nanomaterials
Chapter 7	• Molecular and microbial ecology tools • Enrichment of biodegrading microorganisms • Culture-dependent and culture-independent methods • Identification of biodegrading microorganisms • Specific genes, enzymes, and metabolic pathways • Major techniques in examination of community structure and function • Genetic insights into biodegradation
Chapter 8	• Surface water biotreatment • Groundwater biotreatment • Process configuration • Biofiltration • Biomass accumulation • Rapid rate filters (RRF) • Biological instability • Biodegradable organic matter (BOM) • Specific biodegradable organic compounds (SBOC) • Bench-, pilot-, and demonstration-scale bioremoval
Chapter 9	• Identification of emerging pollutants • Standardization of instrumental methods to quantify emerging pollutants • Emerging instrumental methods for detection and quantification • Alternative methods for testing biodegradation and inhibition • Emerging molecular biology tools for assessment of biodegradation and inhibition • Emerging technologies for removal of hazardous pollutants

Chapter 1: A guide book to studying the fate and effect of hazardous pollutants in biological treatment systems

The present chapter can be read as a stand-alone chapter that provides an insight into the content and rationale of this book. At the same time, this chapter provides an opportunity for a reader to choose among a number of topics.

Chapter 2: Hazardous pollutants in the water environment

This chapter provides a review on the properties of hazardous pollutants in the water environment. It first addresses the underlying historical development that led to the definition of "hazardous pollutants". The chapter differentiates then between toxicity and exposure hazards of these pollutants. The physicochemical properties of chemicals, such as partitioning between phases and degradability that affect their fate, transport, and exposure and toxicity hazards are also discussed. The chapter also focuses on the tools used in estimation of relevant physicochemical properties including molecular, partitioning and degradability properties. In that respect, a large overview is provided on the computational tools EPI (Estimation Program Interface) Suite, Online Chemical Modeling Environment (OCHEM), and the Organisation for Economic Co-operation and Development (OECD) Toolbox, all based on the Quantitative Structure-Activity Relationships (QSAR) technique.

In the chapter, hazardous pollutants in the water environment are classified according to their origins and occurrence. Also discussed are the pathways to reach WWTPs and the role of WWTPs in the fate and transport of hazardous pollutants in the environment. The pollutant groups discussed are persistent organic pollutants (POPs), including many pesticides and specific groups such as polychlorinated biphenyls (PCBs), polychlorinated dibenzo-p-dioxins and polychlorinated dibenzofurans (PCDDs/PCDFs), and also new POPs such as some insecticides, and brominated flame retardants like hexabromocyclododecane (HBCD) and polybrominated diphenyl ethers (PBDEs). Also new POPs used in the manufacturing of industrial products are mentioned. Great emphasis is put on the behavior and fate of PPCPs in WWTPs and in the water environment.

The chapter also discusses the regulatory frameworks that deal with hazardous pollutants. The main focus is on the Toxic Substances Control Act of U.S. EPA (Environmental Protection Agency), the Toxic Substances Management Policy of Environment Canada, and the Regulation for Registration, Evaluation, Authorisation and Restriction of Chemicals (REACH) of the European Union (EU).

Since there are many hazardous pollutants in the water environment nowadays, there is a need to examine them in special groups. Thus, the chapter discusses the groupings of hazardous pollutants as "priority pollutants", "emerging pollutants", "micropollutants", "macropollutants", and "xenobiotic organic compounds (XOCs)". Also mentioned are the conditions and profiles for chemicals to become planetary boundary threats.

Chapter 3: Quantification of hazardous pollutants in biological systems

Due to the presence of excess organics and other constituents, quantification of trace level pollutants in biological systems requires skillful sampling and analysis techniques and advanced analytical instrumentation. Hence, the importance of developing and documenting protocols for accurate and reliable quantification of hazardous pollutants in biological systems couldn't be overemphasized. **Chapter 3** serves the exact purpose by detailing the quantification of hazardous pollutants in biological systems. First, the principles and practices of environmental quantitative analysis are introduced, supported with a case study of quantification of nitrosamines in biological systems. The introductory section also discusses maximum contaminant levels (MCLs) of pollutants and the end users of environmental quantitative analysis. The section emphasizes that the demand for quantitative analysis is largely driven by regulations and concludes with a discussion about the current practices for regulatory method development and list some of U.S. EPA's standard methods. This discussion is then followed by a section on monitoring, sampling, analysis, and quantification practices for hazardous pollutants in biological systems. After listing the basics of good laboratory practices, quality assurance/quality control principles are introduced. Subsequent sections are dedicated to the quantification of key hazardous pollutant groups in biological systems. Quantification covers sample preparation, instrumental determination, analytical challenges, and future improvements. In that respect, the following organic hazardous pollutants are studied in great detail: perfluorinated compounds (PFCs), polybrominated diphenylethers (PBDEs), polychlorinated compounds (PCCs), alkyl phenols (APs), pharmaceuticals and personal care products (PPCPs), estrogens, and disinfection by-products (DBPs). The same discussion is extended for inorganic hazardous pollutants such as carbon-based nanomaterials and metals. Also discussed is the quantification of pathogens in biological media. In the concluding section, the authors emphasize that trace level environmental analysis has to continuously adapt to changing needs as new compounds are released into the environment and environmental regulations are updated.

Chapter 4: Removal and behavior of hazardous pollutants in biological treatment systems

This chapter deals specifically with the removal and behavior of hazardous pollutants in biological treatment systems. The chapter first focuses on the biological mechanisms that are relevant for degradation of hazardous pollutants. A discussion is provided on biotransformation and biodegradation issues as well as the use of hazardous organic pollutants as primary, secondary and cometabolic substrates in biological treatment systems. Since biosorption is the second most important removal mechanism in biological treatment, its quantification and its importance for biodegradation are emphasized too. Other abiotic removal methods in biological treatment systems are also briefly described.

Then, the importance of biomass configuration is discussed in the removal of hazardous pollutants. In that respect, the characteristics of suspended and attached-growth (biofilm) biological reactors are compared. Also included are the impacts of conventional wastewater treatment and Biological Nutrient Removal (BNR). Further emphasized is the effect of redox conditions. In that regard, detailed information is given on biodegradation rates of many hazardous organic pollutants under different redox conditions. Since most hazardous pollutants exert inhibitory effects on biological treatment, either by inhibiting their own removal or the removal of other compounds, another section is devoted to inhibitory effects and inhibition models.

Since most of the hazardous pollutants are not used as a primary substrate, but as a cometabolite, a large section of the chapter deals with the cometabolism which often occurs in biological reactors. In that respect, particular attention is paid to nitrifying systems that enable the cometabolic removal of many hazardous pollutants such as hormones and pharmaceuticals that are otherwise nonbiodegradable under the indicated redox condition. The chapter briefly discusses also the fate of organic pollutants under the anaerobic conditions of a sludge digester. Additionally, to give an idea about the modeling, the chapter provides simple examples of model equations that describe the removal of hazardous organic pollutants by different processes such as biodegradation, biosorption, volatilization and stripping.

Nowadays biological and physicochemical processes are often combined to bring about a more effective removal of organic hazardous pollutants. In another section, the combination of biological treatment with adsorptive processes as well as advanced oxidation processes is briefly mentioned.

Since hazardous inorganic pollutants too are of importance in biological treatment systems, another section is devoted to their fate and effects. In that respect, speciation, biotransformation, biosorption and inhibitory properties of metals are particularly emphasized. This part is also concerned with the behavior and effects of nanometals since such nanomaterials, finding widespread use today, enter wastewater treatment systems. Briefly, also the importance of hazardous ions for biological treatment is discussed in the chapter.

Chapter 5: Experimental assessment of the inhibitory effect and biodegradation of hazardous pollutants

This chapter addresses the experimental methods used in studying biodegradation and inhibitory effects of hazardous pollutants. This chapter is designed in alignment with the key issues discussed in **Chapter 4**.

After an introductory part which provides an insight into conduction of experiments, **Chapter 5** discusses the choice of experimental set-up based on the physicochemical properties of test substances. The importance of reactor configuration such as batch or continuous-flow is also mentioned. Then, in the next section the rationale of inhibition studies is discussed along with the standard tests used for this purpose. Special emphasis is put on inhibition kinetics. A critical appraisal of existing inhibition tests is provided and suggestions are made for their improvement.

Then, the chapter proceeds with standard tests used in biodegradation of hazardous substances. Before a biodegradability test is conducted, it is very useful to acquire information on the biodegradability of a test substance. In that regard, a significant part of **Chapter 5** is allocated to the common databases and computational tools used in prediction of biodegradation and biotransformation pathways. In that content, a large discussion is provided on the database METACyc, and the tools EPISuite BIOWIN module and EAWAG BBD and enviPATH. Detailed examples are given in the chapter on how such databases and tools can be exploited in the prediction of biodegradability. In that respect, the chapter also highlights the link between the experimental and computational tools that are used in the assessment of biodegradability. At the end, the chapter also covers a large discussion about the shortcomings of existing biodegradation tests.

Chapter 6: Removal of hazardous pollutants in full-scale wastewater treatment plants

To cover a complete discussion regarding hazardous pollutants, information on practical experiences is also necessary. Therefore, another chapter, **Chapter 6**, summarizes the removal of hazardous pollutants in full-scale treatment plants as reported in peer-reviewed literature, whereas data are taken mostly from full-scale activated sludge plants. In that respect, pharmaceutically active compounds and personal care products (PPCPs) that are of great concern today in domestic WWTPs receive significant attention. Also discussed are the emissions and fates of Volatile Organic Compounds (VOCs), removal of agricultural pollutants and surfactants, and other more specific groups such as nanomaterials and X-ray contrast agents. Particular emphasis is put on the influent and effluent concentrations of hazardous pollutants that go through wastewater treatment plants.

Chapter 7: Integrating microbial and molecular tools to determine the fate and impact of hazardous pollutants

This chapter covers a comprehensive overview of both conventional and contemporary high-throughput molecular and microbial ecology tools applied in microbiological examination of biological treatment systems for the remediation of hazardous pollutants. Initial subsections of the chapter deal with the enrichment of microorganisms within the microbial community of engineered treatment processes. Then, the chapter addresses the molecular techniques that are used for identification of both individual and consortium level microorganisms capable of biodegradation. Next subsections discuss the methods used for isolation of most competitive species in biotransformation of hazardous pollutants. Subsequently, the chapter covers techniques used for the identification of specific genes, enzymes, and metabolic pathways to provide detailed information on biomolecular mechanisms of biotransformation. The last subsection covers the application of these tools to determine the effects of hazardous pollutants on the microbial community.

Chapter 8: Biological removal of hazardous pollutants in drinking water treatment

While there is a great focus today on the removal of hazardous pollutants from wastewater, **Chapter 8** points to an equally important topic, namely to biological removal of hazardous pollutants in drinking water treatment. The chapter first conveys the message that intentional biological drinking water treatment (BDWT) is not a new topic, but started in the 1800s.

Nowadays biological processes find application in the treatment of surface as well as groundwater. Since different types of hazardous pollutants are present in each case and treatment options vary, the chapter makes a differentiation between the biological processes in surface and groundwater treatment.

In surface water treatment configurations such as rapid rate filters (RRF), slow sand filters (SSF), and river bank filters (RBF) are used, whereas a great emphasis is put on RRF in this chapter. In that respect, the chapter focuses on the importance of biodegradable organic matter (BOM) which includes identifiable specific biodegradable organic compounds (SBOC) as well as non-identifiable BOM. First the biomass accumulation in biofilters as a result of BOM consumption is discussed. Then, the removal of several specific organic pollutants like the disinfection by-products (DBPs) trihalomethanes (THMs) and haloacetic acids (HAAs), endocrine disrupting compounds (EDCs), PPCPs, pesticides, algal metabolites etc. is reviewed. Attention is paid to the design, operation and environmental conditions that impact the effectiveness of such biofilters in the removal of specific organics.

In aerobic biological treatment of groundwater, the removal of relatively common pollutants such as ferrous ion (Fe(II)), manganese (Mn(II)), and ammonia (NH_3) leading to biological instability of water as well as the removal of definitely hazardous pollutants such as the VOCs perchloroethylene (PCE) and trichloroethylene (TCE) and others like methyl tertiary butyl ether (MTBE) are of interest. The chapter first discusses the alternatives for process configuration as well as the factors affecting aerobic treatment of groundwater. In groundwater biotreatment, also anoxic/anaerobic treatment is of great interest to remove a wide range of pollutants such as nitrate (NO_3^-), perchlorate (ClO_4^-), bromate (BrO_3^-), selenate (Se(VI)), hexavalent chromium (Cr (VI)) and VOCs. The process configurations used for that purpose are discussed along with parameters affecting the removal of hazardous pollutants. Examples are given from bench-, pilot-, and full-scale studies. The chapter finally compares testing at the bench, pilot, and demonstration scales and discusses the importance of experimental design, inoculation, start-up and monitoring.

Chapter 9: Future aspects of hazardous pollutants and their biological removal

This final chapter is dedicated to the future aspects of hazardous pollutants and their biological removal. Nowadays, about 7 million chemicals are present in the market and about 100,000 have been detected in the environment. As new chemicals are introduced into the market, the spectrum of chemicals will widen. In this chapter, the role of the NORMAN Network in identification of emerging pollutants and standardization of instrumental methods to quantify those pollutants are critically assessed. Emerging instrumental methods for detection and quantification of hazardous pollutants at nanogram per litre concentrations are evaluated. Limitations of current standard methods for testing biodegradation and inhibition of hazardous pollutants are discussed and alternative methods are suggested. Molecular biology methods should be one of the key tools to assess the biological potential of a plant in removing hazardous pollutants as well as developing new cutting edge advanced treatment technologies. Therefore, this chapter evaluates the emerging molecular biology tools that should be adapted for routine assessment of biodegradation and inhibition of hazardous pollutants in wastewater treatment plants and improvements needed to make them user-friendly and high throughput. In addition, the chapter gives an insight into the emerging technologies, such as immobilized cells and enzymes, that have been newly developed for removal of hazardous pollutants.

1.4 TARGETED READERSHIP

We think that the present book will appeal to all people from science and engineering disciplines. During its preparation, we assumed that the reader is well informed about the fundamentals of environmental science and technology. Particularly, the book necessitates a fundamental knowledge of the principles of biological treatment that are covered in many environmental science and engineering texts.

Principally, the book targets the academic field, but it can also be used by people working in practice. We think that to a researcher or a practitioner who is already involved with hazardous pollutants in biological processes or is intending to be so, the book should appeal in many ways by providing answers to questions such as the following:

- What should be basically known about the properties of hazardous pollutants? Which characteristics make them hazardous?
- How do hazardous pollutants behave and biotransform/biodegrade in biological systems?
- What is the extent of removal in different biological processes?
- How might the presence of hazardous pollutants affect the removal of other pollutants?

- Which experimental methods are used to assess the removal and fate of hazardous pollutants in biological drinking water and wastewater treatment systems?
- How should hazardous pollutants be monitored and measured?
- How can hazardous pollutants affect microorganisms in biological treatment units? What are the major process characterization and microbial ecology tools used in studying hazardous pollutants?
- If an experimental study is conducted on hazardous pollutants, which aspects need to be considered in biological treatment?

1.5 SUGGESTIONS FOR THE READER

While this book tries to be a cohesive one in which all subjects are covered in a unified way, another aim of it is that each chapter should be readable alone. In order to achieve this aim, in each chapter frequent cross-referencing to other chapters is also provided.

However, it is impossible for a reader to have a grasp over all the topics presented in this book. Bearing this in mind, the following suggestions are made for readers to choose from a number of different topics:

In order to acquire a general idea about the presence and importance of hazardous pollutants in biological treatment systems, it might be sufficient to read **Chapter 1**, and eventually the final chapter, **Chapter 9**, which deals with future aspects.

Chapter 2 is comprised of the main sources of hazardous pollutants, their intrinsic properties, hazard effects and transport in the water environment. This chapter may be particularly appealing to a reader who is interested in gaining information on the properties of hazardous pollutants. The reader does not need to have much knowledge on biological treatment. Thus, even for a reader who is not particularly involved with biological treatment, this chapter may be read alone.

Chapter 3 appeals rather to a reader who deals with analytical issues. Although sampling and analysis of different hazardous pollutant groups can be found in other documents too, this chapter has the unique feature that it addresses biological systems.

Chapter 4 is the chapter to be first read by a reader who is interested in the removal and effects of hazardous pollutants in biological treatment systems. It provides a detailed overview on all aspects of biological removal of hazardous pollutants.

Complementary to **Chapter 4**, it is advisable to read **Chapter 5** which, in detail, deals with experimental procedures. Especially for scientists who are involved with laboratory-scale studies, this chapter is a must.

Chapter 6 would be an appealing chapter for all those who are involved with practical applications and want to learn about the extent of removal of hazardous pollutants in full-scale treatment. Although **Chapter 6** may be read alone, **Chapter 4** is suggested as the first chapter to read.

Chapter 7 is a chapter that could also be read alone for a biologist or microbiologist. However, it is best to read **Chapter 4** and **5** before it, for a better understanding of the mechanisms in biological removal of hazardous pollutants.

Chapter 8, dealing with the biological processes in drinking water treatment, is one of the chapters that is independent from the others. However, in the ideal case, to gain an insight into biological processes and experimental studies, the reader is recommended to read also **Chapter 4** and **Chapter 5**, respectively.

Chapter 9 is a conclusive chapter that provides a perspective on future aspects of hazardous pollutants and their biological removal. Along with **Chapter 1**, the reader might read this chapter to acquire a general idea about the subject.

1.6 REFERENCES

Clouzot L., Cloutier F., Love N., Melcer H., Ort C., Patureau D., Plosz B. G., Pomies M., Vanrolleghem P. A. and Choubert J. M. (2012). Are we about to reliably predict the fate of micropollutants through wastewater treatment plants? pp. 299–312, https://hal.archives-ouvertes.fr/hal-00697063.

Forrez I., Boon N., Verstraete W. and Carballa M. (2011). Biodegradation of micropollutants and prospects for water and wastewater biotreatment. In: Comprehensive Biotechnology, Moo-Young M. (ed.), Elsevier, Amsterdam, The Netherlands, pp. 485–494.

Margot J., Rossi L., Barry D. A. and Holliger C. (2015). A review of the fate of micropollutants in wastewater treatment plants. *WIREs Water*, **2015**(2), 457–487. doi: 10.1002/wat2.1090.

Rücker C. and Kümmerer K. (2012). Modeling and predicting aquatic aerobic biodegradation – a review from a user's perspective. *Green Chemistry*, **14**(4), 875.

Sathyamoorthy S., Chandran K. and Ramsburg C. A. (2013). Biodegradation and cometabolic modeling of selected beta blockers during ammonia oxidation. *Environmental Science & Technology*, **47**(22), 12835–12843.

van Leeuwen I. M. and Vermeire T. G. (eds) (2007). Risk Assessment of Chemicals-An Introduction. Springer, Dordrecht, The Netherlands.

Chapter 2

Hazardous pollutants in the water environment

Recep Kaya Göktaş[1] and Matthew MacLeod[2]

[1]Department of Environmental Engineering, Kocaeli University, Umuttepe Yerleşkesi, Mühendislik Fakültesi A Blok, 41380, İzmit, Kocaeli, Turkey, rkgoktas@gmail.com.

[2]Department of Environmental Science and Analytical Chemistry (ACES), Stockholm University, Svante Arrhenius väg 8, SE 11418, Stockholm, Sweden, matthew.macleod@aces.su.se.

2.1 WHAT MAKES A POLLUTANT "HAZARDOUS" IN THE WATER ENVIRONMENT?

The term "hazardous pollutant" is commonly used to refer to a substance that, upon release to the environment, has the potential to degrade the health of humans and the environment they are living in. However, it is not trivial to quantify this potential risk; and it is even more difficult to come to a consensus about how much potential risk should be deemed enough to make a substance "hazardous". The task gets even more complicated by the fact that there are thousands of synthetic chemicals being released to the environment. And, many new chemicals will be produced and will potentially contaminate the environment once they are in use. A systematic methodology is required to determine which of these man-made synthetic chemicals should be defined as "hazardous pollutants".

Rachel Carson's book, Silent Spring, published in 1962 is often used to mark the beginning of the era in which environmental consciousness about chemical pollution became widespread among societies. It was realized that the synthetic

 Hazardous Pollutants in Biological Treatment Systems: Fundamentals and a Guide to Experimental Research
Ferhan Çeçen, Ulaş Tezel
doi: 10.2166/9781780407708_17

chemicals that are deemed vital to the welfare of the society could also have unacceptable negative effects on the environment and the people living in those societies. This was disturbing to many people. Even more alarming was the fact that these negative effects may manifest themselves much later in time and at locations far away from the source of the pollution.

The scientific community had been studying the human and environmental health effects of synthetic chemicals for some time before the publication of Carson's book, and its message depended on the knowledge produced by those studies. As public concern about the problem began to increase, a need emerged to set up regulations to protect the environment and human health from hazardous pollutants. The regulations needed clear-cut definitions and procedures to determine which substances should be identified as "hazardous". As a result, regulatory bodies turned to the scientific community to help derive these definitions and procedures. Thus, scientific studies and the regulations dealing with hazardous substances have been in close interaction since the beginning. Scientific studies have informed regulatory bodies on how to assess the risks associated with different substances whereas the practice and outcomes of the regulations have been guiding the scientific studies.

The United States established a series of environmental regulations that deal with hazardous pollutants beginning in the 1970s: the Clean Air Act (1970), Clean Water Act (1972), Safe Drinking Water Act (1974), Resource Conservation and Recovery Act (1976), Toxic Substances Control Act (1976), and Comprehensive Environmental Response, Conservation and Liability Act (1980). The Clean Water Act (CWA) defines hazardous substances as "such elements and compounds which, when discharged in any quantity into or upon [the water environment] … present an imminent and substantial danger to the public health or welfare, including, but not limited to, fish, shellfish, wildlife, shorelines, and beaches" (Section 311(b)(2)(A)). The CWA also refers to the Toxic Substances Control Act (TSCA) which defines an "imminently hazardous chemical substance or chemical mixture" as "a chemical substance or mixture which presents an imminent and unreasonable risk of serious or widespread injury to health or the environment" (Section 7(f)). Another regulation referred to in the CWA, the Comprehensive Environmental Response, Conservation and Liability Act (CERCLA), defines hazardous substances as "such elements, compounds, mixtures, solutions, and substances which, when released into the environment may present substantial danger to the public health or welfare or the environment" (Section 102(a)).

The definitions provided in these regulations are basically the same as the definition we have provided in the first sentence of this chapter. All of these definitions put a particular focus on the toxic effects of the substance. However, before the toxic effects materialize, exposure to the substance must have occurred. The two key properties of a substance that are used to assess its potential exposure are its persistence and bioaccumulation potential.

In the field of environmental science and environmental chemistry, "persistent, bioaccumulative, and toxic (PBT) chemicals" is used as a more specific term when referring to the hazardous pollutants (e.g. Matthies *et al.* 2016). Scientific approaches that aim to identify hazardous pollutants using PBT assessment have also been adopted by many regulatory agencies. Among others, the United States Environmental Protection Agency (U.S. EPA), European chemicals regulation on the Registration, Evaluation, Authorisation, and Restriction of Chemicals (REACH), and the Stockholm Convention on Persistent Organic Pollutants layout schemes and procedures for assessing substances according to their PBT properties.

Long-range transport potential (LRTP) is another property of a chemical that partially determines its exposure hazard. LRTP indicates the potential for the pollutant to reach remote regions. It is not explicitly included in most national regulations since national regulations deal with pollution problems within a single legal jurisdiction, whereas long-range transport of pollutants implies an international problem that has an impact on the environment in a different legal jurisdiction, and even at the global scale. LRTP is considered in the evaluation of chemicals nominated for addition to the Stockholm Convention, which deals with the global environment as affected by persistent organic pollutants.

In this chapter, we provide a review of the properties of substances that make them hazardous pollutants with a focus on the water environment. In Section 2.2, we differentiate between toxicity hazard and exposure hazard and investigate these two properties separately. The physicochemical properties of chemicals that affect their fate, transport, and exposure and toxicity hazard are discussed. An overview of the available computational tools that can be used to identify the properties, exposure and toxicity, of existing and future chemicals is provided.

In Section 2.3, we investigate the hazardous pollutants that are likely to be present in the water environment by identifying their common properties and their sources. Also, the pathways for these pollutants to reach wastewater treatment plants are discussed.

In Section 2.4, we provide an overview of the regulatory frameworks that deal with hazardous pollutants. The main focus is on the Toxic Substances Control Act of U.S. EPA, the Toxic Substances Management Policy of Environment Canada, and the REACH legislation of the European Chemicals Agency.

Since there are so many hazardous pollutants, special groupings have been devised by researchers and regulatory bodies to facilitate communication and to manage resources more efficiently by focusing on more critical pollutants. In Section 2.5, the terms that are used to define these specific groups of hazardous pollutants, such as "priority pollutants", "emerging pollutants", and "micropollutants" are discussed.

In order to provide an answer to the question "what is a hazardous pollutant?", collaboration between environmental scientists and regulators is necessary. Environmental scientists can devise methods to quantify the risk posed by a chemical by using knowledge of its physicochemical properties and relating them to its toxicity and exposure hazard. Then, regulators can decide on the procedures to apply these methods and on how to use their output when identifying chemicals as hazardous pollutants. In this chapter, we strive to provide the current picture on this process with a focus on the water environment.

2.2 PROPERTIES OF HAZARDOUS POLLUTANTS

The likelihood of a substance to pose harm to human health or the environment is often assessed initially by using its hazard properties, namely its persistence, bioaccumulation potential and toxicity (Mackay *et al.* 2014). Toxicity is the measure of negative health effects expected to be manifested upon exposure to the pollutant. On the other hand, persistence and bioaccumulation potential are properties related to the degree of exposure. Long-range transport potential and the quantity of the substance released to the environment are also important exposure related properties. Characterization of the toxic hazard from a pollutant and the degree of exposure are both required in order to assess the risk of toxic effects associated with a pollutant (Mackay *et al.* 2001).

The toxic hazard and the exposure hazard are determined, to a large extent, by the physicochemical properties of pollutants. Environmental partitioning properties and degradation (or transformation) properties of pollutants affect their environmental fate and thus their exposure pathways. In the case of inorganic chemicals, speciation and complexation reactions may determine the environmental behavior and the availability for exposure. These properties are all related with the molecular structure (molar mass, molar volume, availability of functional groups and branching) of the pollutant.

In this section, the physicochemical properties of chemicals that affect their fate, transport, and exposure and toxicity hazard are reviewed. Then, the toxicity hazard and the exposure hazard properties are discussed. At the end of this section a brief review of some of the available computational tools and databases that can be used to obtain quantitative information about the toxicity, exposure, and physicochemical properties of existing and future chemicals is provided.

2.2.1 Physicochemical properties

It is desirable to describe the toxicity, persistence, bioaccumulation potential and the potential for long-range transport of a pollutant using metrics that are derived from chemical-specific intensive properties (Mackay *et al.* 2001). In the next sections, we will provide the commonly used intensive or quasi-intensive properties for this purpose. Most of these metrics are related with the physicochemical properties of pollutants. For example, in the case of organic pollutants, the partitioning behavior and degradability are key determinants in defining the toxicity, persistence, bioaccumulation potential and the potential for long-range transport. Also, the molecular structures of pollutants determine their partitioning behavior and degradability. In the case of inorganic pollutants, the speciation and complexation properties are critical in determining the environmental fate of the pollutant substance.

Physicochemical properties of pollutants also determine their fate in built environments, such as wastewater/stormwater collection systems and water/wastewater treatment systems. The partitioning properties of a pollutant substance will control the amount that would be lost from the system via volatilization, or the amount that would partition to the solid phase. Its molecular structure will control the chemical and biochemical reactions that it can undergo when traveling through these built environments.

The partitioning behavior of a substance in the air-water-octanol system can be used to model its fate in the natural and built environments. The partitioning behavior of a chemical in this three-phase system can be described by using three partition coefficients: air-water partition coefficient (K_{AW}), octanol-water partition coefficient (K_{OW}) and octanol-air partition coefficient (K_{OA}). Only two of these partition coefficients are required since the third one will be dependent on the other two, such as:

$$K_{OA} = \frac{K_{OW}}{K_{AW}} \tag{2.1}$$

These three partition coefficients can be directly measured or they can be determined by using the physicochemical properties of the pure substance such as vapor pressure, solubility in water, and solubility in octanol.

The octanol phase in this system is a convenient surrogate for organic phases found in the environment such as organic matter in soil particles, sediments, aerosols, and suspended solids in water, and also for lipids and fats in living organisms. There are correlations developed for estimating organic carbon-water, soil-water and biosolids-water partition coefficients of a chemical based on its K_{OW} value. A high K_{OW} value indicates that the chemical has a tendency to partition into organic phases as opposed to the water phase; therefore it is a measure of the **hydrophobicity** of the chemical. A high K_{AW} value indicates the tendency of a chemical to be present in the air phase rather than the water phase; therefore it is a measure of the **volatility** of the chemical. A **hydrophilic** chemical prefers to be in the water phase as opposed to the air and organic phases. A full understanding of the degree of hydrophobicity, hydrophilicity and volatility of a chemical can be obtained by comparing its K_{AW} and K_{OW} values. Figure 2.1 is a diagram of log K_{AW} vs. log K_{OW} produced originally by Gouin *et al.* (2000). The diagonals in this graph are lines of constant log K_{OA} values. A chemical's location in this diagram shows its distribution in the air-water-octanol system, which can be used to assess its behavior in natural and built environmental systems. A volatile chemical will be located at the upper left of this chemical space, a hydrophobic chemical will be located at the lower right, and a hydrophilic chemical will be located at the lower left.

The other key determinant of the environmental behavior and hazard potential of pollutants is their degradability or reactivity. There are many different reactions that a pollutant can undergo in the natural and built environments, such as biodegradation, hydrolysis, photolysis, oxidation and reduction. The potential of a chemical to be susceptible to these reactions is determined by the physicochemical properties of the chemical. However, the rates of the reactions are also very dependent on the availability of reactants and suitable conditions in the ambient environment.

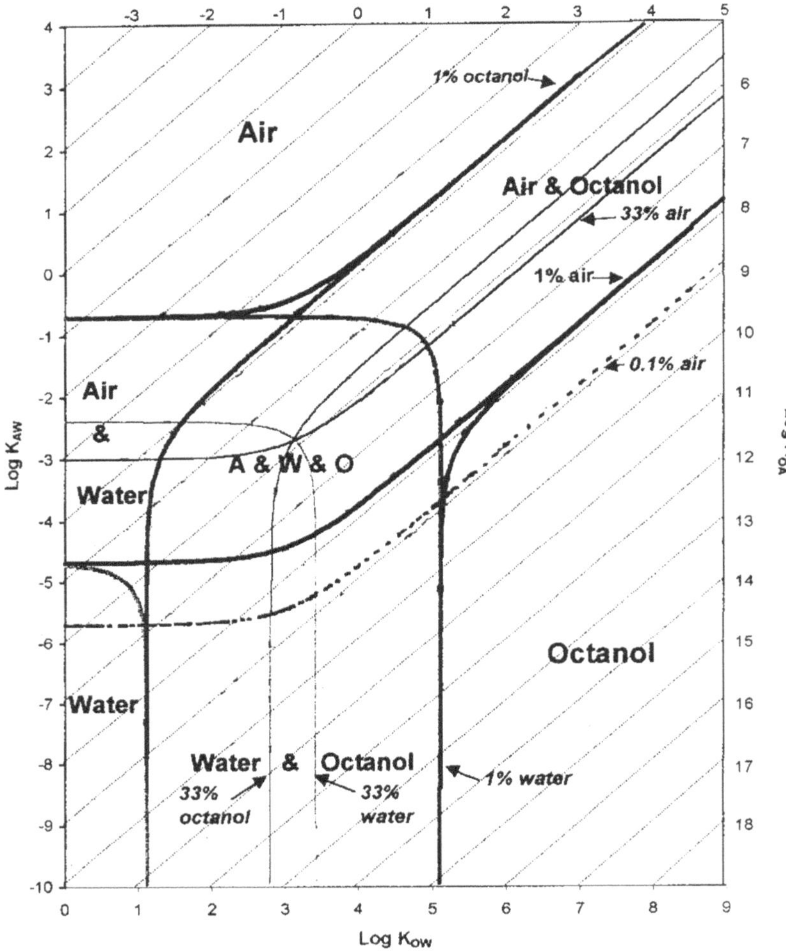

Figure 2.1 A chemical space diagram that can be used to assess the partitioning behavior of a chemical in the air-water-octanol system. The location of a chemical in this diagram is determined by its log K_{AW} and log K_{OW} values. The diagonal lines show constant log K_{OA} values. The thick lines show constant percentages of a chemical present in a specific phase at equilibrium when the ratio of air:water:octanol volumes is 650,000:1300:1. These thick lines can be used as regional boundaries within the chemical space. (Adapted with permission from Gouin, T., Mackay, D., Webster, E. and Wania, F. (2000). "Screening chemicals for persistence in the environment." *Environmental Science & Technology*, 34(5), 881–884. Copyright © 2000 American Chemical Society.)

The molecular structure of the pollutant provides clues about the reactivity of the pollutant substance in certain conditions. For example, according to the rules of thumbs for assessing the biodegradability of organic pollutants, highly halogenated, highly branched, nitroaromatic molecules are expected to be persistent whereas molecules with straight aliphatic chains, esters, acids, hydroxyl functional groups are expected to be biodegradable (Howard & Muir, 2010). These rules of thumbs developed through years of experimental knowledge combined with new controlled laboratory experiments, provide data to be used in developing models that can estimate the reactivity of chemicals based on their molecular structure.

A comprehensive review on the standard test methods and estimation methods for biodegradation can be found in Howard (2000). Raymond *et al.* (2001) and Jaworska *et al.* (2003) provide reviews of biodegradation estimation methods with a focus on aerobic degradation, whereas the works of Rorije *et al.* (1998) and Meylan *et al.* (2007) present studies on estimating anaerobic biodegradability.

Hydrolysis is the reaction of a chemical compound with water or with reactant species provided by water (i.e. H_2O, H^+, OH^-). A comprehensive review of estimation methods for the hydrolysis rate of organic compounds can be found in Wolfe and Jeffers (2000). Photolysis refers to chemical reactions that occur only in the presence of light. Mill (2000) provides a review on this process occurring in surface waters both directly (formation of excited or radical species by absorption of a photon by the chemical compound) and indirectly (formation of oxidants or excited states by the absorption of photons by a compound, which then reacts with other compounds in the water environment). Biodegradation and photolysis reactions are oxidation-reduction (redox) reactions, but there are also abiotic and non-photochemical redox reactions that occur in water environments, as well. Tratnyek and Macalady (2000) provide a review of environmental conditions and chemical molecular structures that affect the kinetics of these reactions for organic pollutants.

There are comprehensive handbooks that compile and organize experimental information on the physicochemical properties of environmental pollutants. These resources also contain data obtained from various estimation methods that relate toxicity and environmental fate properties of chemicals with their intrinsic properties. The handbook of Mackay *et al.* (2006) is an excellent resource on the physicochemical properties and environmental fate for organic chemicals. The handbook of Boethling and Mackay (2000) provides a review of methodologies for estimating chemical properties that are critical for environmental fate assessment.

In recent years, online databases and computational tools that facilitate the extraction and estimation of property information have been available and they are widely used when assessing the risk associated with hazardous pollutants. In Section 2.2.4, we will provide an overview of some of the most popular computational tools currently used for this purpose.

2.2.2 Toxicity hazard

Toxicity hazard manifests itself in different ways that require different methods of measurement and different metrics. A general categorization can be provided as acute toxicity vs. chronic toxicity.

Acute toxicity is usually defined as the exposure concentration or the intake dose that causes death in a target organism. When test results for aquatic toxicity are reported, the test organism and the test duration should be specified. The concentration of a specific substance in water that would kill 50% of the population of a specific aquatic organism (e.g. a specific fish species) after a certain duration of exposure (e.g. 96 hours) would be reported as the 96-hour LC_{50} of that substance for that specific organism, where LC stands for "lethal concentration". A toxic endpoint other than death of the organism may also be the subject of the test. Then, the lowest observed effect concentration (LOEC), the highest concentration causing no effect (no observed effect concentration, NOEC) or the concentration causing 50% of the observed effect (EC_{50}) may be reported (Gerbersdorf *et al.* 2015).

Chronic toxicity is more difficult to measure since the tests must be conducted for a longer duration and the toxicity endpoint is often a more subtle sublethal effect. In the tests for chronic toxicity, the exposure dose is usually reported as the daily dose applied per body mass. Chronic toxicity is often more relevant than acute toxicity for hazardous pollutants since the environmental concentrations are generally too low to cause immediate poisonous effect. Instead, ecological and health effects may manifest themselves after many years of exposure to these low concentrations. In cases where chronic toxicity studies have not been conducted for a specific chemical, the results of short-term acute toxicity testing are sometimes extrapolated to estimate a chronic effect concentration. An acute-to-chronic ratio of 10 has often been used (EPA, 1991; Organisation for Economic Co-operation and Development [OECD], 1992). This ratio is basically a safety factor or an uncertainty factor used to deal with lack of data and to provide a degree of conservatism. Chapman *et al.* (1998) provided a critical review of the use of safety factors in toxicity evaluations for risk assessment.

There are several measures of acute and chronic toxicity as the external dose applied (e.g. LC_{50}, applied dose per day per body mass) that would cause an adverse effect on the test organism during the laboratory test. A variety of test organisms are used and the test durations also differ. It is argued that these differences may complicate the interpretation of the toxicity test results; and the use of an internal dose as the toxicity metric instead of the external dose is suggested (Mackay *et al.* 2001). This internal quantity or concentration may be reported as the critical body burden (CBB), or critical body residue (CBR). When the quantity cannot be applied to the whole body, critical tissue burden (CTB) can be used as the toxicity metric.

Genotoxicity is a type of chronic toxicity where exposure to the hazardous pollutant causes harmful genetic modifications. The harmful effect may be the development of cancer (carcinogenicity) or abnormalities in physiological development (teratogenicity). It is difficult to measure the genotoxicity of a substance because the harmful effect is a result of a complex process that is susceptible to interference from multiple factors.

Some pollutants are able to mimic endogenous hormones; alterations in reproductive development of aquatic organisms have been observed near discharge locations of pollutant chemicals that mimic estrogens and androgens (Tyler *et al.* 1998). This phenomenon is called endocrine disruption and has attracted considerable attention from the public and the environmental scientists since the first discovery of intersex fish at 1978 near a discharge location in an English river (Sumpter & Johnson, 2008).

Antibiotic resistance is a phenomenon closely related to toxicity hazard that has been a subject of serious concern. The release of antibiotic substances from industrial, hospital and household sources provides a low but continuous residue of these chemicals in the environment, and promotes spread of antibiotic-resistant bacteria and antibiotic-resistant genes with potential adverse effects on the ecosystem and human health (Zhang & Li, 2011).

Genotoxicity, endocrine disruption and antibiotic resistance can thus be viewed as different forms of chronic toxicity. Assuming that the long-term effects on ecological systems can be inferred from the short-term effects on single organisms observed in the laboratory is questionable (Johnson & Sumpter, 2016). Adding to the complexity is the fact that individual organisms are being exposed to multiple chemicals concurrently in their environmental habitats. So, it is difficult to single out a chemical as the cause of an observed chronic effect. Also, the exposed mixture of chemicals may have synergistic effects (combined effect being larger than the sum of individual effects) or antagonistic effects (combined effect being less than the sum of individual effects). The review article by Gerbersdorf *et al.* (2015) lists the issues of intra- and interspecies relations, the exposure of individual organisms to a complex mixture of chemicals, and using short-term laboratory tests to determine the long-term chronic effects as complications in assessing the environmental toxic effects of anthropogenic chemicals. They also propose ways to resolve these issues, such as, (i) new test approaches that analyze changes in species composition to account for effects on higher ecological levels; (ii) new studies that investigate combination effects of large numbers of chemicals from pollutant groups such as pharmaceuticals and pesticides; and (iii) more research on the evaluation of chronic toxicity by effect-based tools.

Matthies *et al.* (2016) provide a review of the evolution of assessment criteria of toxicity of chemicals in national and international legislations. They state that the current regulations or conventions do not explicitly consider the synergistic effects since they focus on the regulation of individual chemicals and not the mixture of chemicals found in the environment. Future research studies that aim to improve assessment of toxic effects of chemicals by considering synergistic effects of mixtures of chemicals on ecological communities and human health may soon provide new tools and approaches, which will subsequently be included in the legislations. Also, further research will also probably find completely new toxic effects and the complexity of hazardous substance evaluations will further increase.

2.2.3 Exposure hazard

High toxicity by itself does not immediately imply high risk of adverse effects. In order to assess the environmental and health risk associated with a substance, both its toxic hazard and its exposure hazard must be taken into account (Mackay *et al.* 2001). Exposure hazard is a function of a substance's tendency for accumulation in certain environmental media in which contact with sensitive organisms may occur, and also of the tendency to be uptaken by these organisms that results in an internal dose. A high-toxicity, low-exposure pollutant and a low-toxicity, high-exposure pollutant can pose an equal risk.

Persistence has been called "the most important single criterion affecting chemical exposure and risk via the environment" (Mackay *et al.* 2014). It can be defined as the ability of a substance to remain (in an unchanged form) in the environment (Boethling, 2016). Persistence is usually evaluated as half-lives in specific environmental media such as water, soil, and sediment. For example, the Stockholm Convention sets threshold criteria for persistence as 2 months in water and 6 months in soil and sediments (S.C., 2009). The use of half-lives implies a first-order reaction assumption for the degradation of the pollutant. However, the degradation reactions in the environment are rarely truly first-order as they depend on the availability of another constituent (e.g. oxidants and reductants, specific microbial communities) in addition to the pollutant itself. Pseudo first-order reaction rate constants can be approximated by assuming generic values for the quantity of the other required constituent in cases where it has a relatively constant concentration (Mackay *et al.* 2014). So, the half-life measure for persistence is dependent on the properties of the environment as well as the properties of the pollutant.

When the persistence of a pollutant is evaluated according to media-specific half-lives, its distribution between media in the environment is neglected. Any pollutant that is released into the multi-media environment partitions among different environmental media, such as air, water, sediment, soil and vegetation. The specific medium into which the pollutant enters the environment (i.e. the mode of entry) and the partitioning behavior of the pollutant affect its overall environmental distribution, which in turn affects its overall persistence. Many critiques have been made regarding the use of media-specific half-lives as a measure of persistence, especially in the regulatory context (e.g. Boethling, 2016; Matthies *et al.* 2016). However, the overall environmental persistence of a pollutant in a multi-media environment cannot be directly measured. Instead, methods have been developed that use multi-media mass balance modeling approaches to estimate overall environmental persistence.

Multi-media mass balance models are useful tools in assessing exposure hazard. In the multi-media modeling approach, the environment is conceptualized as a collection of compartments that represent different environmental media. Mathematical descriptions of the inter-media exchanges between the compartments are used to develop mass balances for each of these compartments. With a focus on the characteristics of the pollutant, multi-media mass balance models integrate the knowledge accumulated through studies in multiple scientific disciplines and provide a full picture of the pollutant's fate in the environment (MacLeod *et al.* 2010).

Multi-media mass balance models usually use media-specific half-lives to describe the transformation reactions that result in the loss of the pollutant from the environment. However, by calculating the distribution of the pollutant in a multi-media environment from partition ratios measured in the laboratory and kinetics of transport between compartments, the contributions of degrading reactions in different media to the overall persistence of the pollutant and the influence of mode of entry are taken into account.

In the mass-balance modeling approach, overall persistence is defined as an overall half-life or as the residence time of the pollutant in a multi-media environment due to its degradation to other chemical species under steady-state conditions (Mackay *et al.* 2014). MacLeod and McKone (2004) studied the relationship between overall persistence and population-level intake of pollutants using a multi-media mass balance and multi-pathway exposure model. They showed that for chemicals emitted to air or surface water, the population-level intake is correlated with overall persistence. However, the relationships between population-level intake and media-specific half-lives were not robust.

The **OECD Pov and LRTP Screening Tool** is a modeling tool developed for calculating overall persistence of chemicals for screening-level assessment (Wegmann *et al.* 2009). Site-specific half-lives and overall environmental persistence depend on the properties of the substance, and also on the properties of the environment. However, environmental properties used in the estimation of these quantities are predefined; and persistence can be treated as a quasi-intensive property and can be used in chemical evaluations (Mackay *et al.* 2001). The OECD Tool employs a standardized description of an evaluative environment with properties that represent the entire global environment. The overall persistence of a substance as calculated from its multi-media fate and transport in this generic environment is a quantitative description of persistence and environmental exposure hazard. The OECD Tool is freely available as spreadsheet software (OECD, 2016).

The term "pseudo-persistent" has sometimes been used for non-persistent substances that have a continuous emission to an environmental region such that there is continuous exposure to organisms near the point of release. Pharmaceuticals and personal care products (PPCPs) are of special concern as pseudo-persistent substances since they are continually released to some aquatic environments although they are non-persistent when only their environmental half-lives are taken into account (Daughton, 2002). Mackay *et al.* (2014) proposed to use the term "continuously present" since the risk of continuous exposure to these compounds is not due to an intensive chemical-specific property but due to their continuous release to a certain environment. Such "continuously present" chemicals pose an exposure hazard; however that exposure can be reduced rapidly compared to persistent chemicals if emissions are curtailed.

All other things being equal, ecological communities will be more highly exposed to chemicals that are persistent in the environment than non-persistent pollutants. However, in order for toxic effects to be manifested, the pollutant must enter into and reach high concentrations in the body of the organism, specifically at the site of toxic action in the body. It is possible for some chemical pollutants to reach high concentrations within organisms even when they are present in much lower concentrations in the ambient environment. **Bioaccumulation** is the term used to describe this phenomenon. It can be viewed as persistence within the body of organisms, and the accumulation may continue through the food web resulting in even more magnified concentrations in higher members of the food web.

Bioaccumulation is usually defined and measured for aquatic organisms, and we will follow this convention since our focus is on the water environment. The partition ratio between octanol and water (K_{OW}) is well established as an indicator of potential for chemicals to bioaccumulate in fish and other aquatic organisms since a high K_{OW} value shows a tendency of the chemical to partition into the lipids of fish from the surrounding water (Gobas *et al.* 2009). Bioaccumulation potential of a pollutant chemical may also be directly measured in laboratory conditions by exposing an aquatic organism (e.g. a fish) to the chemical as it is present in known concentrations in the water environment where the fish is being kept. Then, the bioconcentration factor (BCF) can be determined as the ratio of the chemical concentration in the body of the organism (C_B) and the chemical concentration in the water (C_W):

$$\mathrm{BCF} = \frac{C_B}{C_W} \tag{2.2}$$

BCF implies that equilibrium is reached between the organism and the water environment. Also, the uptake mechanism in bioconcentration is assumed to be absorption through the skin and the respiratory surface (Mackay & Fraser, 2000). Bioaccumulation is a more general term that encompasses all the possible exposure routes including dietary uptake, and provides a framework to quantify all possible elimination routes including metabolism and excretion. Mass balance models have been developed for simulating the change of chemical concentration in the body of the organism that takes into account the kinetics of all the uptake and elimination processes acting simultaneously. Hydrophobic organic chemicals that are poorly metabolized and metals that show ligand binding may bioaccumulate in organisms

(Mackay & Webster, 2003) as their elimination kinetics may be slower than their uptake kinetics. The term biomagnification, on the other hand, puts the focus on the dietary uptake route. Biomagnification factor (BMF) can be defined as the ratio of the food uptake of chemicals to their elimination (van der Oost *et al.* 2003). This implies the following relationship at steady-state:

$$\mathrm{BMF} = \frac{k_F}{k_B} = \frac{C_B}{C_F} \tag{2.3}$$

where k_F and k_B are rate constants (T^{-1}) for food uptake and elimination processes, respectively, and C_F is the concentration of chemical in the food.

Biomagnification is the phenomenon of increased concentrations in the body of organisms compared to their food. So, it may lead to food chain accumulation. Trophic magnification factor (TMF), which is defined as the change in the log-concentration of chemical per change in trophic level, is used as a metric to quantify biomagnification through food webs (Walters *et al.* 2016). BMF and TMF provide evidence for bioaccumulation through food webs and they are argued to be the best measures for determining bioaccumulative chemicals at the ecosystem level (Matthies *et al.* 2016; Walters *et al.* 2016). Another metric that has been proposed to quantify the bioaccumulation potential of a chemical is its elimination rate constant (k_B). By the use of the elimination rate constant, an elimination half-life ($EL_{0.5}$) can be defined as a chemical-specific property (Goss *et al.* 2013). $EL_{0.5}$ can be a better measure for categorizing chemicals according to their bioaccumulation potential compared to BMF and TMF since it accounts for chemicals that are taken up by all routes including the dietary route even if their uptake rate is very slow (Matthies *et al.* 2016).

When a PBT chemical released to the environment is transported and reaches remote locations from its source, the spatial scale related to its exposure hazard significantly increases. The **long-range transport potential** (LRTP) of chemical substances is very much dependent on the properties of the chemical. If a chemical is persistent and has a tendency to partition to or transfer to mobile media of air and water, it has a potential to transport to remote regions (Mackay *et al.* 2001). The Stockholm Convention (S.C., 2009) states that if a pollutant has a half-life in air of at least 2 days and is observed in remote regions it can be classified as having LRTP. This criteria excludes pollutants that can be transported in the water environment, also it doesn't provide a definition for remoteness, as well as a threshold concentration to be observed in the remote regions. LRTP analysis in the literature often focuses on transport of pollutants in air although most of the analysis methods can also be applied to transport in water, as well. However, pollutants that are released into the water environment do not necessarily stay in water. They may be transferred to air with subsequent transport to remote regions.

A more refined characterization of LRTP can be obtained from multi-media mass balance models. Several metrics have been developed that assume generic environmental conditions and use the properties of the pollutant in a mass balance modeling context to define LRTP. A widely used metric is the characteristic travel distance (CTD), which is defined as the distance at which the concentration of a chemical present in a mobile media of air or water has dropped to 1/e (where e is the base of the natural logarithm) of its initial value due to degrading reactions in the mobile medium and irreversible transfers to immobile media it interacts with during transport (Beyer *et al.* 2000). CTD is not dependent on the quantity of emissions and not related to environmental concentrations, which are required characteristics for a metric that is a pseudo-intensive chemical property (Mackay *et al.* 2001). However, CTD is not applicable to pollutants that are discharged to an immobile medium with only a small fraction being transferred to the mobile medium. In this case, Beyer *et al.* (2000) propose the use of an effective travel distance (ETD), defined as the distance that a smaller percentage (e.g. 1%) of the initially discharged pollutant reaches. CTD and ETD are transport-oriented metrics for LRT. There are also target-oriented metrics that focus on the fraction of the pollutant that is transported to remote regions and deposited there (Scheringer, 2009). Arctic Contamination Potential (ACP) is a target-oriented metric that focuses on the potential of a persistent organic pollutant to be transported to the Arctic region, which is among the most remote regions on the globe. ACP was defined as the ratio of the amount of pollutant present in the surface media of the Arctic region to the total amount of global emissions (Wania 2003, 2006). A global scale multi-media mass balance model, Globo-POP (Wania & Mackay, 1995), has been used in most ACP calculations to date. When ACP is being determined, all transport media are taken into account including the water environment. ACP analysis for a large set of hypothetical chemicals with a broad range of partitioning properties yielded a categorization of chemicals based on the mechanism by which they are transported to the Arctic regions (Wania, 2006; Gouin & Wania, 2007). This categorization classifies chemicals as "flyers", "swimmers", "single-hoppers" and "multiple-hoppers". "Swimmers" are those chemicals that are partitioned to the water environment and transported to the Arctic by ocean currents. They have $\log K_{AW} < -2$ and $\log K_{OA} < 10$.

Transfer efficiency (TE) is another target-oriented LRTP metric that is calculated by using multi-media mass balance models. It is defined as the ratio of the rate of contaminant flux to a target compartment or compartments in a remote region to the rate of emissions in a source region (MacLeod & Mackay, 2004). Recently, it has been used to define a metric for remoteness from sources of chemicals in the global environment, which may facilitate the identification of remote regions so that environmental measurements can be better interpreted and also better designed for improved evaluation of LRTP (Göktaş & MacLeod, 2016).

The OECD Pov and LRTP Screening Tool, mentioned before, calculates CTD and TE for organic chemicals, besides their overall persistence, using a generic global multi-media mass balance model (Wegmann *et al.* 2009). The output of the OECD Tool is used to screen chemicals for their persistence and LRTP to facilitate chemical management.

Zarfl *et al.* (2011) derived half-life criteria for LRTP of organic pollutants transported in the water environment that would correspond to the 2-day half-life criteria in air for airborne pollutants. They proposed half-lives of 10 days and 90 days for freshwater and seawater environments, respectively. They also calculated CTD values for transport in air and water separately for a large selection of hazardous pollutants. It was shown that pollutants that have a large half-life in water and a low air-water partition coefficient have higher CTDs for transport in water than for transport in air.

Ranking of pollutants according to their persistence, bioaccumulation potential and potential for long-range transport is an effective method for identifying their exposure hazard. However, the **quantity** released to the environment should be considered as an additional factor (Mackay *et al.* 2001). Environmental emission rates of pollutants must be known to estimate the actual degree of exposure. The term "high production volume (HPV) chemicals" is used to describe chemicals that are produced in large amounts, and as a result that have a potential for high emission rates to the environment and a potential for high degree of exposure. The European chemicals legislation, REACH, requires industrial chemicals that are manufactured or imported in amounts that exceed 10 tonnes per year to be assessed for their persistence, bioaccumulation potential and toxicity (Strempel *et al.* 2012). Besides production volume, the knowledge of how and where the chemical is used (use profile) is another criteria that is related to the quantity of chemical release to the environment (Muir & Howard, 2006).

In order to include the property of quantity into the characterization of hazardous pollutants, however, accurate and reliable inventory data for the production, transport, and usage of chemicals is required. Incomplete inventory data is a common and serious problem worldwide, the problem being more pronounced in the case of developing countries. In the work of Arnot *et al.* (2012), where a model-based exercise was conducted for ranking chemicals according to their exposure potential, it was revealed that the uncertainty in the actual emission rates dominates the uncertainty in human exposure estimates. They state that in order to reduce this key source of uncertainty, coordinated efforts of the regulatory, manufacturing and use communities are required.

2.2.4 Tools for estimating the physicochemical properties, exposure and toxicity hazard of pollutants

It is desirable to estimate the physicochemical properties of pollutants, and also their toxicity and exposure hazard through the knowledge of their molecular structure, since experimental determination is a costly and time-consuming process. We have provided some reference information on the estimation methods for key properties of hazardous pollutants in the previous sections. In this section, we will provide an overview of some of the available computational tools that can be used to obtain chemical property estimations. Our focus will be on the tools that use the Quantitative Structure-Activity Relationships (QSAR) technique. A description of the QSAR approach can be given as: obtaining relationships "based on mathematical dependencies between the variance in molecular structures, encoded by so-called molecular descriptors (e.g. number of atoms, dipole moment etc.), and the variance in a given physicochemical property (e.g. log K_{OA}) within a set of similar compounds" (Puzyn, 2011).

It is generally recommended to prefer empirical property data over estimated data. However, empirical databases have failed to keep up with the rapid pace at which the chemical industry develops new chemicals, and measured property data are also lacking for many chemicals currently in use in commerce. Also, the available empirical data need to be checked for potential errors and inconsistencies. So, the use of computational tools for chemical property estimation is inevitable and necessary.

The United States Environmental Protection Agency's (EPA) **EPI (Estimation Program Interface) Suite** is one of the most widely used QSAR tools (EPA, 2017g). It is freely available through U.S. EPA's website. EPI Suite includes a number of physicochemical property and environmental fate estimation programs and an aquatic toxicity estimation program (Table 2.1). EPI Suite has been developed by U.S. EPA's Office of Pollution Prevention and Toxics in collaboration with Syracuse Research Corporation. The developers recommend it to be used as a screening-level tool. They also recommend to use experimentally measured values when available, and in order to facilitate this, EPI Suite includes a database of chemical properties. The estimation programs in EPI Suite are developed for organic chemicals, and inorganic and organometallic chemicals are generally outside the applicability domain of the programs. It is critical to be familiar with the underlying procedures in the estimation programs included in EPI Suite when using their outputs. EPI Suite comes with comprehensive documentation that explains the theory and applicability of the individual programs together with a list of reference publications.

The only input required by EPI Suite is the chemical structure of the compound. This input is entered into the program as a SMILES notation. SMILES is an acronym for Simplified Molecular Input Line Entry System, which is a notation system to represent the molecular structure as a string of symbols (Weininger, 1988). The SMILES notation can be directly entered into the program, but there are several alternative ways that may be more practical for the user. The Chemical Abstract Service (CAS) Registry number or the name of a chemical can be searched within a database included with EPI Suite; and, if a match is found, the SMILE notation can be retrieved. EPI Suite includes a structure drawing module that can be used to draw the molecular structure of the chemical and then to generate SMILES notation based on the drawing. Also, chemical structures drawn or obtained outside EPI Suite can be imported in and converted to the corresponding SMILES notation.

Table 2.1 The individual programs included in EPI Suite (EPA, 2017g).

Program Name	Estimated Property
Physicochemical Properties	
KOWWIN	Octanol-water partition coefficient (K_{OW})
HENRYWIN	Air-water partition coefficient (K_{AW})
KOCWIN	Organic carbon-water partition coefficient (K_{OC})
KOAWIN	Octanol-air partition coefficient (K_{OA})
MPBPWIN	Melting point, boiling point, vapor pressure
WSKOWWIN	Water solubility (using the estimated K_{OW} value)
WATERNT	Water solubility (using a "fragment constant" method)
BCFBAF	Fish bioconcentration factor (BCF)
AEROWIN	Fraction of airborne substances sorbed to airborne particulates
Reactivities	
AOPWIN	Gas-phase reaction rate
BIOWIN	Aerobic and anaerobic biodegradability
BioHCwin	Biodegradation half-life for hydrocarbons
HYDROWIN	Aqueous hydrolysis rate constant
Fate and Transport Properties	
WVOLWIN	Rate of volatilization from rivers and lakes
STPWIN	Removal in a typical activated sludge-based sewage treatment plant
LEV3PI	Fate in a default model "environment" using a steady-state multimedia model
Toxicity	
ECOSAR	Acute and chronic toxicity to aquatic organisms

After the chemical structure is provided, EPI Suite can be applied to automatically execute each individual estimation program and report their output. The output report can be selected to be a summary of the results or can be selected to include the details of the calculations. It is also possible to omit the outputs of selected estimation programs if they are not of interest to the user.

Puzyn (2011) compared the performance of the OECD multimedia mass balance model (The OECD Pov and LRTP Screening Tool) using different combinations of experimentally derived and estimated values of partition coefficients and half-lives. The estimated physicochemical data were obtained by three different estimation tools one of which was EPI Suite. They showed that replacing empirical data with estimated data did not significantly decrease the model performance.

Several studies have used property data estimations from EPI Suite in large-scale chemical screening exercises. These studies were conducted using large numbers of chemicals and many chemicals lacked empirically derived property data, which needed to be replaced by EPI Suite estimations. Strempel *et al.* (2012) and Scheringer *et al.* (2012) assessed PBT chemicals using aerobic biodegradation half-lives estimated by BIOWIN, BCF values estimated by BCFBAF, and EC_{50} and LC_{50} values estimated by ECOSAR. Arnot *et al.* (2012) conducted an exposure assessment study for priority ranking of chemicals using a multimedia fate model, for which they obtained partitioning property and reaction half-life values of most of the chemicals using EPI Suite.

EPI Suite is a collection of already available property estimation programs developed using the QSAR approach. The freely available web-based platform, called the **Online Chemical Modeling Environment (OCHEM)**, on the other hand, is developed for facilitating QSAR modeling so that new property estimation models can be created and shared with the research community (Sushko *et al.* 2011). It is available online at http://www.ochem.eu. OCHEM actually has two interconnected subsystems: a chemical database of experimental measurements and the modeling framework. The chemical database is user-oriented and created by entries from the user community. There are protocols that ensure the reliability of the data entered and the original source information (scientific article or book) is always provided. The database can be reached by using a guest account, but registration is required to access all the available features.

The OCHEM developers' aim is to facilitate developing predictive models of physicochemical and biological properties of chemical compounds. It is also a project that encourages and enables the sharing, reuse and reconstruction of models; so that the developed models can be used in predictions for brand new chemicals or they can be improved by other researchers. The article by Sushko *et al.* (2011) is the main introductory source for the OCHEM platform. Additionally, the project's wiki (http://wiki.ochem.eu) contains the latest developments and extensive help documentation.

OCHEM has been used in several studies for developing QSARs. For example, the study by Vorberg and Tetko (2014) used OCHEM to develop a predictive model for the biodegradability of chemicals. With the tools provided by OCHEM, they compiled a large dataset of empirical data and used different modeling techniques to develop several different QSARs using different sets of molecular descriptors. From the individual models, they developed a consensus model that can provide confidence intervals for predictions. All the datasets used in the study and the consensus model are publicly available through the OCHEM platform at https://ochem.eu/article/31660.

The **OECD QSAR Toolbox** is another example of a computational tool that can be used in obtaining chemical property information. The OECD Toolbox is intended to be used for deriving (eco)toxicity data by filling the data gaps in the available experimental data within a group of similar chemicals (Dimitrov *et al.*

2016). The Toolbox is a software application that can be freely downloaded from the OECD website (OECD, 2017). The OECD Toolbox has a library of QSAR models; however it guides the user to obtain predictions of toxicity hazard of chemicals through implementing simpler approaches called read-across and trend analysis. Read-across is a very simple method of data gap filling. In this method, the key characteristics of the target chemical are identified and chemicals with the same characteristics are grouped together. Then, the lacking data for the target chemical is directly replaced with the available data for the chemicals within the same group. In trend analysis, the same type of a chemical categorization is followed by a regression analysis (usually linear) used to obtain a correlation equation that relates the selected properties of the group of chemicals with the target unknown property. A summary of these methods based on chemical categorization and their implementation in the OECD Toolbox are provided in van Leeuwen *et al.* (2009) and Dimitrov *et al.* (2016).

The characteristic feature of the OECD Toolbox is its integrated design that computerizes the application of the read-across and trend analysis approaches by combining chemical categorization, data collection and data estimation steps in a logical workflow. The reliability of the data obtained from the OECD Toolbox is very much dependent on the modules that categorize chemicals according to specified toxicity end points. Mombelli (2012) evaluated the estrogen receptor profiler of the OECD Toolbox that categorizes chemicals into different classes based on their estrogen receptor binding potentials. The categorization is accomplished by a set of criteria derived on the basis of scientific knowledge of the phenomenon, which is organized into a rule-based system. The study of Tebby *et al.* (2011) applied multiple categorization modules in the Toolbox for exploring acute aquatic toxicity of chemicals. Evaluations of some of the QSAR models available through the Toolbox can be found in the works of Devillers and Mombelli (2010a, b) and Mombelli and Devillers (2010).

Complementary to this chapter, **Chapter 5** provides a detailed discussion on the use of databases and computational methods prior to conduction of standard biodegradability tests.

2.3 ORIGINS AND OCCURRENCE OF HAZARDOUS POLLUTANTS IN WATER ENVIRONMENT

The introduction of hazardous pollutants to the water environment can occur directly or indirectly. Direct release occurrences consist of wastewater effluents produced during the production phase or during and after the use of products. Indirect routes to the water environment consist of various environmental transport processes after a pollutant's release to non-aquatic media in the environment. The fate of the pollutant in the water environment depends on the characteristics of both the chemical and the environment as we have discussed in the previous sections.

Wastewater treatment plants (WWTPs) are built water environments that have a critical role in the fate of waterborne hazardous pollutants. They may be the final destination for the pollutant chemical if its degradation occurs within the WWTP; or, they may serve as a temporary residence for the pollutant as it continues its journey after it leaves the WWTP. Pollutants may leave the WWTP by way of volatilization, through the effluent or by partitioning to the WWTP sludge. Figure 2.2, which was originally published in Gerbersdorf *et al.* (2015), summarizes the transport pathways of anthropogenic pollutants into the water environment. This figure presents a comprehensive picture that takes into account all the modes of entry into the environment based on the production and usage patterns of the pollutants including the reuse of biosolids. The role of stormwater and surface runoff as potential source of pollutants to the WWTPs and surface waters is also considered.

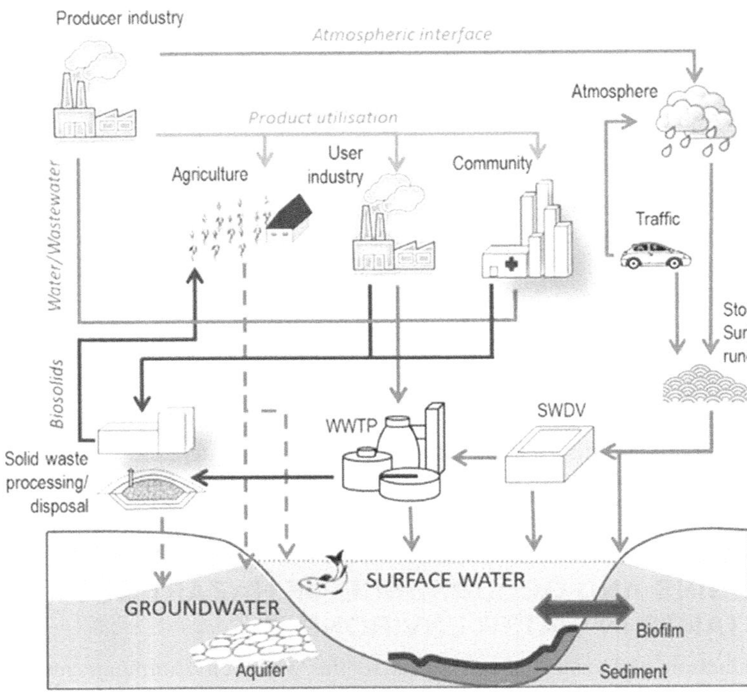

Figure 2.2 Transport pathways of anthropogenic pollutants to the water environment (WWTP: Wastewater treatment plant; SWDV: Stormwater Detention Vault). Biosolids represent both WWTP sludge and biowaste. (Graph by Demet Antakyali (Grontmij GmbH) published in Gerbersdorf *et al.* (2015) (http://doi.org/10.1016/j.envint.2015.03.011), an open access article under the CC BY-NC-ND license (http://creativecommons.org/licenses/by-nc-nd/4.0/)).

The classification of hazardous pollutants may be accomplished by taking into account where and how they are used, since this will be closely related to their mode of entry into the environment. However, their actual behavior in the environment, which depends on their physicochemical properties, is also very important since it determines if it will end up in WWTPs or in natural water environments (Gerbersdorf *et al.* 2015). In this section we will classify hazardous pollutants that may occur in the water environment according to their sources and common properties. We will discuss the role of WWTPs in their overall fate and transport in the environment.

2.3.1 Persistent organic pollutants

Persistent organic pollutants (POPs) are persistent, bioaccumulative and toxic (PBT) chemicals with a potential for long-range transport. They include both intentionally produced chemicals and non-intentional by-products, for example from combustion processes. POPs pose environmental problems at a global scale. The Stockholm Convention deals with the identification, reduction and elimination of POPs (S.C., 2009). Initially 12 chemicals have been recognized as POPs (the so-called "dirty dozen"), many of which are highly chlorinated compounds (Table 2.2). Most of these legacy POPs are **pesticides**: aldrin, chlordane, DDT, dieldrin, endrin, heptachlor, hexachlorobenzene, mirex, toxaphene. Other POPs include polychlorinated biphenyls (PCBs), which are industrial chemicals, and polychlorinated dibenzo-p-dioxins and polychlorinated dibenzofurans (PCDDs/PCDFs), which are unintentional by-products of combustion processes and production processes of chlorinated substances. Since the Stockholm Convention's adoption in 2001, new POPs have been added to the list (Table 2.2). Most of the new POPs are also highly chlorinated compounds. Some of them are pesticides (e.g. hexachlorocyclohexane isomers). However, there are also brominated compounds such as hexabromobiphenyl, hexabromocyclododecane (HBCD) and **polybrominated diphenyl ethers (PBDEs)**, which are used as flame retardants. Another group of chemicals that are among the new POPS are **per- and polyfluoroalkyl substances (PFASs)** such as perfluorooctanoic acid (PFOA) and perfluorooctyl sulfonate (PFOS), which have been used in the manufacture of many industrial products.

POPs may follow different routes to enter the WWTPs or natural waters depending on their mode of entry to the environment and their physicochemical properties (Figure 2.2). Many POPs are highly hydrophobic (e.g. PCBs, PCDD/PCDFs, PBDEs) and they are thus associated with the particles present in the environmental medium and travel with these particles. In WWTPs, they will largely be present in the wastewater sludge; and, in surface water, they will be sorbed to the sediments and also bioaccumulate in the aquatic organisms. However, some of the new POPs (hexachlorocyclohexane isomers, chlordecone and PFOS) have considerable tendency to partition to the water phase and can be transported via water flows (Zarfl *et al.* 2011).

Table 2.2 Initial and new persistent organic pollutants (POPs) and chemicals proposed for listing under the Stockholm Convention (S.C., 2017).

POP	Production/Usage
Initial POPs	
Aldrin	Pesticide
Chlordane	Pesticide
Dichlorodiphenyltrichloroethane (DDT)	Pesticide
Dieldrin	Pesticide
Endrin	Pesticide
Hexachlorobenzene (HCB)	Pesticide, industrial chemical, by-product
Mirex	Pesticide
Toxaphene	Pesticide
Polychlorinated biphenyls (PCBs)	Industrial chemical
Polychlorinated dibenzo-p-dioxins (PCDD)	By-product
Polychlorinated dibenzofurans (PCDFs)	By-product
New POPs	
Alpha hexachlorocyclohexane (α-HCH)	Insecticide
Beta hexachlorocyclohexane (β-HCH)	Insecticide
Chlordecone	Agricultural pesticide
Hexabromobiphenyl	Flame retardant
Hexabromocyclododecane (HBCD)	Flame retardant additive
Hexabromodiphenyl ether (hexaBDE) and Heptabromodiphenyl ether (heptaBDE)	Flame retardants
Hexachlorobutadiene (HCBD)	Industrial chemical (solvent)
Lindane (γ-HCH)	Insecticide
Pentachlorobenzene (PeCB)	Industrial chemical, by-product
Pentachlorophenol (PCP)	Pesticide, disinfectant
Perfluorooctane sulfonic acid (PFOS) and perfluorooctane sulfonyl fluoride (PFOS-F)	Industrial chemicals
Polychlorinated naphtalenes (PCNs)	Industrial chemicals
Endosulfan	Insecticide
Tetrabromodiphenyl ether (tetraBDE) and pentabromodiphenyl ether (pentaBDE)	Flame retardants
Chemicals proposed for listing as POPs	
Decabromodiphenyl ether (c-decaBDE)	Flame retardant
Dicofol	Pesticide
Short-chained chlorinated paraffins	Industrial chemicals (flame retardants)
Pentadecafluorooctanoic acid (PFOA) and related compounds	Industrial chemicals (surfactants)

Hexachlorocyclohexane (HCH) isomers and chlordecone are pesticides and their entry to the wastewater streams may be during the production phase or if they are used in urban environments such as parks and lawns. On the other hand, PFASs have been detected in freshwaters near industrial manufacturing plants, military fire training areas, airports, and wastewater treatment plants (Hu *et al.* 2016). They have been identified as important drinking water contaminants (Sun *et al.* 2016). By using a newly developed river-basin-scale multimedia fate and transport model (Lindim *et al.* 2016b), Lindim *et al.* (2016a) found that major rivers in Europe are highly contaminated with PFASs. They recommended installing effective treatment technologies at WWTPs as an effective way of reducing PFAS pollution in European rivers.

2.3.2 Pesticides

The group of chemicals broadly called pesticides are usually divided into many subgroups according to their targeted usage, such as herbicides, fungicides, insecticides, plant growth regulators, bactericides, and defoliants (Houtman, 2010). Some of the most persistent and problematic pesticides such as DDT, HCB, HCHs, and endosulfan are recognized as POPs (Section 2.3.1). Many newer pesticides are less persistent, but they still may pose a threat to the environment due to their high volume of production and usage, as well as their inherent property of being biologically active.

Apart from the pesticides recognized as POPs that have long-range transport potential, international trade of pesticides as commercial products and also the international trade of agricultural products that may contain residues of these pesticides, make the production, trade and usage of pesticides a global problem. In order to regulate the international trade of hazardous chemicals and pesticides, the United Nations Environment Programme (UNEP) and the Food and Agriculture Organization of the United Nations (FAO) developed a legally binding text, which is now called the Rotterdam Convention since it was first adopted in Rotterdam, the Netherlands in 1998 (UNEP & FAO, 2015). At the heart of the Rotterdam Convention is the Prior Informed Consent (PIC) procedure that facilitates information exchange between countries about dangerous pesticides in the context of international trade (Galt, 2008). Hazardous pesticides and industrial chemicals are included in Annex III of the convention, which is also called the PIC list. A procedure is specified in the convention text on how a chemical is included in the PIC list. The procedure is triggered by the ban or severe restriction of a pesticide or industrial chemical by two or more countries. Then, a formal review process is conducted and a decision is made on the inclusion of the chemical to the PIC list. The decision document is circulated among all the countries and each country makes a decision on whether or not they will continue to import the chemical. Currently, there are a total of 47 chemicals in the PIC list: 33 pesticides (including 3 severely hazardous pesticide formulations) and 14 industrial chemicals (UNEP & FAO, 2017). Table 2.3 contains the chemicals in the PIC list and the candidate chemicals that are under consideration to be included in the list.

Table 2.3 Chemicals (pesticides and industrial chemicals) listed in Annex III of the Rotterdam Convention (PIC list). Candidate chemicals that are under consideration to be included in the list are also given.

Chemicals in the PIC List	
Pesticides	
2,4,5-T and its salts and esters	Endosulfan
Alachlor	Ethylene dichloride
Aldicarb	Ethylene oxide
Aldrin	Fluoroacetamide
Azinphos-methyl	HCH (mixed isomers)
Binapacryl	Heptachlor
Captafol	Hexachlorobenzene (HCB)
Chlordane	Lindane (γ-HCH)
Chlordimeform	Mercury compounds
Chlorobenzilate	Methamidophos
DDT	Monocrotophos
Dieldrin	Parathion
Dinitro-ortho-cresol (DNOC) and its salts	Pentachlorophenol and its salts and esters
Dinoseb and its salts and esters	Toxaphene (Camphechlor)
EDB (1,2-dibromoethane)	Tributyl tin compounds
Severely Hazardous Pesticide Formulation	
Formulations containing a combination of benomyl, carbofuran and thiram	
Methyl-parathion	
Phosphamidon	
Industrial Chemicals	
Actinolite asbestos	Polybrominated biphenyls (PBBs)
Anthophyllite	Polychlorinated biphenyls (PCBs)
Amosite asbestos	Polychlorinated terphenyls (PCTs)
Crocidolite	Tetraethyl lead
Tremolite	Tetramethyl lead
Commercial octabromodiphenyl ether	Tris(2,3 dibromopropyl)phosphate
Commercial pentabromodiphenyl ether	
PFOS, perfluorooctane sulfonates, perfluorooctane sulphonamides and perfluorooctane sulfonyls	
Candidate Chemicals to be Included in the PIC List	
Pesticides	
Carbofuran	
Carbosulfan	
Trichlorfon	
Severely Hazardous Pesticide Formulation	
Fenthion	
Paraquat dichloride	
Industrial Chemicals	
Chrysotile asbestos	
Tributylin compounds	
Short-chained chlorinated paraffins	

Source: UNEP and FAO (2017).

Older and persistent organochlorine pesticides, aldrin, chlordane, DDT, dieldrin, endosulfan, HCH isomers, HCB, toxaphene, and pentachlorophenol are included in both the PIC list (Table 2.3) and the Stockholm Convention's POPs list (Table 2.2). In the 1980s, there was a shift from organochlorine pesticides to organophosphate and carbamate pesticides, which are less persistent but much more acutely toxic (Galt, 2008). Parathion, methyl-parathion, malathion, chlorpyrifos and diazinon are among the most commonly used organophosphate pesticides. Parathion is included in the PIC list (Table 2.3). Aldicarb, which is in the PIC list, and carbofuran, which is a candidate PIC chemical (Table 2.3), are among the most common carbamate pesticides. Currently the use of organophosphates and carbamates are in decline and they are being replaced by many new pesticides. Neo-nicotinoids (e.g. imidacloprid) and fipronil are among the most widely used pesticides today (Bonmatin *et al.* 2005; Bonmatin *et al.* 2015). New pesticides are constantly being developed by the chemical companies due to increased pest resistance to the current pesticides, but the need for companies to sustain their profits as the patents on the older pesticides expire is an additional driver of innovation in the pesticide industry (Galt, 2008).

Pesticide application on agricultural fields is a significant diffuse source of hazardous pollution to groundwater and surface waters through soil-water leaching and runoff. However, pesticides can also reach WWTPs; and, WWTPs can become a point source for pesticide pollution in surface waters. Gerecke *et al.* (2002) found that WWTPs are a significant source of pesticides in surface waters of Switzerland. They monitored atrazine, which is an agricultural pesticide, and mecoprop, which is used in both agriculture and urban areas (lawns), in WWTP effluents and detected high concentrations. They argue that the presence of pesticides in WWTPs is probably due to improper operations at farmyards and urban gardens and parks (e.g. washing of equipments, disposal of unused products, accidental releases). Combined sewer systems (combined discharge of domestic wastewater and stormwater) are also a source of pesticide input to WWTPs as pesticides applied on lawns and road embankments are flushed into the stormwater drainage during rain events. Urban use of pesticides also cover antifouling agents in paints and coatings (e.g. diuron, irgarol), and agents to protect construction materials (e.g. mecoprop), which are also prone to stormwater flushing.

2.3.3 Pharmaceuticals and personal care products

Pharmaceuticals and personal care products (PPCPs) are a diverse group of chemicals that include medical drugs and the active ingredients or preservatives in cosmetics and fragrances. They are used in large quantities throughout the world, some being in comparable amounts to the agricultural chemicals, and new PPCPs are constantly entering the marketplace (Daughton & Ternes, 1999). PPCPs are also called "down the drain" chemicals since their main route of environmental release is via sewerage systems through excretion and bathing (Daughton, 2016a). Intentional or accidental release to the sewerage system is also an important source (Petrie *et al.* 2016).

Collectively large contributions from human end-users to sewerage systems make WWTPs the most important sources of PPCPs to water bodies. Some of the PPCPs are persistent (e.g. fragrances). However, PPCPs with short environmental half-lives may behave as "pseudo-persistent" or "continuously present" chemicals in the hot spots near WWTP effluent discharges due to their continuous release (Daughton, 2003).

Pharmaceuticals are chemicals used in medicine to diagnose, treat and prevent diseases, to maintain or alter the health condition or structure/function of the human body (Daughton & Ternes, 1999). Veterinary drugs and illicit drugs should also be considered in this group. The diverse group of pharmaceutical chemicals can be classified into subgroups of antibiotics, hormones, anti-inflammatory drugs, antiepileptic drugs, blood lipid regulators, β-blockers, contrast media, and cytostatic drugs (Liu & Wong, 2013). Table 2.4 provides a list of pharmaceutical subgroups and representative chemical compounds in these subgroups.

Personal care products are chemicals used in cosmetics, toiletries or fragrances, as active ingredients or preservatives (Daughton & Ternes, 1999). Main subgroups of personal care products are given in Table 2.5. Although personal care products are produced and used in higher amounts and detected in higher concentrations than pharmaceuticals, their environmental occurrence and toxicity have been the focus of much less research (Daughton & Ternes, 1999; Brausch & Rand, 2011).

Steroid estrogens in the pharmaceuticals group have attracted considerable attention because of their endocrine disruption effects. An **endocrine disruptor** is defined by the World Health Organization (WHO) as "an exogenous substance or mixture that alters function(s) of the endocrine system and consequently causes adverse health effects in an intact organism, or its progeny, or (sub) populations" (WHO, 2013). Among the personal care products, triclosan and parabens, which are used as antimicrobial agents, and some widely used sunscreen agents have estrogenic effects (Brausch & Rand, 2011). Other than PPCPs, some *industrial chemicals* such as *nonylphenol* and *bisphenol-A* can act as estrogen mimics and add to the estrogenic chemicals present in the aquatic environment (Runnalls *et al.* 2010).

The steroid estrogens, estrone (E1), estradiol (E2), and estriol (E3), are naturally produced and excreted by humans and animals, whereas ethinylestradiol (EE2) is a synthetic estrogen used as an oral contraceptive (Liu & Wong, 2013). EE2 is found in trace concentrations in WWTP effluents, but due to its high potency, and in combination with the natural estrogens, it has been shown to cause adverse effects on fish populations (Daughton & Ternes, 1999; Runnalls *et al.* 2010). Runnalls *et al.* (2010) provided a review on the occurrence of steroidal and anti-steroidal drugs in the aquatic environment with an aim for prioritizing research needs. They state that EE2 is not the only steroidal drug in widespread use. Androgens, progestogens, and glucocorticoids are used in greater amounts than estrogens. Also, there is a significant use of anti-estrogens, anti-androgens, and anti-mineralocorticoids. These pharmaceuticals have the potential to be constantly entering the aquatic environment via WWTPs. Although these compounds are expected to be present in very low concentrations, they may have adverse effects on aquatic organisms.

Table 2.4 Pharmaceutical subgroups and representative chemicals within these subgroups.

Subgroup	Representative Chemicals	Comment
Hormones and hormone mimics	Steroid estrogens: Estrone (E1) Estradiol (E2) Estriol (E3) Ethinyl estradiol (EE2)	Endocrine disrupting effects.
Antibiotics	Clarithromycin Erythromycin Sulfamethoxazole Sulfadimethoxine Ciprofloxacin Norfloxacin Chloramphenicol Ofloxacin Amoxicillin Tetracycline Azithromycin	Promotion of antibiotic resistance in pathogens, genotoxicity. Pose a high risk to aquatic organisms (Verlicchi *et al.* 2012).
Blood lipid regulators	Clofibric/fenofibric acids Fenofibrate Gemfibrozil	High usage. Ubiquitous, persistent pollutants. High risk to aquatic organisms (Verlicchi *et al.* 2012).
Analgesics and anti-inflammatory drugs	Diclofenac Ibuprofen Acetaminophen Naproxen Acetylsalicylic acid Mefenamic acid	Most prevalent drugs other than clofibric acids. High mass loads in WWTP effluents and high risk to aquatic organisms (Verlicchi *et al.* 2012).
Antihypertensives	Diltiazem Enalapril Hydrochlorothiazide	Detected in WWTP effluents (Verlicchi *et al.* 2012).
β-blockers	Metoprolol Propanolol Atenolol	Highly prescribed drugs resistant to removal in WWTPs (Verlicchi *et al.* 2012; Filipe *et al.* 2017).
Antidepressants	Venlafaxine Citalopram Fluoxetine (Prozac) Fluvoxamine (Luvox)	Worldwide usage. Potential for multiple biological effects on aquatic organisms. (Fong & Ford, 2014)
Antiepileptics	Carbamazepine Primidone	Ubiquitous and prevalent due to poor removal at WWTPs.

(*Continued*)

Table 2.4 Pharmaceutical subgroups and representative chemicals within these subgroups (*Continued*).

Subgroup	Representative Chemicals	Comment
Antineoplastics (cytostatic drugs)	Ifosfamide Cyclophosphamide Carboplatin Cisplatin	Highly (geno)toxic compounds released primarily from hospitals where they are used for chemotherapy. Poor removal at WWTPs.
Impotence drugs	Sildenafil citrate	Widespread, non-prescribed usage.
Tranquilizers	Diazepam	High risk to aquatic organisms (Verlicchi *et al.* 2012)
Retinoids	Tretinoin	High usage rates.
Diagnostic contrast media	Diatrizoate Iopromide	Very high usage rates. Persistent, but no bioaccumulation potential, no acute toxicity.

Source: Daughton and Ternes (1999); Liu and Wong (2013).

Table 2.5 Personal care product subgroups and representative chemicals within these subgroups.

Subgroup	Representative Chemicals	Comment
Fragrances/ Synthetic musks	Galaxolide (HHCB) Toxalide (AHTN)	Used in large quantities. Persistent and bioaccumulative.
Preservatives	Parabens (alkyl-*p*-hydroxybenzoates)	Potential to cause estrogenic effects.
Disinfectants (antiseptics)	Triclosan Triclocarban	Triclosan has potential to promote antibiotic resistance in bacteria, is weakly estrogenic.
Sunscreen agents	2-ethyl-hexyl-4-trimethoxycinnamate Methyl benzilidene camphor Benzophenone-3	Bioaccumulative. Potential for estrogenic activity. More environmental and toxicological studies needed.
Insect repellants	*N,N*-diethyl-*m*-toluamide (DEET) 1,4-dichlorobenzene	Relatively persistent in aquatic environment. Low BCF.

Source: Daughton and Ternes (1999); Brausch and Rand (2011); Liu and Wong, (2013).

Since the contribution of human end-users through sewerage systems is the main source of occurrence of PPCPs in water environments, aquatic concentrations may be estimated using population data, usage information, and rates of metabolic excretion, per-capita wastewater production and WWTP removal for individual PPCPs. Meyer *et al.* (2016) present such a study that evaluates the impact of WWTP discharges on water quality of German rivers. They compare the measured and predicted aquatic concentrations of carbamazepine (antiepileptic), diclofenac (analgesic) and metopropol (β-blocker). Their study demonstrated the impact of WWTP discharges on increased pharmaceutical concentrations and that simple models can provide conservative estimates of aquatic concentrations of these "down-the-drain" pollutants. The review of Verlicchi *et al.* (2012) provides information on the removal of various pharmaceuticals in WWTPs using a large data set and also evaluates mass loadings to the environment with the aim of determining environmental risk. Their analysis shows that highest discharge amounts are observed for an antihypertensive drug (hydrochlorothiazide), several β-blockers and analgesics/anti-inflammatories. The highest risk was estimated to be posed by antibiotics (i.e. erythromycin, ofloxacin, sulfamethoxazole, clarithromycin, amoxicillin, tetracycline and azithromycin), several psychiatric drugs (i.e. fluoxetine and diazepam), analgesics/anti-inflammatories (i.e. ibuprofen and mefenamic acid), and lipid regulators (i.e. fenofibric acid, fenofibrate and gemfibrozil).

PPCPs in WWTP influents may be incorporated into the sludge produced at various units of the treatment plant. When this sludge is spread on land in order to enhance agricultural productivity of the soil, an environmental exposure route is created for the PPCPs present in the sludge. It is possible for the sludge-incorporated PPCPs to pollute surface water or groundwater as they are released from the sludge or from the soil to which they are applied. Verlicchi and Zambello (2015) provide a comprehensive review on the occurrence and environmental risk of PPCPs in sewage sludge. They found that the highest concentrations detected in secondary sludge were those of fragrances, antiseptics (disinfectants) and antibiotics. They determined that, in sludge-amended soil, the occurrence of hormones (i.e. estradiol), antibiotics (i.e. ciprofloxacin, ofloxacin, tetracycline), caffeine, and disinfectants (i.e. triclosan and triclocarban) posed critical environmental risk.

2.4 REGULATORY FRAMEWORKS

In this section, an overview of selected established regulatory frameworks that deal with hazardous pollutants is provided. The aim is to lay out the main approaches adopted by these frameworks in defining, classifying, prioritizing and regulating hazardous substances. The selected regulatory frameworks to be discussed are the Toxic Substances Control Act (TSCA) of the U.S., the Toxic Substances Management Policy (TSMP) of Canada, and the Regulation for Registration, Evaluation, Authorisation and Restriction of Chemicals (REACH) of the European

Union (EU). These legislations were selected based on their representativeness of the approach applied in most parts of the world.

2.4.1 Toxic Substances Control Act (TSCA) of the United States

The **TSCA**, which was put into force in 1976 in the U.S., is the oldest of the three regulatory frameworks discussed here. The authority responsible for its implementation is the U.S. Environmental Protection Agency (EPA). Since 1976, it has been subjected to several updates, the last of which was signed into law in 2016 (EPA, 2017i). The TSCA does not regulate substances that fall entirely into categories such as drugs, cosmetics, pesticides, food additives, nuclear materials, munitions and tobacco. In the U.S., pesticides are regulated under the Federal Insecticide, Fungicide, and Rodenticide Act (FIFRA), and drugs, hormones, and pharmacological chemicals are regulated under the Federal Food, Drug, and Cosmetic Act (FFDCA).

The TSCA requires pre-manufacture notification (PMN) for "new chemical substances" before their manufacture or import. A "new chemical substance" is defined as any substance that is not on the TSCA chemical substance inventory. Currently, the TSCA inventory lists about 85,000 chemicals (EPA, 2017a). The TSCA inventory has been assembled from the reports submitted by manufacturers, processors and importers of chemical substances. There are certain exemptions that can be granted so that the chemical is not subject to reporting and not added to the inventory. Any substance that is included in the inventory is considered to be an "existing" chemical substance for the purposes of regulation under the TSCA. When an existing chemical's production volume, its use, or its manufacturing and processing methods significantly changes, EPA may require a "significant new use" notice that is prepared in a similar way to the PMN reports.

The PMN reports for new chemical substances should include all the available data on the physicochemical properties of the chemical or the chemical mixture, and also the available information on its production volume, use, environmental release, human exposure and toxicity (EPA, 1997). The PBT properties of the new chemicals are assessed during the review process of the PMN reports (EPA, 1999a). The PBT criteria established by EPA have two tiers (Table 2.6). For the substances with environmental half-lives of more than 2 months, and fish BCF or BAF values of more than 1000, moderate control actions may be implemented with specific limits on exposures, releases and uses. However, substances with environmental half-lives of more than 6 months, and fish BCF and BAF values of more than 5000 are subjected to more stringent control actions that may result in a production ban. In each tier, the toxicity data is reviewed and further toxicity testing may be required if the chemical is confirmed to be persistent and bioaccumulative.

After the review of the PMN report, the substance is included in the TSCA inventory and becomes an "existing chemical". At the time of the enactment of the

TSCA in 1976, all the chemical substances that were already in commerce were put into the TSCA inventory as "existing" chemicals. This process of "grandfathering" existing chemicals at the time of implementation of a new legislation has also been applied in other countries (Muir & Howard, 2006). About 50,000 substances were included in the initial TSCA inventory and were exempted from the legislation by being grandfathered in, which was criticized by Lohmann *et al.* (2013).

Table 2.6 PBT criteria used in different regulatory frameworks.

		Persistence (Half-Lives in Days[a])	**Bioaccumulation**	**Toxicity**
EPA[b]	Moderate action	>60	BCF/BAF > 1000	Develop data where necessary
	High Action	>180	BCF/BAF > 5000	Develop data where necessary
CEPA[c]		air: ≥ 2 water: ≥ 152 sediments: ≥ 365 soil: ≥ 182	BAF ≥ 5000 BCF ≥ 5000 log K_{OW} ≥ 5	"CEPA toxic" or "CEPA toxic" equivalent
REACH[d]	PBT	mw: > 60 fw: > 40 mw sediment: > 180 fw sediment: > 120 soil: > 180	BCF/BAF > 2000	NOEC or EC10 < 0.01 mg/L CMR[e] or chronic toxicity
	vPvB	water: > 60 sediment: > 180 soil: > 180	BCF/BAF > 5000	

[a]mw: marine water; fw: fresh or estuarine water.
[b](EPA, 1999a).
[c](Canada, 2000).
[d](EU, 2006).
[e]CMR: carcinogenic, mutagenic, or toxic to reproduction.

Under the TSCA, another reporting requirement is the Chemical Data Reporting (CDR) Rule (formerly Inventory Update Rule) (EPA, 2017d). CDR requires manufacturers and importers to report production volume and use information if the annual production volume of a chemical at a single site is above 25,000 lbs. Chemicals that are subject to certain TSCA regulatory actions may be required to be reported under CDR if production volume exceeds 2500 lbs. CDR information is collected every 4 years.

EPA's Toxic Release Inventory (TRI) program is another mandatory reporting program related to toxic chemicals although it is not a part of the TSCA (EPA, 2017k). TRI requires industries that belong to a list of specific sectors, employ

10 or more full-time equivalent employees, and manufacture, process or use chemicals in the TRI list, to submit environmental release information for these chemicals (EPA, 2017c). It is aimed to create incentives for the industries to reduce their environmental release of hazardous pollutants by making the submitted information public. The chemicals are included in the TRI list considering their adverse environmental effects, and their acute and chronic human health effects, including carcinogenicity. PBT chemicals included in the TRI list generally have very low manufacture, process or use threshold amounts for reporting (EPA, 1999b). These PBT chemicals include lead and lead compounds, mercury and mercury compounds, dioxin and dioxin-like compounds, polycyclic aromatic hydrocarbons, hexabromocyclododecane (HBCD) and pesticides.

EPA uses industry submitted data through CDR and TRI for preparing a list of existing chemicals to be further assessed under the TSCA. This list is called the TSCA Work Plan for Chemical Assessments. The screening criteria for chemicals to be included in the TSCA Work Plan considers the chemical's use in children's products and children's health effects, neurotoxic effects, PBT properties, carcinogenicity, and the results of biomonitoring programs (EPA, 2012). The current TSCA Work Plan was updated in 2014 and contains 90 chemicals (EPA, 2014). The most up-to-date status of EPA's ongoing assessments of these chemicals can be found in EPA's TSCA website (EPA, 2017b). The latest update to the TSCA in 2016 identifies clear and enforceable deadlines for EPA's chemicals assessments and risk management actions, and requires EPA to implement a new risk-based safety standard (EPA, 2017h). Under the updated TSCA, EPA will establish a risk-based screening process and criteria to identify chemicals as either "high" or "low" priority for assessment. Recently, EPA has designated an initial list of 10 chemicals among the TSCA Work Plan Chemicals that it will evaluate for potential risk to human health and the environment (EPA, 2016). These chemicals are 1,4-dioxane, 1-bromopropane, asbestos, carbon tetrachloride, cyclic aliphatic bromide cluster (HBCD), methylene chloride, n-methylpyrrolidone (NMP), pigment violet 29, trichloroethylene (TCE), and tetrachloroethylene (PCE). EPA will designate additional chemicals for evaluation as the risk evaluations for these chemicals are completed; and, eventually all the remaining TSCA Work Plan chemicals will have been reviewed for their potential toxicity and exposure hazards.

If a risk is identified by the TSCA chemical assessments, EPA is required to take action to reduce the risk associated with the particular use of that chemical (EPA, 2017e). The potential risk reduction actions are stricter reporting and use requirements, or limitations on the manufacture, processing and use of the chemical. In some cases, EPA may be involved in promoting voluntary actions by the industry.

Recently, EPA has implemented a new database called ChemView, which is available online through https://java.epa.gov/chemview. ChemView contains information on chemical health and safety data EPA collects through CDR and TRI for chemicals regulated under the TSCA. It also makes available to the public EPA's chemical assessments and regulatory actions. Confidential business information is not included in ChemView.

2.4.2 Toxic Substances Management Policy (TSMP) of Canada

The TSMP of Canada was published in 1995. The key management objectives of the policy are: (i) "virtual elimination from the environment of toxic substances that result predominantly from human activity and that are persistent and bioaccumulative (Track 1 substances)"; and (ii) "management of other toxic substances and substances of concern, throughout their entire life cycles, to prevent or minimize their release into the environment (Track 2 substances)" (Environment Canada [EC] 1995).

The criteria to identify Track 1 substances are given in Table 2.6. Different half-life thresholds are set for different environmental media. A substance is considered persistent if it meets the criterion in any one medium. Also, if it is shown that a substance is subjected to long-range transport in air to remote regions such as the Arctic, it is considered persistent in air. The bioaccumulation potential of a substance is determined using BAF, BCF and log K_{OW} values. BAF is preferred over BCF; and log K_{OW} is used in the absence of BAF or BCF data. Official definitions of these persistence and bioaccumulation measures and also the methods that should be used to determine them are included in the annexes of the Canadian Environmental Protection Act (CEPA) (Canada, 2000). The toxicity of a substance is required to be determined through a systematic and risk-based assessment of the exposure properties and environmental and human health effects of the substance. If, at the end of the assessment, the substance is shown to meet or shown to be equivalent to the toxicity definition given in the CEPA, it is considered toxic. The toxicity definition in the CEPA is stated as "a substance is toxic if it is entering or may enter the environment in a quantity or concentration or under conditions that (a) have or may have an immediate or long-term harmful effect on the environment or its biological diversity; (b) constitute or may constitute a danger to the environment on which life depends; or (c) constitute or may constitute a danger in Canada to harm life or health" (EC, 1995). Thus, the toxicity definition of the CEPA includes both the toxicity hazard and the exposure hazard properties of the pollutant. Track 1 substances also need to satisfy the criterion of being predominantly anthropogenic. This is determined based on expert judgment and serves the purpose of excluding elements and naturally occurring inorganic substances such as minerals from the virtual elimination list.

According to the TSMP, virtual elimination from the environment of Track 1 substances means eliminating their measurable release to the environment (EC, 1995). The use of a Track 1 substance is allowed if its release is eliminated by implementing pollution prevention measures. This measure is less protective than the total elimination of the use of the specific substance, and has been criticized

(Canada Environmental Law Association [CELA], 2000). The TSMP states that the phase-out of the substance should be targeted if pollution prevention measures fail. For Track 1 substances that are already present in the environment, remediation is considered after taking into account the clean-up costs and the projected benefits to the ecosystem and human health.

If a pollutant does not satisfy all of the four criteria used to determine Track 1 substances, it is specified as a Track 2 substance. The TSMP emphasizes risk assessment and risk management throughout the full life-cycle of Track 2 substances to prevent or minimize their release to the environment (EC, 1995). Federal government is responsible for carrying out the risk assessment and management activities, setting long-term environmental goals, and determining time lines and the means for achieving these goals.

The Canadian Environmental Protection Act (CEPA) is the main regulation for managing hazardous pollutants in Canada. It was put into force in 1988. In 1999, it was updated based on the strategies and approaches identified in the TSMP of 1995. The most current version of the act and the list of regulations made under the act can be found in Government of Canada's Justice Laws Website (Canada, 2017).

Several lists of chemicals have been established through the CEPA for regulatory purposes (EC, 2017a). Among these lists, the Domestic Substances List (DSL) includes all chemicals that were commercially used, manufactured, or imported within the period 1984–1986. A total of 23,000 substances were initially added to the DSL list without having been assessed. Thus they were "grandfathered in" as was done when the U.S. established the TSCA inventory. Similar to the TSCA inventory, the DSL is used in Canada to define "existing" substances. Currently, 1954 substances have been added to the original DSL after being assessed as new substances, and they became "existing" substances. The substances that are in commercial use internationally but that are not included in the DSL are included in the Non-Domestic Substances List. The Non-Domestic Substances List chemicals are subject to notification before use but the information requirements are lower compared to totally new chemicals. The Non-Domestic Substances List is based on the TSCA inventory of the U.S. and it contains more than 58,000 substances.

The Priority Substances List (PSL) is used as a tool in determining the chemicals that need to be assessed as a priority and in detail. Chemicals on the DSL can be put onto the PSL after a screening and categorization process that uses the persistence, bioaccumulation potential and toxicity criteria in the regulation (Table 2.6) (EC, 2017a). The first PSL was published in 1989 and included 44 substances. The assessments of these first set of priority substances resulted in the declaration of 25 of the listed substances as toxic. The second PSL list published in 1995 added 25 substances to the list. Currently, the assessments of 23 substances in the second PSL have been completed and 18 substances have been determined to be toxic (EC, 2017b). When a substance is determined to be toxic after detailed scientific assessments, it is recommended to be included into the Toxic Substances List. The Government of Canada is responsible for the regulation of the substances in the

Toxic Substances List and for implementing pollution prevention and remediation activities. Currently there are 137 pollutants or pollutant groups listed in the Toxic Substances List (EC, 2017c). The substances in the Toxic Substances List can be further included in the Virtual Elimination List if they are determined to be persistent, bioaccumulative and anthropogenic under the criteria set in the TSMP (Table 2.6). Currently, there are only two pollutant substances included in the Virtual Elimination List (EC, 2017d). The first substance to be added into the list was hexachlorobutadiene (HCBD). The level of quantification for HCBD was set as 0.06 ng/mL in a chlorinated solvent. Thus, HCBD releases into the environment have to be below this level. The second substance added to the list for virtual elimination from the environment was perfluorooctane sulfonate (PFOS) and its salts. No level of quantification was determined for PFOS.

2.4.3 Regulation for Registration, Evaluation, Authorisation and Restriction of Chemicals (REACH) of the European Union

REACH entered into force in 2007 as the common legislation for managing chemicals among the member states of the EU (EU, 2006). The European Chemicals Agency (ECHA) was created as the responsible agency to implement the requirements of REACH. The system of chemical management through REACH has four elements: registration, evaluation, authorisation, and restriction.

Manufacturers and importers of chemicals are required to submit a registration dossier for any chemical produced or imported in amounts of more than 1 tonne per year per producer or importer whether they are hazardous or not. There are certain exemptions to the registration requirement, such as for substances that pose minimum risk because of their intrinsic properties (e.g. stearic acid), that are naturally occurring (e.g. water), and for polymers and for substances that are already regulated under other legislations (i.e. food or feed additives, medicinal products). Annexes IV and V of REACH provide details of the exemptions from the registration obligation. ECHA's guidance document on registration provides further details on the registration procedure (ECHA, 2016).

Certain substances are regarded as registered; and no new registration under REACH is required for these substances for their registered uses. These substances include biocidal products and plant protection products whose active substances have been approved according to the requirements of the relevant EU regulations. Also, substances that have been notified according to the Dangerous Substances Directive (Directive 67/547/EEC) and included in the European List of Notified Substances (ELINCS) are regarded as registered under REACH since the notification process requires the submission of a similar set of information required by REACH's registration procedure. The Dangerous Substances Directive did not require the submission of a notification for the chemicals that were on the European Community market on 1981, which were listed in the European Inventory of

Existing Commercial Chemical Substances (EINECS). Thus, like the TSCA of the U.S. and CEPA of Canada, European legislations on hazardous pollutants also had adopted the approach of exempting existing chemicals from assessment. However, the Dangerous Substances Directive had been replaced by the CLP (Classification, Labelling and Packaging) Regulation of the EU in 2008, which had been prepared according to the requirements of REACH. When REACH came into force in 2007, chemicals listed in EINECS were regarded as phase-in substances, which were required to be registered within a specified transition period. Thus, by requiring the registration of existing substances listed in EINECS, REACH has adopted a more stringent and comprehensive approach to chemicals management compared to similar legislations applied in North America. Phase-in substances were grouped according to their production volumes; and, higher production volume substances had earlier registration deadlines within this transition period. Phase-in substances manufactured or imported in quantities of more than 1000 tonnes per year were required to be registered by 2010. The same deadline also applied to substances that are carcinogenic, mutagenic, or toxic to reproduction with production or import quantities of more than 1 tonne, and to substances that are very toxic to aquatic organisms with production or import quantities of more than 100 tonnes. Thus, substances that pose higher risk to human health and the environment were required to be registered earlier. Phase-in substances that are manufactured or imported in quantities between 100 and 1000 tonnes per year were required to be registered by 2013. For the remaining phase-in substances with a production volume of more than 1 tonne, registration has to be completed by 2018.

The registration dossiers are required to include information on the physicochemical properties, toxicology and ecotoxicology of the chemicals. The minimum information requirements are given in Annexes VII to X of REACH. The information requirements increase in a stepwise fashion as the manufactured or imported quantity of the chemical corresponds to a higher tonnage band.

For chemicals that are being manufactured or imported in quantities of more than 10 tonnes per year, the submission of a Chemical Safety Report is required. The Chemical Safety Report should document a chemical safety assessment performed by the registrant. The aim is to classify and label the substance according to its human health, physicochemical and environmental hazards. The human health hazard assessment should include the determination of the derived no-effect level (DNEL), which defines the upper level of human exposure that would avoid adverse health effects. The physicochemical hazard assessment should include the potential for explosivity, flammability and oxidizing potential of the substance. The environmental hazard assessment should include the determination of the predicted no-effect concentration (PNEC), which is the environmental concentration below which adverse environmental effects are not expected to occur.

A critical part of the chemical safety assessment is the PBT/vPvB assessment. Annex XIII of REACH defines the criteria for identifying a substance as a PBT substance or as a vPvB (very persistent, very bioaccumulative) substance (Table

2.6). If at the end of the chemical safety assessment, the substance is evaluated to belong to any of the hazard classes as set out in Article 14(4) (based on the hazard categorization in Annex I of CLP regulation), or it is evaluated to be a PBT or vPvB substance, an exposure assessment is required. The exposure assessment should include all the life-cycle stages of the chemical and should determine the human and environmental exposure levels. The exposure assessment is followed by a risk characterization step where the outcome of the exposure assessment is compared with the results of the hazard assessments. This includes the comparison of DNELs and PNELs with the calculated exposure concentrations estimated in the exposure assessment. Moermond *et al.* (2012) provided a detailed overview of the PBT/vPvB assessment procedure and the risk management follow-up applied in REACH and compared it with other regulatory frameworks. This is important since a group of chemical substances including medicinal products, biocides, plant protection products, and food and feed additives are not regulated by REACH but other EU legislations are in force to regulate their marketing and use. Moermond *et al.* (2012) also criticize the exemption of chemical substances that are produced within or imported into the EU in amounts below 10 tonnes from the PBT assessment.

The evaluation of the submitted registration dossiers is conducted by ECHA and the EU member states. The evaluation phase includes the compliance check of the submitted dossier, the examination of testing proposals if submitted by the registrant in case of insufficient substance data, and the substance evaluation (ECHA, 2017c).

The evaluation phase may result in the identification of the registered chemical to be a substance of very high concern (SVHC). A chemical substance is identified as a SVHC if it is determined to be carcinogenic, mutagenic or toxic to reproduction, or if it is determined to have PBT or vPvB properties, or if it raises an equivalent concern to either of these cases (Zarfl & Matthies, 2013). SVHCs are included in the "candidate list for substances of very high concern (SVHCs) for authorisation" and become a part of the authorisation phase. The candidate list is published on the ECHA website (ECHA, 2017b). The substances in the candidate list may be recommended by ECHA to be included in the Authorisation List (Annex XIV of REACH). The authority to decide on the inclusion of a substance in the Authorisation List is the European Commission. Currently, there are 31 unique substances in the Authorisation List (ECHA, 2017a). Substances in the Authorisation List are subject to authorisation before they can be produced, imported or used. Authorisation is given if it can be demonstrated by the industry that the risks from the use of the substance are adequately controlled or that the socio-economic benefits outweigh the risks (Moermond *et al.* 2012). Hansen *et al.* (2007) criticized the authorization process in REACH for being risk-based rather than hazard-based although the SVHC identification is based on a hazard assessment. They argued that the evaluation of risk control measures for their adequacy and the evaluation of the balance between socio-economic benefits and risks are not trivial and not clearly specified in the regulation, raising concerns about the incorporation of the precautionary principle by REACH.

As a part of REACH, restrictions may also be applied to those substances for which it is demonstrated that the risk management options are not sufficient for the safe production and use of the substance. These restrictions may be in the form of enforced limits or a ban on the production or use. Currently, the list of substances restricted under REACH (Annex XVII) include 64 unique substances (ECHA, 2017d).

The commentary article by Tarazona (2013) uses REACH as an example to lay out the challenges and opportunities in regulatory risk assessments. It is argued that the large regulatory databases formed as a part of the legislations like REACH provide opportunities in new scientific developments in chemical management and control. The registration dossiers of REACH contain information on chemical structure, physical-chemical, environmental fate, and toxicological and ecotoxicological properties of a very large number of chemicals in a standardized format. This information can be used in the development of QSARs and related tools. Furthermore, when the georeferenced information regarding the use, marketing and production of chemicals, which is also available through regulatory databases is considered together with these property data, completely new approaches, paradigms and models can be developed that provide a holistic assessment of the risk posed by chemicals on individuals and ecosystems.

2.5 SELECTED GROUPS OF HAZARDOUS POLLUTANTS

As of March 2017, there are more than 129 million unique organic and inorganic chemical substances included in the CAS Registry of the American Chemical Society's Chemical Abstract Service (CAS, 2017a). About 15,000 substances are being added to the registry every day. On the other hand, only a very small portion of these substances is regulated by international, national or state agencies, as reflected by the regulated chemicals list (CHEMLIST) of Chemical Abstract Service, which currently contains about 347,000 substances (CAS, 2017b). CHEMLIST is also being expanded (50 substances added per week), but at a much slower rate compared to the CAS Registry. When the regulatory chemical management practices (Section 2.4) and the vastness of the chemical universe is considered, it is apparent that the list of regulated chemical substances does not cover all the hazardous pollutants present in the environment. Moreover, the vastness of the chemical universe is dwarfed when the universe of potential organic chemicals (chemicals that can be synthetically produced and chemicals that are already present but not yet been identified) is considered (Daughton, 2004). While scientific researchers and regulatory bodies are constantly working on assessing the hazards related to the known chemicals and also on anticipating and identifying new hazardous pollutants, new terminologies and expressions are emerging. Special groupings of hazardous pollutants are devised to facilitate communication and to manage resources more efficiently by focusing on more critical pollutants. In this section, we will try to identify some of these terms used to specify groupings of hazardous pollutants.

The assessment of PBT properties of chemicals has been used by the scientific community in identifying the hazardous pollutants. The PBT assessment provides a sound scientific basis for determining the toxicity and exposure hazards of pollutants. Many regulatory frameworks have also adopted this approach and, mostly by using threshold criteria, categorized chemicals according to their PBT properties so that they can put more effort into managing risks associated with the production and use of more critical pollutants. The production volume is also considered in this screening process. "**Priority pollutants**" is the term used to identify these more critical pollutants and many regulatory agencies have produced official lists of priority pollutants. The TSCA Work Plan for Chemicals Assessment of EPA, Priority Substances List (PSL) of Environment Canada, and Substances of Vey High Concern (SVHCs) of EU's REACH are examples of priority pollutants lists (see Section 2.4 for details).

The Clean Water Act (CWA) of EPA maintains two lists of pollutants that are used in water quality regulatory programs (EPA, 2017j). The Toxic Pollutants List contains aquatic pollutants that are selected based on their toxicity properties as defined in the CWA. The Priority Pollutants List is derived from the Toxic Pollutants List after considering the pollutant's frequency of occurrence in water environments, its production volume and the availability of a chemical standard for testing. Currently there are 126 entries in the Priority Pollutants List of EPA.

The Water Framework Directive (WFD) of the EU requires the creation of a priority substances list (Annex X of WFD) in a similar way to the Priority Pollutants List of the CWA (EU, 2000). Environmental quality standards are determined for these priority substances and used in determining the chemical status of water bodies. The latest amendment proposal to the directive in 2012 stressed improving the efficiency of monitoring of PBT substances (European Commission [E.C.], 2012). The latest list contains 48 substances or substance groups (E.C., 2017).

Priority pollutants has been the primary focus of the studies on hazardous pollution since the 1970s although these conventional, regulated pollutants represent only a tiny piece of the vast universe of potential pollutants (Daughton, 2004). The term "**emerging pollutants**" is being widely used to refer to the group of pollutants that have attracted less attention as hazardous pollutants, that are unregulated, but that are being recognized as potential concern. PPCPs are among the most recognized emerging pollutants and they possess all the characteristics mentioned in the previous sentence. Probably for this reason, sometimes the topic of emerging pollutants has been narrowed down to include only PPCPs (Daughton, 2016b). However, there is some vagueness in the term "emerging". The "emergence" of a new pollutant substance may refer to the fact that the substance is a newly synthesized chemical, which was previously non-existent in the environment. Or, it may refer to the fact that the existence of the pollutant in environmental matrices has just recently become known due to recent advances in analytical techniques. PPCPs contain many totally new substances as well as substances that have been present but not considered as pollutants (Daughton, 2004). The Network of Reference Laboratories, Research Centres and Related Organisations for Emerging Environmental Substances (NORMAN) publishes a list of emerging substances that are potential emerging pollutants on their website (NORMAN, 2017). The list contains surfactants, flame retardants, PPCPs, gasoline additives and their degradation products, biocides, polar pesticides and their degradation products, and proven or suspected endocrine disrupting compounds (EDCs).

EPA maintains a list titled Drinking Water Contaminant Candidate List (CCL), which includes chemical and microbial contaminants that are found in drinking water but that are not currently regulated in drinking water regulations (EPA, 2017f). The contaminants listed in the CCL may be regulated in the future under the Safe Drinking Water Act (SDWA) of the U.S. The latest CCL (CCL4) was announced by EPA in 2016 and includes 97 chemicals or chemical groups and 12 microbial contaminants. The chemical pollutants in the list include pesticides, PPCPs, and industrial chemicals.

The term "**micropollutants**" is used with a similar meaning to the emerging pollutants, but it puts more emphasis on the fact that these pollutants are detected at extremely low levels in the environment, which is made possible by the advances in analytical chemistry (Daughton, 2016b). "**Macropollutants**" is a related term that is used to refer to the mostly inorganic pollutants occurring at much higher (in the order of milligrams per litre) levels (Schwarzenbach *et al.* 2010).

The sources of micropollutants that are found in the water environment are diverse, including PPCPs, pesticides, and industrial chemicals and products introduced mostly through the municipal and industrial wastewaters (Schwarzenbach *et al.* 2006). The term "**anthropogenic trace compounds**" (ATC) is also used to refer to micropollutants as they are mostly originated from human activities (Gerbersdorf *et al.* 2015).

"**Xenobiotics**" is another term that is used to define substances that are released to the environment by human activities. It refers to the reality that these substances are exogenous or foreign to the natural environmental and biological systems. **Xenobiotic organic compounds (XOCs)** include organic ATCs such as industrial (e.g. PCBs) and agricultural chemicals (e.g. pesticides), and also organic chemicals released as unintentional by-products of human activities (e.g. PAHs, PCDDs) (van der Oost *et al.* 2003). Additionally, PPCPs, bisphenols, phthalates, and alkylphenolic compounds are XOCs that are of particular concern since they are present in municipal wastewater (Fatta-Kassinos *et al.* 2011). Conventional wastewater treatment systems are not very effective in removing many of the XOCs and they may be present in treated wastewaters and in wastewater sludge.

Anthropogenic trace compounds or xenobiotics represent a very large number of chemicals that are in production and use throughout the world. It is not possible to monitor and study all the anthropogenic substances emitted to the environment. So, a prioritization process is required to manage the risks posed by hazardous pollutants. Many regulatory bodies and scientific organizations have developed such lists of priority chemicals and emerging chemicals using science-based approaches as mentioned above. However, it is not possible to list all the potential "hazardous pollutants". Sobek *et al.* (2016) discusses the dilemma in prioritizing chemicals for environmental risk assessment. They reviewed the available data obtained through environmental monitoring studies conducted in the Baltic Sea and showed that the majority of the studies were on a small number of already regulated chemicals, which are known to be hazardous. They state that there may be a bias for the prioritization of environmental studies based on the "knowns of the past" and they call for revised strategies in prioritizing environmental chemical studies.

The planetary boundaries framework proposes to define a safe operating space for humanity by identifying and quantifying boundaries of human interference with vital earth-system processes (Rockström *et al.* 2009a; Rockström *et al.* 2009b; Steffen *et al.* 2015). One of the boundaries is defined by novel entities introduced to the environment by anthropogenic activities. Hazardous pollutants are among these novel entities and they may become a planetary boundary threat when they exhibit persistence, widespread distribution, and "potential impacts on vital Earth-system processes or subsystems" (Steffen *et al.* 2015). However, a single planetary boundary for chemical pollution has not been quantified because of the complexity of the problem. Since the proposal of the planetary boundaries approach, there have been several studies that focus on approaches to determine the chemical pollution boundary. Notably, Persson *et al.* (2013) laid out three conditions that must be met for a chemical to pose a "**planetary boundary threat**": (i) the chemical has a disruptive effect on a vital Earth-system process; (ii) the disruptive effect is not discovered before it is a planetary scale problem; and (iii) the effect cannot be readily reversed. In a further study, MacLeod *et al.* (2014) developed scenarios that fulfill these conditions and then defined chemical profiles that fit each scenario as summarized in Table 2.7. The profiles are defined based on either the effect of the chemical on a vital Earth-system process or the exposure of the environment to the chemical. They proposed to use these profiles to determine chemicals that are potential planetary boundary threats. There are currently considerable uncertainties for defining criteria to screen chemicals against these chemical profiles. Prioritizing hazardous pollutants against currently unknown disruptive effects is a great challenge and it is more difficult to develop criteria for effect-based profiles such as 1.1, 2.2, 2.3, 2.4, and 3.2 (Table 2.7). However, it may be feasible to prioritize chemicals based on their exposure properties (profiles 2.1 and 3.1) with the current state of knowledge and using insights from chemical fate and transport modeling. Further research on the interaction between hazardous pollutants and critical biogeochemical processes can identify currently unknown adverse effects of pollutants, leading to a better hazardous pollution management that contributes to the efforts for maintaining the Earth within the safe operating space for humanity.

Table 2.7 Conditions and profiles for chemicals to be a planetary boundary threat.

Conditions (Persson *et al.* 2013)	Chemical Profiles (MacLeod *et al.* 2014)
1. The chemical or mixture of chemicals has a disruptive effect on a vital Earth System Process	1.1. Unknown disruptive effect on a vital Earth-system process
2. The disruptive effect is not discovered until it is, or inevitably becomes, a problem at a planetary scale	2.1. Nearly homogeneous environmental exposure at the global scale 2.2. Rapid global distribution of effects 2.3. Time delay between exposure and effects 2.4. Effects are only manifested on a global scale
3. The effects of the pollutant in the environment cannot be readily reversed	3.1. Environmental exposure is poorly reversible 3.2. Effects are poorly reversible

2.6 REFERENCES

Arnot J. A., Brown T. N., Wania F., Breivik K. and McLachlan M. S. (2012). Prioritizing chemicals and data requirements for screening-level exposure and risk assessment. *Environmental Health Perspectives*, **120**(11), 1565–1570.

Beyer A., Mackay D., Matthies M., Wania F. and Webster E. (2000). Assessing long-range transport potential of persistent organic pollutants. *Environmental Science & Technology*, **34**(4), 699–703.

Boethling R. (2016). Present, bioaccumulative, and toxic: sufficient for PBT? *Integrated Environmental Assessment and Management*, **12**(1), 204–205.

Boethling R. and Mackay D. (2000). Handbook of Property Estimation Methods for Chemicals: Environmental and Health Sciences. CRC Press, Boca Raton, FL, USA.

Bonmatin J. M., Marchand P. A., Charvet R., Moineau I., Bengsch E. R. and Colin M. E. (2005). Quantification of imidacloprid uptake in maize crops. *Journal of Agricultural and Food Chemistry*, **53**(13), 5336–5341.

Bonmatin J. M., Giorio C., Girolami V., Goulson D., Kreutzweiser D. P., Krupke C., Liess M., Long E., Marzaro M., Mitchell E. A. D., Noome D. A., Simon-Delso N. and Tapparo A. (2015). Environmental fate and exposure; neonicotinoids and fipronil. *Environmental Science and Pollution Research*, **22**(1), 35–67.

Brausch J. M. and Rand G. M. (2011). A review of personal care products in the aquatic environment: environmental concentrations and toxicity. *Chemosphere*, **82**(11), 1518–1532.

Canada (2000). Canadian Environmental Protection Act, 1999: Persistence and Bioaccumulation Regulations. Registration SOR/2000-107 March 23, 2000.

Canada (2017). Justice Laws Website: Canadian Environmental Protection Act, 1999 (S.C. 1999, c. 33). http://laws-lois.justice.gc.ca/eng/acts/c-15.31/ (Last accessed: 16.03.2017)

CAS (2017a). CAS REGISTRY – The gold standard for chemical substance information. http://www.cas.org/content/chemical-substances (Last accessed: 27.03.2017)

CAS (2017b). Regulated Chemicals – CHEMLIST – Find whether a substance is regulated and by what agency. http://www.cas.org/content/regulated-chemicals (Last accessed: 27.03.2017)

CELA (2000). Toxic Substances. Environmental Standard Setting and Children's Health. Canada Environmental Law Association & Ontario College of Family Physicians Environmental Health Committee. Chapter 6, Toronto, Ontario.

Chapman P. M., Fairbrother A. and Brown D. (1998). A critical evaluation of safety (uncertainty) factors for ecological risk assessment. *Environmental Toxicology and Chemistry*, **17**(1), 99–108.

Daughton C. G. (2002). Environmental stewardship and drugs as pollutants. *Lancet*, **360**(9339), 1035–1036.

Daughton C. G. (2003). Cradle-to-cradle stewardship of drugs for minimizing their environmental disposition while promoting human health. I. Rational for and avenues toward a green pharmacy. *Environmental Health Perspectives*, **111**(5), 757–774.

Daughton C. G. (2004). Non-regulated water contaminants: emerging research. *Environmental Impact Assessment Review*, **24**(7–8), 711–732.

Daughton C. G. (2016a). In response: government perspective. *Environmental Toxicology and Chemistry*, **35**(2), 266–268.

Daughton C. G. (2016b). Pharmaceuticals and the Environment (PiE): evolution and impact of the published literature revealed by bibliometric analysis. *Science of the Total Environment*, **562**, 391–426.

Daughton C. G. and Ternes, T. A. (1999). Pharmaceuticals and personal care products in the environment: agents of subtle change? *Environmental Health Perspectives*, **107**, 907–938.

Devillers J. and Mombelli, E. (2010a). Evaluation of the OECD QSAR Application Toolbox and Toxtree for estimating the mutagenicity of chemicals. Part 1. Aromatic amines. *Sar and Qsar in Environmental Research*, **21**(7–8), 753–769.

Devillers J. and Mombelli E. (2010b). Evaluation of the OECD QSAR application toolbox and toxtree for estimating the mutagenicity of chemicals. Part 2. – unsaturated aliphatic aldehydes. *Sar and Qsar in Environmental Research*, **21**(7–8), 771–783.

Dimitrov S. D., Diderich R., Sobanski T., Pavlov T. S., Chankov G. V., Chapkanov A. S., Karakolev Y. H., Temelkov S. G., Vasilev R. A., Gerova K. D., Kuseva C. D., Todorova N. D., Mehmed A. M., Rasenberg M. and Mekenyan O. G. (2016). QSAR Toolbox – workflow and major functionalities. *Sar and Qsar in Environmental Research*, **27**(3), 203–219.

EC (1995). Toxic Substances Management Policy. *En40-499/1-1995*, Environment Canada.

E.C. (2012). Proposal for a Directive of the European Parliament and of the Council amending Directives 2000/60/EC and 2008/105/EC as regards priority substances in the field of water policy. European Commission.

E.C. (2017). Priority substances under the Water Framework Directive. http://ec.europa.eu/environment/water/water-dangersub/pri_substances.htm (Last accessed: 27.03.2017)

EC (2017a). A Guide to Understanding the Canadian Environmental Protection Act, 1999. https://www.ec.gc.ca/lcpe-cepa/default.asp?lang=En&n=E00B5BD8-1 (Last accessed: 16.03.2017)

EC (2017b). Priority Substances List. https://www.ec.gc.ca/lcpe-cepa/default.asp?lang=En&n=C6C230D5-1 (Last accessed: 17.03.2017)

EC (2017c). Toxic Substances List. https://www.ec.gc.ca/lcpe-cepa/default.asp?lang=En&n=0DA2924D-1 (Last accessed: 17.03.2017)

EC (2017d). Virtual Elimination List. https://www.ec.gc.ca/lcpe-cepa/default.asp?lang=En&n=2F918AD5-1 (Last accessed: 17.03.2017)

ECHA (2016). Guidance on registration, Version 3.0, November 2016. European Chemicals Agency.

ECHA (2017a). Authorisation List. https://echa.europa.eu/addressing-chemicals-of-concern/authorisation/recommendation-for-inclusion-in-the-authorisation-list/authorisation-list (Last accessed: 26.03.2017)

ECHA (2017b). Candidate List of substances of very high concern for Authorisation. https://echa.europa.eu/candidate-list-table (Last accessed: 26.03.2017)

ECHA (2017c). Evaluation. https://echa.europa.eu/regulations/reach/evaluation (Last accessed: 26.03.2017)

ECHA (2017d). Substances restricted under REACH. https://echa.europa.eu/addressing-chemicals-of-concern/restrictions/substances-restricted-under-reach (Last accessed: 26.03.2017)

EPA (1991). Technical Support Document for Water Quality-based Toxics Control. EPA/505/2-90-001. Washington, DC.

EPA (1997). Chemistry Assistance Manual for Premanufacture Notification Submitters. *EPA 744-R-97-003*, United States Environmental Protection Agency, Office of Pollution Prevention and Toxics.

EPA (1999a). Category for Persistent, Bioaccumulative, and Toxic New Chemical Substances. E. P. Agency, ed., Federal Register, 60194-60204.

EPA (1999b). Persistent Bioaccumulative Toxic (PBT) Chemicals; Lowering of Reporting Thresholds for Certain PBT Chemicals; Addition of Certain PBT Chemicals; Community Right-to-Know Toxic Chemical Reporting. Federal Register, 58666–58753.

EPA (2012). TSCA Work Plan Chemicals: Methods Document. U.S. Environmental Protection Agency, Office of Pollution Prevention and Toxics.

EPA (2014). TSCA Work Plan for Chemical Assessments: 2014 Update. U.S. Environmental Protection Agency, Office of Pollution Prevention and Toxics.

EPA (2016). Designation of Ten Chemical Substances for Initial Risk Evaluations Under the Toxic Substances Control Act. E. P. Agency, ed., Federal Register, 91927–91929.

EPA (2017a). About the TSCA Chemical Substance Inventory. https://www.epa.gov/tsca-inventory/about-tsca-chemical-substance-inventory (Last accessed: 14.03.2017)

EPA (2017b). Assessing and Managing Chemicals under TSCA. https://www.epa.gov/assessing-and-managing-chemicals-under-tsca (Last accessed: 14.03.2017)

EPA (2017c). Basics of TRI Reporting. https://www.epa.gov/toxics-release-inventory-tri-program/basics-tri-reporting (Last accessed: 14.03.2017)

EPA (2017d). Chemical Data Reporting under the Toxic Substances Control Act. https://www.epa.gov/chemical-data-reporting (Last accessed: 14.03.2017)

EPA (2017e). Current Chemical Risk Reduction Activities. https://www.epa.gov/assessing-and-managing-chemicals-under-tsca/current-chemical-risk-reduction-activities (Last accessed: 14.03.2017)

EPA (2017f). Drinking Water Contaminant Candidate List (CCL) and Regulatory Determination. https://www.epa.gov/ccl (Last accessed: 27.03.2017)

EPA (2017g). Estimation Programs Interface Suite for Microsoft Windows, v 4.11. United States Environmental Protection Agency, Washington, DC, USA.

EPA (2017h). Highlights of Key Provisions in the Frank R. Lautenberg Chemical Safety for the 21st Century Act. https://www.epa.gov/assessing-and-managing-chemicals-under-tsca/highlights-key-provisions-frank-r-lautenberg-chemical (Last accessed: 14.03.2017)

EPA (2017i). Summary of Toxic Substances Control Act. https://www.epa.gov/laws-regulations/summary-toxic-substances-control-act (Last accessed: 09.03.2017)

EPA (2017j). Toxic and Priority Pollutants Under the Clean Water Act. https://www.epa.gov/eg/toxic-and-priority-pollutants-under-clean-water-act (Last accessed: 27.03.2017)

EPA (2017k). Toxics Release Inventory (TRI) Program. https://www.epa.gov/toxics-release-inventory-tri-program (Last accessed: 14.03.2017)

EU (2000). Directive 2000/60/EC of the European Parliament and of the Council of 23 October 2000 establishing a framework for Community action in the field of water policy. *The European Parliament and the Council of the European Union, Official Journal of the European Communities*, OJ L327, 22.12.2000.

EU (2006). Regulation (EC) No 1907/2006 of the European Parliament and of the Council of 18 December 2006 concerning the Registration, Evaluation, Authorisation and Restriction of Chemicals (REACH). The European Parliament and the Council of the European Union.

Fatta-Kassinos D., Kalavrouziotis I. K., Koukoulakis P. H. and Vasquez M. I. (2011). The risks associated with wastewater reuse and xenobiotics in the agroecological environment. *Science of The Total Environment*, **409**, 3555–3563.

Filipe O. M. S., Mota N., Santos S. A. O., Domingues M. R. M., Silvestre A. J. D., Neves M. G. P. M. S., Simões M. M. Q. and Santos E. B. H. (2017). Identification and characterization of photodegradation products of metoprolol in the presence of natural fulvic acid by HPLC-UV-MSn. *Journal of Hazardous Materials,* **323**(Part A), 250–263.

Fong, P. P. and Ford A. T. (2014). The biological effects of antidepressants on the molluscs and crustaceans: a review. *Aquatic Toxicology*, **151**, 4–13.

Galt, R. E. (2008). Beyond the circle of poison: significant shifts in the global pesticide complex, 1976–2008. *Global Environmental Change-Human and Policy Dimensions*, **18**(4), 786–799.

Gerbersdorf S. U., Cimatoribus C., Class H., Engesser K. H., Helbich S., Hollert H., Lange C., Kranert M., Metzger J., Nowak W., Seiler T. B., Steger K., Steinmetz H. and Wieprecht S. (2015). Anthropogenic Trace Compounds (ATCs) in aquatic habitats – Research needs on sources, fate, detection and toxicity to ensure timely elimination strategies and risk management. *Environment International*, **79**, 85–105.

Gerecke A. C., Scharer M., Singer H. P., Muller S. R., Schwarzenbach R. P., Sagesser M., Ochsenbein U. and Popow, G. (2002). Sources of pesticides in surface waters in Switzerland: pesticide load through waste water treatment plants-current situation and reduction potential. *Chemosphere*, **48**(3), 307–315.

Gobas F. A. P. C., de Wolf W., Burkhard L. P., Verbruggen E. and Plotzke K. (2009). Revisiting bioaccumulation criteria for POPs and PBT assessments. *Integrated Environmental Assessment and Management*, **5**(4), 624–637.

Göktaş R. K. and MacLeod, M. (2016). Remoteness from sources of persistent organic pollutants in the multi-media global environment. *Environmental Pollution*, **217**, 33–41.

Goss K. U., Brown T. N. and Endo S. (2013). Elimination half-life as a metric for the bioaccumulation potential of chemicals in aquatic and terrestrial food chains. *Environmental Toxicology and Chemistry*, **32**(7), 1663–1671.

Gouin T. and Wania F. (2007). Time trends of arctic contamination in relation to emission history and chemical persistence and partitioning properties. *Environmental Science & Technology*, **41**, 5986–5992.

Gouin T., Mackay D., Webster E. and Wania F. (2000). Screening chemicals for persistence in the environment. *Environmental Science & Technology*, **34**(5), 881–884.

Hansen S. F., Carlsen L. and Tickner J. A. (2007). Chemicals regulation and precaution: does REACH really incorporate the precautionary principle. *Environmental Science & Policy*, **10**(5), 395–404.

Houtman C. J. (2010). Emerging contaminants in surface waters and their relevance for the production of drinking water in Europe. *Journal of Integrative Environmental Sciences*, **7**(4), 271–295.

Howard P. (2000). Biodegradation. In: Handbook of Property Estimation Methods for Chemicals: Environmental and Health Sciences, Boethling R. and Mackay D. (eds), CRC Press, Boca Raton, FL, USA.

Howard P. H. and Muir D. C. G. (2010). Identifying new persistent and bioaccumulative organics among chemicals in commerce. *Environmental Science & Technology*, **44**(7), 2277–2285.

Hu X. D. C., Andrews D. Q., Lindstrom A. B., Bruton T. A., Schaider L. A., Grandjean, P., Lohmann, R., Carignan, C. C., Blum, A., Balan, S. A., Higgins, C. P. and Sunderland, E. M. (2016). Detection of Poly – and Perfluoroalkyl Substances (PFASs) in US drinking water linked to industrial sites, military fire training areas, and wastewater treatment plants. *Environmental Science & Technology Letters*, **3**(10), 344–350.

Jaworska J. S., Boethling R. S. and Howard P. H. (2003). Recent developments in broadly applicable structure-biodegradability relationships. *Environmental Toxicology and Chemistry*, **22**(8), 1710–1723.

Johnson A. C. and Sumpter J. P. (2016). Are we going about chemical risk assessment for the aquatic environment the wrong way? *Environmental Toxicology and Chemistry*, **35**(7), 1609–1616.

Lindim C., van Gils J. and Cousins I. T. (2016a). Europe – wide estuarine export and surface water concentrations of PFOS and PFOA. *Water Research*, **103**, 124–132.

Lindim C., van Gils J. and Cousins I. T. (2016b). A large-scale model for simulating the fate & transport of organic contaminants in river basins. *Chemosphere*, **144**, 803–810.

Liu J. L. and Wong M. H. (2013). Pharmaceuticals and personal care products (PPCPs): a review on environmental contamination in China. *Environment International*, **59**, 208–224.

Lohmann R., Stapleton H. M. and Hites R. A. (2013). Science should guide TSCA reform. *Environmental Science & Technology*, **47**(16), 8995–8996.

Mackay D. and Fraser A. (2000). Bioaccumulation of persistent organic chemicals: mechanisms and models. *Environmental Pollution*, **110**(3), 375–391.

Mackay D. and Webster E. (2003). A perspective on environmental models and QSARs. *Sar and Qsar in Environmental Research*, **14**(1), 7–16.

Mackay D., McCarty L. S. and MacLeod M. (2001). On the validity of classifying chemicals for persistence, bioaccumulation, toxicity, and potential for long-range transport. *Environmental Toxicology and Chemistry*, **20**(7), 1491–1498.

Mackay D., Shiu W. Y., Ma K.-C. and Lee S. C. (2006). Handbook of Physical-Chemical Properties and Environmental Fate for Organic Chemicals, 2nd edn, CRC Press, Boca Raton, FL, USA.

Mackay D., Hughes D. M., Romano M. L. and Bonnell M. (2014). The role of persistence in chemical evaluations. *Integrated Environmental Assessment and Management*, **10**(4), 588–594.

MacLeod M. and Mackay D. (2004). Modeling transport and deposition of contaminants to ecosystems of concern: a case study for the Laurentian Great Lakes. *Environmental Pollution*, **128**(1–2), 241–250.

MacLeod M. and McKone T. E. (2004). Multimedia persistence as an indicator of potential for population-level intake of environmental contaminants. *Environmental Toxicology and Chemistry*, **23**(10), 2465–2472.

MacLeod M., Scheringer M., McKone T. E. and Hungerbuhler K. (2010). The state of multimedia mass-balance modeling in environmental science and decision-making. *Environmental Science & Technology*, **44**(22), 8360–8364.

MacLeod M., Breitholtz M., Cousins I. T., Wit, C. A. d., Persson, L. M., Rudén, C. and McLachlan, M. S. (2014). Identifying chemicals that are planetary boundary threats. *Environmental Science & Technology*, **48**(19), 11057–11063.

Matthies M., Solomon K., Vighi M., Gilman A. and Tarazona J. V. (2016). The origin and evolution of assessment criteria for persistent, bioaccumulative and toxic (PBT) chemicals and persistent organic pollutants (POPs). *Environmental Science: Processes & Impacts*, **18**(9), 1114–28.

Meyer W., Reich M., Beier S., Behrendt J., Gulyas H. and Otterpohl R. (2016). Measured and predicted environmental concentrations of carbamazepine, diclofenac, and metoprolol in small and medium rivers in northern Germany. *Environmental Monitoring and Assessment*, **188**(8), 487.

Meylan W., Boethling R., Aronson D., Howard P. and Tunkeli J. (2007). Chemical structure-based predictive model for methanogenic anaerobic biodegradation potential. *Environmental Toxicology and Chemistry*, **26**(9), 1785–1792.

Mill T. (2000). Photoreactions in surface waters. In: Handbook of Property Estimation Methods for Chemicals: Environmental and Health Sciences, Boethling R. S. and Mackay D. (eds), CRC Press, Boca Raton, FL, USA.

Moermond C. T. A., Janssen M. P. M., de Knecht J. A., Montforts M. H. M. M., Peijnenburg W. J. G. M., Zweers P. G. P. C. and Sijm, D. T. H. M. (2012). PBT assessment using the revised annex XIII of REACH: a comparison with other regulatory frameworks. *Integrated Environmental Assessment and Management*, **8**(2), 359–371.

Mombelli E. (2012). Evaluation of the OECD (Q)SAR Application Toolbox for the profiling of estrogen receptor binding affinities. *Sar and Qsar in Environmental Research*, **23**(1–2), 37–57.

Mombelli E. and Devillers J. (2010). Evaluation of the OECD (Q)SAR Application Toolbox and Toxtree for predicting and profiling the carcinogenic potential of chemicals. *Sar and Qsar in Environmental Research*, **21**(7–8), 731–752.

Muir D. C. G. and Howard P. H. (2006). Are there other persistent organic pollutants? A challenge for environmental chemists. *Environmental Science & Technology*, **40**(23), 7157–7166.

NORMAN (2017). Emerging substances. http://www.normandata.eu/?q=node/19 (Last accessed: 27.03.2017)

OECD (1992). Report of the OECD Workshop on the Extrapolation of Laboratory Aquatic Toxicity Data to the Real Environment. Environmental Monograph No. 59. Organisation for Economic Co-Operation and Development, Paris.

OECD (2016). OECD Pov and LRTP Screening Tool. http://www.oecd.org/chemicalsafety/risk-assessment/oecdpovandlrtpscreeningtool.htm (Last accessed: 29.12.2016)

OECD (2017). The OECD QSAR Toolbox. http://www.oecd.org/chemicalsafety/risk-assessment/oecd-qsar-toolbox.htm (Last accessed: 27.01.2017)

Persson L. M., Breitholtz M., Cousins I. T., de Wit C. A., MacLeod M. and McLachlan M. S. (2013). Confronting unknown planetary boundary threats from chemical pollution. *Environmental Science & Technology*, **47**(22), 12619–12622.

Petrie B., Youdan J., Barden R. and Kasprzyk-Hordern B. (2016). New framework to diagnose the direct disposal of prescribed drugs in wastewater – a case study of the antidepressant fluoxetine. *Environmental Science & Technology*, **50**(7), 3781–3789.

Puzyn T. (2011). On the replacement of empirical parameters in multimedia mass balance models with QSPR data. *Journal of Hazardous Materials*, **192**(3), 970–977.

Raymond J. W., Rogers T. N., Shonnard D. R. and Kline A. A. (2001). A review of structure-based biodegradation estimation methods. *Journal of Hazardous Materials*, **84**(2–3), 189–215.

Rockström J., Steffen W., Noone K., Persson A., Chapin F. S., Lambin E., Lenton T. M., Scheffer M., Folke C., Schellnhuber H. J., Nykvist B., de Wit C. A., Hughes T., van der Leeuw S., Rodhe H., Sorlin S., Snyder P. K., Costanza R., Svedin U., Falkenmark M., Karlberg L., Corell R. W., Fabry V. J., Hansen J., Walker B., Liverman D., Richardson K., Crutzen P. and Foley J. (2009a). Planetary boundaries: exploring the safe operating space for humanity. *Ecology and Society*, **14**(2), 32.

Rockström J., Steffen W., Noone K., Persson A., Chapin F. S., Lambin E. F., Lenton T. M., Scheffer M., Folke C., Schellnhuber H. J., Nykvist B., de Wit C. A., Hughes T., van der Leeuw S., Rodhe H., Sorlin S., Snyder P. K., Costanza R., Svedin U., Falkenmark M., Karlberg L., Corell R. W., Fabry V. J., Hansen J., Walker B., Liverman D., Richardson K., Crutzen P. and Foley J. A. (2009b). A safe operating space for humanity. *Nature*, **461**(7263), 472–475.

Rorije E., Peijnenburg W. and Klopman G. (1998). Structural requirements for anaerobic biodegradation of organic chemicals: a fragment model analysis. *Environmental Toxicology and Chemistry*, **17**(10), 1943–1950.

Runnalls T. J., Margiotta-Casaluci L., Kugathas S. and Sumpter J. P. (2010). Pharmaceuticals in the aquatic environment: steroids and anti-steroids as high priorities for research. *Human and Ecological Risk Assessment*, **16**(6), 1318–1338.

S.C. (2009). Stockholm Convention on Persistent Organic Pollutants (POPs) as amended in 2009: Text and Annexes. UNEP.

S.C. (2017). The POPs. http://chm.pops.int/TheConvention/ThePOPs/tabid/673/Default.aspx (Last accessed: 31.01.2017)

Scheringer M. (2009). Long-range transport of organic chemicals in the environment. *Environmental Toxicology and Chemistry*, **28**(4), 677–690.

Scheringer M., Strempel S., Hukari S., Ng C. A., Blepp M. and Hungerbuhler K. (2012). How many persistent organic pollutants should we expect? *Atmospheric Pollution Research*, **3**(4), 383–391.

Schwarzenbach R. P., Escher B. I., Fenner K., Hofstetter T. B., Johnson C. A., von Gunten, U. and Wehrli, B. (2006). The challenge of micropollutants in aquatic systems. *Science*, **313**(5790), 1072–1077.

Schwarzenbach R. P., Egli T., Hofstetter T. B., von Gunten U. and Wehrli B. (2010). Global water pollution and human health. *Annual Review of Environment and Resources*, **35**, 109–136.

Sobek A., Bejgarn S., Rudén C. and Breitholtz M. (2016). The dilemma in prioritizing chemicals for environmental analysis: known versus unknown hazards. *Environmental Science: Processes & Impacts*, **18**(8), 1042–1049.

Steffen W., Richardson K., Rockström J., Cornell S. E., Fetzer I., Bennett E. M., Biggs R., Carpenter S. R., de Vries W., de Wit C. A., Folke C., Gerten D., Heinke J., Mace G. M., Persson L. M., Ramanathan V., Reyers B. and Sörlin S. (2015). Planetary boundaries: guiding human development on a changing planet. *Science*, **347**(6223), 1259855.

Strempel S., Scheringer M., Ng C. A. and Hungerbuehler K. (2012). Screening for PBT Chemicals among the "Existing" and "New" Chemicals of the EU. *Environmental Science & Technology*, **46**(11), 5680–5687.

Sumpter J. P. and Johnson A. C. (2008). 10th Anniversary perspective: reflections on endocrine disruption in the aquatic environment: from known knowns to unknown unknowns (and many things in between). *Journal of Environmental Monitoring*, **10**(12), 1476–1485.

Sun M., Arevalo E., Strynar M., Lindstrom A., Richardson M., Kearns B., Pickett A., Smith C. and Knappe D. R. U. (2016). Legacy and emerging perfluoroalkyl substances are important drinking water contaminants in the cape fear river watershed of North Carolina. *Environmental Science & Technology Letters*, **3**(12), 415–419.

Sushko I., Novotarskyi S., Korner R., Pandey A. K., Rupp M., Teetz W., Brandmaier S., Abdelaziz A., Prokopenko V. V., Tanchuk V. Y., Todeschini R., Varnek A., Marcou G., Ertl P., Potemkin V., Grishina M., Gasteiger J., Schwab C., Baskin II, Palyulin V. A., Radchenko E. V., Welsh W. J., Kholodovych V., Chekmarev D., Cherkasov A., Aires-de-Sousa J., Zhang Q. Y., Bender A., Nigsch F., Patiny L., Williams A., Tkachenko V. and Tetko I. V. (2011). Online chemical modeling environment (OCHEM): web platform for data storage, model development and publishing of chemical information. *Journal of Computer-Aided Molecular Design*, **25**(6), 533–554.

Tarazona J. V. (2013). Use of new scientific developments in regulatory risk assessments: challenges and opportunities. *Integrated Environmental Assessment and Management*, **9**(3), e85–e91.

Tebby C., Mombelli E., Pandard P. and Pery A. R. R. (2011). Exploring an ecotoxicity database with the OECD (Q)SAR Toolbox and DRAGON descriptors in order to prioritise testing on algae, daphnids, and fish. *Science of the Total Environment*, **409**(18), 3334–3343.

Tratnyek P. G. and Macalady, D. L. (2000). Oxidation-reduction reactions in the aquatic environment. In: Handbook of Property Estimation Methods for Chemicals: Environmental and Health Sciences, Boethling R. S. and Mackay D. (eds), CRC Press, Boca Raton, FL, USA.

Tyler C. R., Jobling S. and Sumpter J. P. (1998). Endocrine disruption in wildlife: a critical review of the evidence. *Critical Reviews in Toxicology*, **28**(4), 319–361.

UNEP and FAO (2015). Rotterdam Convention on the Prior Informed Consent Procedure for Certain Hazardous Chemicals and Pesticides in International Trade, Texts and Annexes (Revised in 2015).

UNEP and FAO (2017). Rotterdam Convention: Share Responsibility. www.pic.int (Last accessed: 08.02.2017)

van der Oost R., Beyer J. and Vermeulen N. P. E. (2003). Fish bioaccumulation and biomarkers in environmental risk assessment: a review. *Environmental Toxicology and Pharmacology*, **13**(2), 57–149.

van Leeuwen K., Schultz T. W., Henry T., Diderich B. and Veith G. D. (2009). Using chemical categories to fill data gaps in hazard assessment. *Sar and Qsar in Environmental Research*, **20**(3–4), 207–220.

Verlicchi P. and Zambello E. (2015). Pharmaceuticals and personal care products in untreated and treated sewage sludge: occurrence and environmental risk in the case of application on soil – a critical review. *Science of the Total Environment*, **538**, 750–767.

Verlicchi P., Al Aukidy M. and Zambello E. (2012). Occurrence of pharmaceutical compounds in urban wastewater: removal, mass load and environmental risk after a secondary treatment – a review. *Science of The Total Environment*, **429**, 123–155.

Vorberg S. and Tetko I. V. (2014). Modeling the biodegradability of chemical compounds using the Online CHEmical Modeling Environment (OCHEM). *Molecular Informatics*, **33**(1), 73–85.

Walters D. M., Jardine T. D., Cade B. S., Kidd K. A., Muir D. C. G. and Leipzig-Scott P. (2016). Trophic magnification of organic chemicals: a global synthesis. *Environmental Science & Technology*, **50**(9), 4650–4658.

Wania F. (2003). Assessing the potential of persistent organic chemicals for long-range transport and accumulation in polar regions. *Environmental Science & Technology*, **37**(7), 1344–1351.

Wania F. (2006). Potential of degradable organic chemicals for absolute and relative enrichment in the arctic. *Environmental Science & Technology*, **40**(2), 569–577.

Wania F. and Mackay D. (1995). A global distribution model for persistent organic-chemicals. *Science of the Total Environment*, **160–61**, 211–232.

Wegmann F., Cavin L., MacLeod M., Scheringer M. and Hungerbuehler K. (2009). The OECD software tool for screening chemicals for persistence and long-range transport potential. *Environmental Modelling & Software*, **24**(2), 228–237.

Weininger, D. (1988). SMILES, a chemical language and information system. 1. Introduction to methodology and encoding rules. *Journal of Chemical Information and Computer Sciences*, **28**(1), 31–36.

WHO (2013). State of the Science of Endocrine Disrupting Chemicals – 2012: An Assessment of the State of the Science of Endocrine Disruptors Prepared by a Group of Experts for the United Nations Environment Programme and World Heath Organization. United Nations Environment Programme (UNEP) and the World Health Organization (WHO), Geneva, Switzerland.

Wolfe N. L. and Jeffers P. M. (2000). Hydrolysis. In: Handbook of Property Estimation Methods for Chemicals: Environmental and Health Sciences, Boethling R. S. and Mackay D. (eds), CRC Press, Boca Raton, FL, USA.

Zarfl C. and Matthies M. (2013). PBT borderline chemicals under REACH. *Environmental Sciences Europe*, **25**(1), 11.

Zarfl C., Scheringer M. and Matthies M. (2011). Screening criteria for long-range transport potential of organic substances in water. *Environmental Science and Technology*, **45**(23), 10075–10081.

Zhang T. and Li B. (2011). Occurrence, transformation, and fate of antibiotics in municipal wastewater treatment plants. *Critical Reviews in Environmental Science and Technology*, **41**, 951–998.

Chapter 3

Quantification of hazardous pollutants in biological systems

Lokesh P. Padhye[1], Peizhe Sun[2] and Ching-Hua Huang[2]*

[1]The University of Auckland, Department of Civil and Environmental Engineering, Auckland, New Zealand

[2]Georgia Institute of Technology, School of Civil and Environmental Engineering, Atlanta, GA, U.S.A.

**Corresponding author: l.padhye@auckland.ac.nz*

3.1 INTRODUCTION

3.1.1 Environmental quantitative analysis

The most common purpose for sampling and analysis of environmental pollutants is to quantify their levels for compliance check with regulatory requirements. However, other objectives may exist such as: to monitor ambient background concentrations of pollutants for studying their occurrence and fate in the environment; to detect their accidental releases; to evaluate their risks and toxicity to biota; and to evaluate efficiency of remediation systems in removing these contaminants from the environment (Zhang, 2007). Since the demand for environmental quantitative analysis is largely driven by regulations, the principal end users are industries which have to demonstrate their effluents/emissions fall within the maximum contaminant levels (MCLs) for the kinds of pollutants that are released (Loconto, 2006). In some cases, end users are researchers seeking answers or working on method developments in the private sector or federal, state, or academic laboratories.

 Hazardous Pollutants in Biological Treatment Systems: Fundamentals and a Guide to Experimental Research
Ferhan Çeçen, Ulaş Tezel
doi: 10.2166/9781780407708_69

Across the globe, there exist numerous environmental regulations that govern discharge of hazardous pollutants in the environment, and usually these laws also specify MCLs for targeted pollutants. For example, in the U.S., Title 40 Code of Federal Regulations (40 CFR) is the ultimate authority for environmental compliance. The CFR annual edition is the codification of the general and permanent rules published in the Federal Register by the departments and agencies of the Federal Government and is updated every year (CFR, 2017). Table 3.1 summarizes MCLs for primary drinking water pollutants from the U.S. Environmental Protection Agency (EPA).

To set an MCL for a pollutant, regulatory agencies first need to determine how much of the pollutant can be present in the environment without any adverse health effects. Although the health criterion guides MCL values, for many pollutants, environmental quantitative analysis becomes a critical step in setting their MCLs because of difficulties in measuring small quantities of these pollutants in complex matrices. It highlights the need for accuracy and reliability of environmental quantitative analysis through established methods.

Regulatory approved method development was the norm in the last century. However, getting method approval through regulatory agencies is a time-consuming process. Especially, in the current age of rapid technology developments, a long wait can make a method obsolete even before its passage through all the paperwork. Hence, in late 1990s, the U.S. EPA adopted the Performance Based Measurement System (PBMS) where it placed more of the responsibility for quality assurance on the individual analyst and the laboratory rather than on the method itself. Performance-based methods have enabled analysts and the laboratories that they work in to become more creative and have encouraged attempts to develop alternative methods to perform environmental quantitative analysis (Loconto, 2006). Through the U.S. EPA's review of PBMS, it became evident that while it may be possible to specify performance criteria, independent of methods, techniques, or instruments, it may not be possible to develop a single protocol for the validation of all current and future environmental quantitative analyses. Hence, in 2008, PBMS evolved further to a more flexible approach where additional emphasis was placed on flexibility in choosing sampling and analytical approaches to meet regulatory requirements for measurements (U.S. EPA, 2017).

Once an analytical method is developed, a validation process should provide evidence that quality requirements are achieved in quantification. Two key steps of validating the method which are important for scientific and legal defensibility are: providing evidence on general performance of a quantification method on a range of materials that define a matrix class; and demonstrating that the requirements for a specific use are met through the proposed quantitative analysis, a step called verification (U.S. EPA, 2017). Accuracy and precision of the method are principal data quality indicators. Accuracy describes closeness of the agreement between the quantification results of an analyte and its true value, while precision describes the

Table 3.1 U.S. EPA's national primary drinking water pollutants (MCL in mg/L).

Disinfection By-Products	Inorganics	Microorganisms	Semivolatile Organic Compounds (SVOCs)	Volatile Organic Compounds (VOCs)
Bromate (0.01)	Antimony (0.006)	Cryptosporidium	Adipates (0.4)	Benzene (0.005)
Chlorite (1)	Arsenic (0.01)	Giardia lamblia	Alachlor (0.002)	Carbon tetrachloride (0.005)
Haloacetic acids (0.06)	Asbestos	Legionella	Atrazine (0.003)	Chlorobenzene (0.1)
Total Trihalomethanes (0.08)	Barium (2)	Total Coliforms	Benzo(a)pyrene (PAHs) (0.0002)	o-Dichlorobenzene (0.6)
	Beryllium (0.004)	Viruses	Carbofuran (0.04)	p-Dichlorobenzene (0.075)
	Cadmium (0.005)		Chlordane (0.002)	1,2-Dichloroethane (0.005)
	Chromium (0.1)		Dalapon (0.2)	1,1-Dichloroethylene (0.007)
	Copper (1.3)		Dibromochloropropane (0.0002)	Cis-1,2-dichlorethylene (0.07)
	Cyanide (0.2)		2,4-dichlorophenoxy-acetic acid (0.07)	Trans-1,2-dichloroethylene (0.1)
	Fluoride (4)		Dinoseb (0.007)	Dichloromethane (0.005)
	Lead (0.015)		Diquat (0.02)	1,2-Dichloropropane (0.005)
	Mercury (0.002)		Endothall (0.1)	Ethylbenzene (0.7)
	Nickel (1.1)		Endrin (0.002)	Styrene (0.1)
	Nitrate-N (10)		Ethylene dibromide (0.0005)	Tetrachloroethylene (PCE) (0.005)
	Nitrite-N (1)		Glyphosate (0.7)	Toluene (1)
	Selenium (0.05)		Heptachlor (0.0004)	1,1,1-Trichloroethane (0.2)
	Thallium (0.002)		Hexachlorobenzene (0.001)	1,1,2-Trichloroethane (0.005)

(*Continued*)

Table 3.1 U.S. EPA's national primary drinking water pollutants (MCL in mg/L) (*Continued*).

Disinfection By-Products	Inorganics	Microorganisms	Semivolatile Organic Compounds (SVOCs)	Volatile Organic Compounds (VOCs)
			Hexachlorocyclopentadiene (0.05)	Trichloroethylene (TCE) (0.005)
			Lindane (0.0002)	Vinyl chloride (0.002)
			Methoxychlor (0.04)	Total xylene (10)
			Oxamyl (0.2)	
			PCBs (0.0005)	
			Pentachlorophenol (0.001)	
			Phthalates (0.006)	
			Picloram (0.5)	
			Simazine (0.004)	
			Toxaphene (0.003)	
			2,3,7,8-TCDD (dioxin) (0.00000003)	
			2,4,5-TP (Silvex) (0.05)	

agreement of a set of quantifications among themselves. A sound quantification technique should demonstrate high accuracy and precision.

It's important for analytical methods to be able to detect environmental pollutants at trace levels even in complex matrices such as wastewater influent. The method detection limit (MDL) is defined by the U.S. EPA as "the minimum concentration that can be measured and reported with 99% confidence that the analyte concentration is greater than zero". The MDLs are dependent on method, sample matrix, analytical instrument, and analytical techniques used for quantification but need to be less than the MCLs. During a routine laboratory analysis, MDL is rarely achieved and use of a more realistic detection limit called the "practical quantitation limit" (PQL) is practiced. It is defined as "the lowest concentration that can be reliably achieved within specified limits of precision and accuracy during routine operating condition". Laboratories set their PQL to the concentration of the lowest standard they run for which they feel comfortable signing their name to, confidently, on a daily basis. It is matrix dependent and is usually 2–10 times higher than the MDL (Zhang, 2007).

3.1.2 Hazardous pollutants in biological systems

Hazardous pollutants are those which are known to cause, or suspected to cause, health impacts on aquatic and terrestrial organisms, including humans. Emerging contaminants, in turn, are those for which their presence and health effects in the environment are becoming of concern. Hence, for this chapter, emerging contaminants are treated as a subset of hazardous pollutants. Accordingly, 10 major classes of hazardous pollutants have been identified for review of their quantification methods in biological systems. These are the following:

- perfluorinated compounds (PFCs)
- polybrominated diphenylethers (PBDEs)
- polychlorinated compounds (PCCs)
- alkyl phenols (APs)
- pharmaceuticals and personal care products (PPCPs)
- estrogens
- disinfection by-products (DBPs)
- nanomaterials
- metals
- pathogens

Since many of the above hazardous pollutants are released into the environment as complex mixtures and are present at trace levels, their quantification in environmental samples is challenging. This is particularly true for samples taken from biological systems as, invariably for such systems, the sampling matrix is much more complex due to the presence of suspended organics in addition to the aqueous phase. Hence, to quantify these pollutants in biological systems, innovative sampling methodologies need to be coupled with robust analytical techniques which can confirm the identity of targeted and unknown chemicals at trace concentrations. This chapter discusses common techniques which can be used to address the issues of sampling and analysis for quantification of the hazardous pollutants mentioned above.

BOX 3.1 CASE STUDY ON QUANTIFICATION OF NITROSAMINES IN BIOLOGICAL SYSTEMS

We start by introducing a case study of successful quantification of nitrosamines, an emerging class of highly toxic DBPs, carried out in municipal sludge and anaerobic digestion systems. The occurrence and fate of six nitrosamines in municipal wastewater treatment plants' primary sludge (PS), waste-activated sludge (WAS), and anaerobic digester mixed liquor (ADML) was investigated by Padhye *et al.* (2009). Samples were collected at three municipal wastewater treatment plants (WWTPs) to provide variability in terms of geographical location, size, biological treatment system configuration, and industrial discharge contribution. After arriving in the laboratory, the PS and WAS samples were stored at 4°C whereas the ADML samples were transferred to glass reactors flushed with helium and incubated at 35°C with continuous mixing. Samples were analyzed within 48 hours of receiving them.

For nitrosamine extraction from WAS and ADML, an aliquot of 200 mL of each sludge filtrate was fortified with three representative deuterated compounds at 250–500 ng/L, expected range of target analytes, as surrogates in an amber glass bottle. Then, 200 mg of resin beads of adsorbent were added into each bottle for solid-phase extraction of nitrosamines and the content was agitated at 200 rpm for 2 hours. The content was filtered through a glass-fiber filter. The resins collected on the filter were air-dried at room temperature for 3–6 hours and transferred to a 2-mL amber vial and immersed in 500 μL of dichloromethane. The solvent layer was analyzed by gas chromatography-mass spectrometry (GC-MS) in the electron ionization (EI) mode using selected ion monitoring (SIM).

Detection of the most abundant nitrosamine, *N*-nitrosodimethylamine (NDMA), in some PS filtrates could not be achieved by the above-mentioned procedure due to the complexity of the sample matrix. In such cases, a new extraction protocol was developed. The PS filtrates were first extracted twice with 25 mL of hexane to remove hydrophobic constituents in the matrix. The extracted sample was then subjected to liquid-liquid extraction with 50 mL of dichloromethane. The dichloromethane extract was then evapo-concentrated to 0.5 mL and the concentrate was extracted with 5 ml of deionized (DI) water. The DI water extract was back extracted with 5 mL of clean dichloromethane. The extracted dichloromethane was concentrated to 0.5 mL and analyzed for NDMA using GC-MS in the EI mode using SIM.

Samples were analyzed by an Agilent GC-MS (6890/5973) system equipped with a HP-5MS column (30 m × 250 μm × 0.25 μm) and a large volume injector. Solvent vent mode injections of 5 μL were used with an initial injection port temperature of 0°C, which was rapidly increased to 220°C after 0.3 sec. The temperatures of the MS ion source and transfer line were 150 and 260°C, respectively. GC oven temperatures were 35°C held for 1 min followed by ramping at 10°C min^{-1} to 170°C, then ramping at 15°C min^{-1} to a final temperature of 240°C held for 2 min.

This was one of the first studies to report nitrosamines and their precursor concentrations in municipal wastewater sludges. NDMA was detected at significant levels, with mean concentrations of 680, 390, and 270 ng/L in PS, WAS, and ADML samples, respectively. *N*-Nitrosopyrrolidine was the other nitrosamine detected in sludge samples but at about an order of magnitude lower concentrations than NDMA. Overall, anaerobic sludge digestion was found to be a sink for nitrosamines. Quantification of nitrosamines from the study proved highly relevant for understanding the fate of these hazardous pollutants through these widely adopted biological treatment processes in WWTPs.

Similar to the above case study, many laboratories develop their own analytical protocols for environmental quantification of hazardous pollutants. The U.S. EPA has published a number of methods over the past 40 years and, with its new flexible approach, does not mandate a review to modify approved methods as long as modifications are documented as described in 40 CFR 136.6 (CFR, 2017). Such modifications lower the costs of measurements, overcome matrix interferences, or otherwise improve the analysis. All the EPA methods are numbered and may have modification letters appended at the end, signifying a newer version of the method. Three-digit methods refer to methods for the chemical analysis of water and wastewater. The four-digit series, methods 0000 to 9999, are for the analysis of solid wastes. Table 3.2 lists some of the most commonly used EPA methods for quantification of hazardous pollutants in biological systems.

Table 3.2 U.S. EPA's methods relevant to quantification in biological systems.

Method Series	Description
100 Series	General chemistry of water (drinking water and wastewater)
200 series	Metals in water
300 series	Inorganics and nonmetals in water
400 series	Organics in wastewater
500 series	Trace organics in drinking water
600 series	Trace organics in wastewater
900 series	Radioactivity in drinking water
1000 series	Hazardous waste and microbiological pollutants
2000 series	Bioassays
3000 series	Sample preparation for inorganics and organics
4000 series	Immunoassays
5000 series	Organic hazardous waste
6000 & 7000 series	Metals
8000 series	Organics – semivolatiles and volatiles
9000 series	Inorganics (nonmetals) and oil and grease

Due to a paucity of fresh water resources in many regions of the world, the need for reuse of wastewater is growing every year. Hence, it is becoming increasingly more important to understand trace level concentrations of hazardous pollutants in wastewater and sludges. Since biological systems represent complex matrices for environmental analysis due to the presence of excess organics (Table 3.3) and other constituents, wastewater and sludge samples require skillful sampling and analysis techniques and advanced analytical instrumentation. Typical environmental analysis for biological system matrices involves extraction, concentration, clean-up, and quantification. Presence of background noise (excess organics) often makes these steps complicated. Hence, the importance of developing protocols for accurate and reliable quantification of hazardous pollutants in biological systems couldn't be overemphasized.

Table 3.3 Typical water quality parameters (in mg/L) of common biological system matrices (high-strength).

Parameter	Raw Wastewater (High-Strength)	Septic Sludge (High-Strength)	Leachate (High-Strength)	Digester Supernatant (High-Strength)
COD	1,200	90,000	16,000	9,000
BOD	560	30,000	12,000	4,000
N Total	100	1,500	500	800
P Total	25	300	10	300
TSS	600	100,000	500	10,000
VSS	480	60,000	300	6,000

Note: COD – Chemical Oxygen Demand; BOD – Biochemical Oxygen Demand; N – Nitrogen; P – Phosphorus; TSS – Total Suspended Solids; VSS – Volatile Suspended Solids.
Source: Adapted from Henze *et al.* (2008).

3.2 ENVIRONMENTAL ANALYSIS FOR BIOLOGICAL SYSTEMS

3.2.1 Good laboratory practices

Data obtained through environmental analysis need to be scientifically reliable and legally defensible. Scientific reliability means that validated methods for sampling and analysis are followed so that both, accuracy and precision for measurements are achieved. This section introduces practices that help in ensuring reliability of the data during multiple stages of environmental analysis (Figure 3.1).

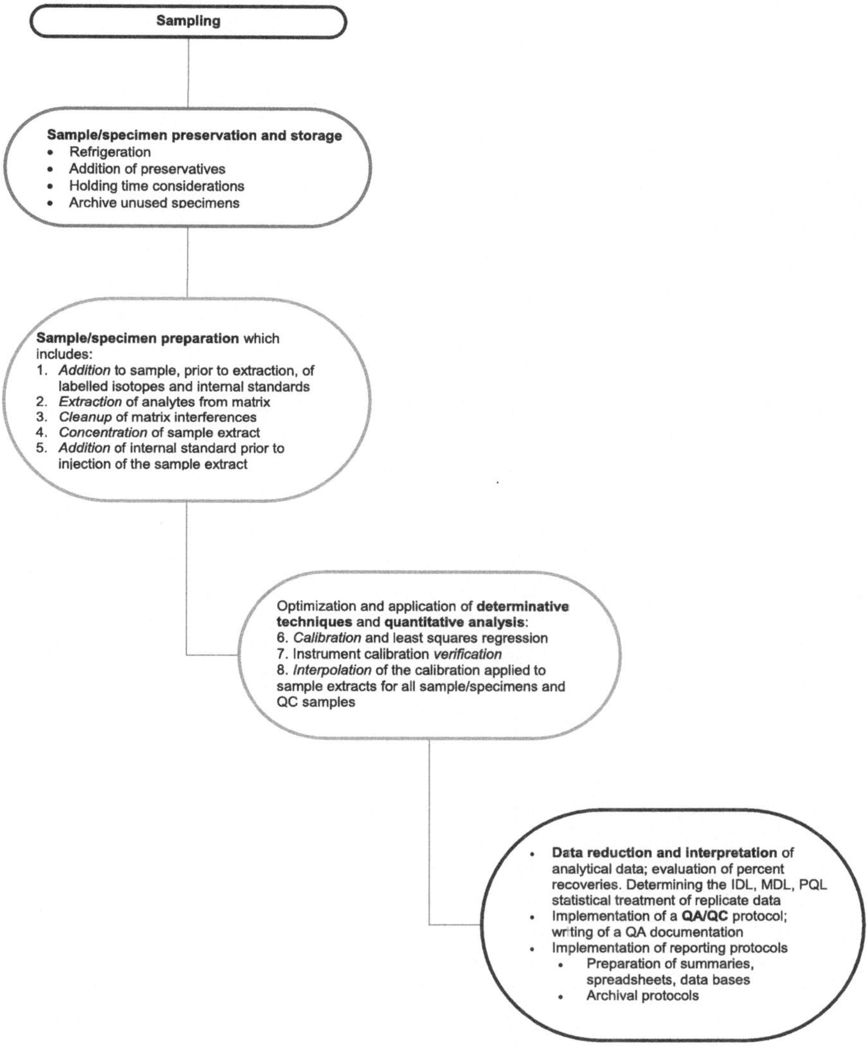

Figure 3.1 Generic protocol for environmental analysis of hazardous pollutants (adapted from Loconto (2006)).

Good laboratory practice is a laboratory quality system covering all aspects of laboratory operations, including management, sampling, method development, analytical quality control, quantification, measurement traceability, and reporting results. Good laboratory practice for a laboratory striving to achieve scientifically reliable analysis of hazardous pollutants in biological systems requires the following considerations:

(1) Storing the samples at appropriate temperatures (usually 4°C for wastewater samples) as soon as they are received in the laboratory and if the analysis is not planned immediately. This is particularly relevant for samples from biological systems as microbes in samples can degrade some of the pollutants at room temperatures before analysis is performed, and hence result in underestimation of their true concentrations. Storing the samples away from the light is also required to prevent photolysis. It is also recommended to perform analysis within 48 hours of receiving samples.
(2) Deciding whether the matrix spike or isotope dilution approach will be suitable for the intended quantitative analysis application. For the latter, a representative surrogate/internal standard should be chosen which behaves similarly to the analyte during sample preparation, extraction, and analytical stages.
(3) Selecting a validated method for sample preparation and analysis which has demonstrated accuracy and precision for the given analyte.
(4) Establishing a standard calibration curve in the range of expected analyte concentrations. Such a curve relates instrument response to concentration of the analyte by establishing a relationship between instrument responses and standards of known concentrations of the analyte. It is recommended to run the standard calibration curve in the same batch as that of samples.
(5) Determining the instrument detection limits (IDLs), MDL, and PQL. MDL and PQL rely on the sample matrices. For biological systems, PQL is usually higher than MDL.
(6) Running samples in duplicates. The EPA methods recommend running all samples in duplicate as a minimum. However, given the complexity of matrices for samples from biological systems, triplicate samples are recommended for ensuring effective data validation.
(7) Preparing and running a minimum of two reference standards during a batch of samples to monitor instrument precision and accuracy. A batch is defined in the EPA methods to be approximately 20 samples.
(8) Preparing and running quality control samples to correct for matrix effects such as matrix spikes, standard spikes, and surrogates to assess extraction efficiency where applicable.
(9) Preparing and running blanks and control samples from the laboratory as well as analyzing field and trip blanks to rule out contamination during sampling and sample transport.
(10) Documenting method validation, proficiency testing, quality control measures, instrument calibration, and quantification.

As evident above, good laboratory practices need to incorporate quality assurance (QA) and quality control (QC). QA refers to a plan for maintaining quality in all aspects of analysis, equivalent to a management strategy. It is an integrated system providing confidence that quality requirements will be fulfilled to the needs of customers, government agencies, regulators, certifiers, and/or third parties. It involves planning, implementation, assessment, reporting, and quality improvement. As a part of good laboratory practices, it is recommended that each laboratory should maintain a document titled "Quality Assurance" in its possession (Loconto, 2006).

Quality Control (QC) consists of the steps a laboratory needs to take to determine the validity of specific sampling and analytical procedures. QA is a broad management plan to ensure quality in all aspects of the operation, whereas QC can be viewed as a task associated with laboratory practices for sampling and analysis. A record of certified reference standards incorporated in analysis is a key component of QC. It is necessary to include reference standards for analysis of hazardous pollutants in biological systems for data accuracy and validation. A brief description of common QC reference standards used in such analysis is provided below.

Field Blanks – A field blank (also known as a trip blank) is a deionized water sample prepared during the trip from the field to the laboratory. It is used to detect any possible contamination arising from sample collection and transport to the laboratory.

Field Duplicates – A field duplicate is a duplicate field sample collected at the same place and at the same time as that of the original sample. It is required to verify precision of the analytical protocol.

Laboratory Method Blanks – A laboratory method blank is a deionized water sample that goes through the entire sample preparation and analysis protocol, along with rest of the field samples. It is used to detect the extent of laboratory contamination in samples.

Lab Replicates – Lab replicates are subsamples prepared in the laboratory from the same sample to test the precision of the analytical protocol for the laboratory method.

Matrix Spike Samples – A known concentration of the analyte being measured is spiked in the sample. This increases the concentration of the analyte in the sample by a predictable amount. It is used to quantify the effect of the sample matrix on percent recovery of the analyte. For methods using mass spectrometry, isotope dilution is commonly used to quantify matrix effect where an unknown sample is spiked with an enriched isotope of the analyte.

Calibration Blanks – A calibration blank is a sample prepared in the laboratory to calibrate the instrument to "zero". It should not have any contamination. It is used to check the analytical instrument performance and to recalibrate it, if needed.

Calibration Standards – Calibration standards consist of multiple "standard concentrations" (prepared through a serial dilution or independently of each other) of the analyte being measured. Calibration standards are used for calibration of instrument and quantification of the analyte.

The above reference standards are important to prevent errors in environmental analysis. When a "false positive" is reported, it means that a sample is tested positive in analysis even though no analyte is present in the sample. It is analogous to a Type I error. On the contrary, a "false negative", analogous to a Type II error, is reported when no analyte is detected even though the analyte is present in a sample. There can be many reasons for false positive and false negative readings. However, incorporating the above-mentioned reference standards, as a part of "Good Laboratory Practices", ensures that these errors get identified and corrected during analysis.

Thus, environmental monitoring is the process of measuring or collecting environmental data by following well documented QA/QC protocols. Environmental monitoring programs are usually implemented through regulatory agencies to collect the environmental data necessary for decision making. One of the most important aims of environmental monitoring for hazardous pollutants in biological systems is to provide spatial and temporal trends of their concentrations, which can only be achieved with sophisticated sampling, analysis, and quantification using state-of-the-art analytical instruments. Technology in environmental monitoring is continuously advancing, and hence, it is important to highlight contemporary techniques which are widely practiced for sampling, analysis, and quantification of hazardous pollutants in biological systems.

3.2.2 Sampling

Obtaining representative samples of the biological system being investigated is of primary importance for a relevant description of the studied environment. This section introduces established sampling practices, possible formation of artifacts during sampling of biological systems, and sample storage protocols. This section also discusses stability issues associated with hazardous pollutants during the sample storage. The QA/QC samples, which must be procured, prepared, and analyzed in the same manner as the environmental samples, are also listed.

Sampling is a vital component of any environmental monitoring program. The validity of the final results depends on proper collection of samples. Any error in sampling can nullify all efforts invested in subsequent analysis. Hence, it is critical to confirm that a collected sample represents the matrix of the system and is devoid of any external contamination arising during sampling. The process of sampling starts in the office or laboratory with the training of personnel and formulation of a sampling plan, followed by field sampling as per the sampling plan. The process ends with the shipment of the sample to the laboratory for analysis and quantification (Alvarez & Jones-Lepp, 2010).

A successful sampling plan should include: a) detailing objectives for sampling; b) choosing the appropriate sampling method for procuring a representative sample; c) identifying QA/QC samples needed for validation of representativeness of a sample; d) determining the minimum quantity of sample needed for sample preparation, analysis, and quantification; and e) detailing transport and storage requirements of samples. A sampling trip and on-site parameter measurements during sampling should be documented for data validation. Sample collection can be judgmental, where sampling is carried out near a predetermined spot such as a known point source, or it can be systematic, where sampling is carried out at a consistent grid pattern. Sometimes random sampling, where a high number of samples are taken to understand a pollutant profile in the region, is practiced (Keith, 1992).

For sampling of biological systems, added precautions are needed to maintain the sample quality and prevent loss of any analytes. Samples should preferably be collected in containers made up of stainless steel, aluminum, glass, or fluorocarbon polymers. Materials such as polyethylene, rubber, Tygon®, or other plastics should be avoided due to the possibility of these materials adsorbing or desorbing targeted chemicals from/into the collected sample. Also, plastic containers should not be used for sampling if analysis of plasticizers and flame retardants is desired as plastic may contain high levels of these compounds from the manufacturing process (Alvarez & Jones-Lepp, 2010). A container or equipment blank should be included as a QA/QC sample during analysis.

Grab or spot sampling is still the most common method for collecting samples from biological systems. Samples are collected by hand directly into the shipping sample container. However, such sampling only provides a snapshot of pollutants, and thus may not capture their periodic variations. Hence, composite sampling is preferred since it is prepared by thoroughly mixing several grab samples over a predetermined time window (usually 24 hours for biological systems to account for diurnal variation). Recently, passive sampling techniques which provide time-weighted average (TWA) concentrations of pollutants are gaining popularity due to advances in passive samplers. These devices rely on physicochemical processes such as adsorption, partition, absorption, etc. Their versatility in terms of deployment period, deployment site, power requirements, handling and storage of samples makes them an ideal choice for modern day sampling of pollutants from biological systems.

Sometimes chemicals are added to samples from biological systems for sample preservation (Table 3.4). For example, samples containing residual disinfectants, such as chlorine and chloramines, need to be quenched to prevent degradation of pollutants or further formation of DBPs during the holding time between sample collection and analysis. Sodium thiosulfate and ascorbic acid are the commonly used quenching agents. For most of the aqueous samples taken from the biological systems, acidification of samples is usually sufficient to preserve the sample. Acidification of aqueous samples to $pH < 2$ is practiced to preserve trace metals and reduce precipitation, microbial activity, and sorption losses to container walls. An acid blank should be included as a QA/QC sample to confirm any contamination arising from the acid used for acidification. Once collected and amended to preserve, samples from biological systems are usually shipped chilled (<4°C) via an overnight carrier to the laboratory in amber or opaque containers to prevent biodegradation and photodegradation.

Table 3.4 Preservation needs and methods for sampling different types of analytes.

Analytes	Change During Storage	Preservation
Metals (M)	Adsorption to glass wall Precipitation in the form of MO, $M(OH)_2$	Use plastic bottle Add HNO_3 to pH < 2
Phthalate ester	Diffusion from plastics	Use Teflon or glass bottle
Oil	Adsorption to plastics	Use glass bottle
VOCs	Volatilization	Avoid headspace
NH_3	Volatilization	Add H_2SO_4 to pH < 2
S^{2-}	Volatilization	Add zinc acetate and NaOH to pH > 9
CN^-	Volatilization	Add NaOH to pH > 12
Polycyclic aromatic hydrocarbons (PAH)	Photochemical degradation	Use amber glass container
DBPs	Formation of more DBPs if residual disinfectant is present	Use quenching agent – sodium thiosulfate or ascorbic acid
Organics in wastewater	Biodegradation	Low pH and temperature; add $HgCl_2$ to kill bacteria
Total phenolics	Biodegradation	Add H_2SO_4 to stop biodegradation

Source: Adapted from Zhang (2007).

3.2.3 Sample preparation

The section introduces sample preparation and discusses objectives of sample preparation for quantification of hazardous pollutants in biological systems. Different sample preparation techniques for effective analysis, specifically focusing on biological system matrices, are introduced. In addition, common protocols of sample preparation for various pollutant classes are described.

Due to trace levels of pollutants in complex matrices, it is important to extract, pre-concentrate, and clean-up of environmental samples before using detection techniques to quantify pollutants. Collectively, all three steps can be grouped under "sample preparation". Although a few biological system samples can be directly injected into instruments, most samples need all three or at least two of the above-mentioned sample preparation steps. It is the most labor-intensive and time-consuming stage of the quantitative analysis of pollutants in biological systems (Zhang, 2007).

The purpose of environmental sample preparation can be one of the following:

- *To concentrate or dilute the analyte* – Most biological system samples usually go through a concentration step to increase the concentration of an analyte for its detection. However, in some cases where analyte is present at very high concentrations, dilution is required to bring down analyte concentrations within the calibration range of the instrument.
- *To homogenize the sample* – For better representativeness, samples may need to be homogenized. This can be achieved either through mechanical or ultrasonic methods.
- *To remove interferences* – For biological system samples, presence of excess organic matter poses a challenge in detection of trace levels of pollutants. Hence, it's often recommended to clean the matrix of interferences arising due to presence of organic matter found in biological systems.
- *To extract analytes from solid or aqueous phase into solvents* – Since instruments such as GC-MS cannot handle solid or aqueous samples, analytes of interest need to be collected from the sample and transferred to the phase suitable for instrumental analysis using different extraction techniques.
- *To change the chemical structure of analytes* – Some analytes need to be chemically modified to make them suitable for instrumental analysis. Derivatization of secondary amines to make them volatile for GC-MS detection is one such example.

Liquid-liquid extraction (LLE) or solvent extraction is a method to transfer analytes from one phase (usually aqueous) to the other (usually an organic solvent) based on their partitioning coefficients in two different immiscible liquids. It is an extraction of an analyte from one liquid into another liquid, and hence the name. Even though it is one of the oldest methods of extraction, the significant quantity of toxic solvents involved in the extraction makes it unappealing. Regardless, it is still commonly used for certain aqueous pollutants. Such analysis is usually coupled with derivatization as most of the LLE methods are developed for GC-MS quantification. After LLE, analytes are usually subjected to further preconcentration through evaporation of solvent.

Solid-phase extraction (SPE) is an extraction method in which the analytes in a liquid sample are immobilized on a solid-phase adsorbent bed and then eluted off. The beds are made up of either a polymeric or a silica-based stationary phase in a cartridge, a small disk, or 96-well plates. The adsorbent bed can be silica based (e.g. C18), hydrophilic lipophilic balanced (HLB), of mixed cation exchange (MCX), or of mixed anionic exchange (MAX). SPE is preferred over LLE for many hydrophilic pollutants which do not partition well into an organic solvent thus resulting in poor extraction efficiencies. SPE involves a lower quantity of toxic solvents, shorter analysis time, simpler process, and better adaptability for automation (Alvarez & Jones-Lepp, 2010). The traditional SPE involves four steps: conditioning the stationary phase, running samples through the stationary phase, rinsing the stationary phase, and eluting the analytes from the stationary phase, using a solvent, into collection vials.

Other extraction techniques include equilibrium liquid-liquid-liquid microextraction (LLLME) which incorporates a disposable polypropylene hollow fiber to extract analytes into a few μL of an organic phase and subsequently into another phase before analysis. In addition, use of the solid-phase microextraction (SPME) technique, which uses hollow fibers to extract compounds from an aqueous sample by absorption in the case of liquid coatings, or adsorption in the case of solid coatings, is widely used for extraction of pharmaceuticals from aqueous media (Alvarez & Jones-Lepp, 2010). For VOCs in aqueous media, purge-and-trap and SPME are the most commonly used methods of extraction and sample preconcentration (Chary & Fernandez-Alba, 2012). Figure 3.2 lists common sample preparation techniques used for generic classes of pollutants.

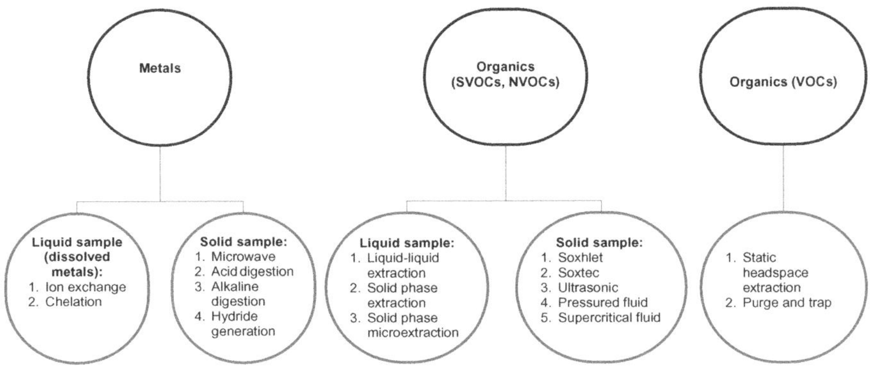

Figure 3.2 Overview of sample preparation (adapted from Loconto (2006)). *Note*: NVOCs – Non-Volatile Organic Compounds

Sample clean-up is often essential for isolating trace levels of pollutants from biological system matrices. Co-extracted compounds from samples of biological systems may share a similar structure to analytes of interest, thus not only causing interference in analysis but also posing a challenge for their removal from the matrix. The common techniques adopted for sample clean-up involve some modifications and combinations of SPE, LLE, and dialysis (Alvarez & Jones-Lepp, 2010). To determine recovery of analytes through multiple steps of sample clean-up and extraction, it is recommended to spike surrogates, such as isotopically-labeled analogs of targeted pollutants, at known concentrations as a QA/QC measure for data validation.

3.2.4 Quantification

This section introduces the detection aspect of environmental quantitative analysis. A verified calibration is a necessity in acquiring data from biological system samples for quantitative analysis. The section discusses the internal standard mode of validation of calibration. The section also defines the instrument detection limit and discusses its relationship with MDLs. Finally, common detection techniques for quantification of pollutants in biological systems are presented, along with their pre-quantification analyses.

Most detection techniques for trace level quantification of hazardous pollutants in biological systems are based on mass spectrometry, which has become the preferred method in environmental analysis due to the inherent complexity of most biological system samples. Different types of mass spectrometers (quadrupole, ion traps (ITMS), time-of-flight (TOF), triple quadrupole (QqQ), magnetic sector, and orbitrap are currently in use as detectors coupled to either GC or liquid chromatography (LC). Their selection depends on the chromatography technique used (GC or LC), objectives of quantification, mass accuracy required, and specificity desired by regulations (Alvarez & Jones-Lepp, 2010). A better understanding of mass spectrometry and its application for environmental analysis can be gained from the literature (Santos & Galceran, 2003; Petrović *et al.* 2005). Figure 3.3 provides established protocols of environmental quantification of organic compounds.

Instrument calibration and verification are critical steps for precise quantification of analytes. Calibration is the process of testing and adjusting an instrument output signal to establish a correlation between the instrument's measurement of the analyte and the actual concentration of the analyte while verification means the testing of analytes of known concentrations in the same manner as samples to assure the instrument is accurately measuring values throughout the quantification. For details about various modes of instrument calibration and verification, please refer to Loconto (2006). For biological system samples, the internal standard (IS) mode of calibration is preferred. In this mode, a compound of similar chemical structure, as that of analyte, is spiked in samples at known concentrations. Both the analyte and the IS behave similarly during quantification as they generate a similar response from an instrument. Hence, the ratio of analyte response to IS response remains constant, and a calibration curve can be plotted for different ratios of analyte concentrations to a constant spiked IS concentration.

The IDL for a specific analyte is defined as the lowest possible concentration that can be reliably measured with that instrument. Usually, it is the smallest concentration of analytical grade standard of an analyte that is barely detectable on the instrument. The steps for determining exact IDL are listed in the U.S. EPA guidance document for environmental analysis (CFR, 2017). The only situation whereby an IDL is the same as an MDL is when there is no sample preparation step involved and where the sample is directly injected into the instrument. However,

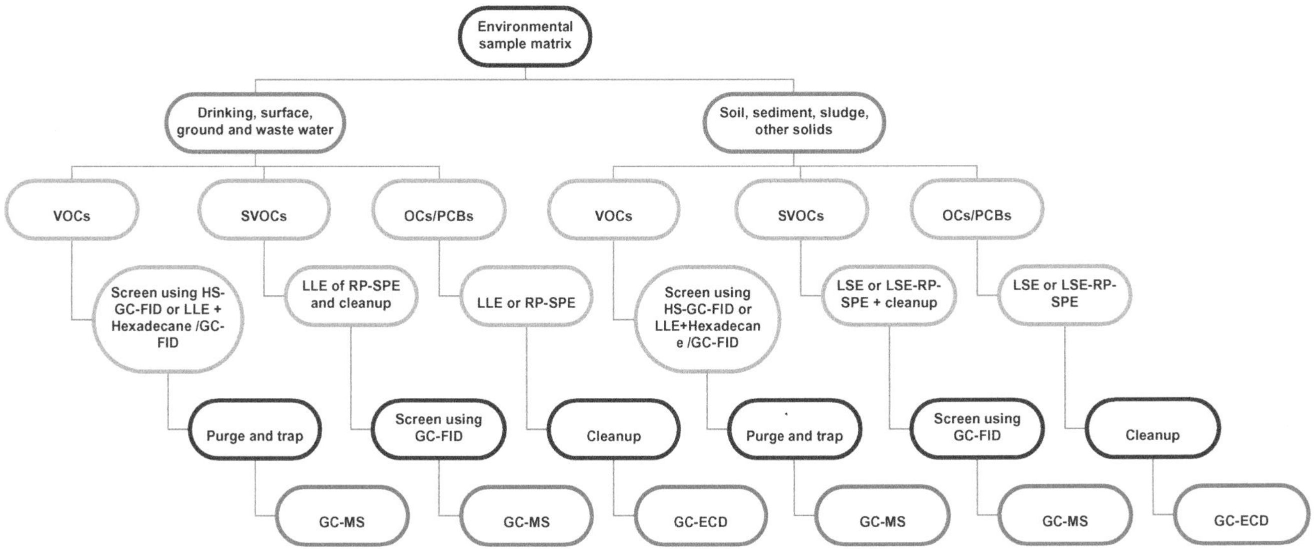

Acronyms defined that are used in the scheme:
VOCs – volatile organics
SVOCs – semi-volatile organics
OCs/PCBs – organochlorine pesticides/polychlorinated biphenyls
HS-GC-FID – static headspace coupled to gas chromatography incorporating flame ionization detection
LLE – liquid-liquid extraction (separatory funnel, countercurrent apparatus)
LSE – liquid-solid extraction (Soxhlet apparatus) or solvent leaching from solid matrices
RP-SPE – reversed-phase-solid-phase extraction using chemically bonded silica sorbents
LSE-RP-SPE – liquid-solid leaching followed by isolation on reversed-phase sorbents
GC-ECD – gas-chromatography incorporating electron-capture detection
GC-MS – gas chromatography interfaced to low-resolution quadruple mass spectrometry

Figure 3.3 Environmental analytical protocol for organics (adapted from Loconto (2006)).

for biological system samples, such instances are rare. MDLs are significantly lower than IDLs because sample preparation methods are designed to concentrate the analyte from the sample to the solvent. Although recoveries of analytes from biological system samples are significantly lower than 100%, the large concentration factors involved (100–1000), lower the MDLs considerably.

3.3 QUANTIFICATION OF HAZARDOUS POLLUTANTS

3.3.1 Perfluorinated compounds

A perfluorinated compound (PFC) is an organofluorine compound with all hydrogen atoms replaced by fluorine on a carbon chain, but the molecule also contains at least one different atom or functional group. The major application of PFCs is to make stain/water repellents that are widely applied to coat fabrics, carpets, metals, and electronics, as well as grease-proof food packages. PFCs have unique chemical properties – both hydrophobic and lipophobic, which make them excellent repellents. In addition, because they consist of the strongest chemical bonds (C-F), PFCs are also used in fire extinguishers. PFCs were produced in very large quantities worldwide. The major compounds in the PFC group are 8-carbon PFCs, including perfluorooctanesulfonate (PFOS) and perfluorooctanoic acid (PFOA). However, in the early 2000s, PFCs brought great public concern because they were frequently detected in the environment and even in humans. The U.S. EPA reported that the potential health risks of exposure to PFCs may include cancer, developmental effects, bioaccumulation, and immunotoxicity. As a result, PFOS has been ceased to use in North America since 2002. In 2016, the U.S. EPA established the health advisory level of 70 parts per trillion for PFOS and PFOA. In addition to PFOS and PFOA, which have been listed in the U.S. EPA's CCL-4 (https://www.epa.gov/ccl/chemical-contaminants-ccl-4), another four PFCs (perfluorononanoic acid (PFNA), perfluorohexanesulfonic acid (PFHxA), perfluoroheptanoic acid (PFHpA), and perfulorobutanesulfonic acid (PFBS)) are of public concern as well and are listed on the U.S. EPA's UCMR-3 (https://www.epa.gov/dwucmr/third-unregulated-contaminant-monitoring-rule). This section focuses on the quantification methods for the six PFCs in water and wastewater matrices.

3.3.1.1 Sample preparation for perfluorinated compounds

Several major precautions should be taken for sampling and sample pre-treatment of PFCs. First, the sample containers should be carefully selected to avoid adsorption of PFCs. Fluorine-coated bottles should also be avoided because they may introduce PFCs into water samples. Before sampling, rinsing sample containers with (semi-) polar solvents is suggested. Second, acidification of samples for preservation is discouraged because it will increase the volatilization and adsorption of PFCs. Third, to remove particulates from the water matrix, centrifugation is preferred over filtration, because the contents of some types of filters also contain PFCs.

A great number of studies have developed methods for extracting and concentrating PFCs from the aqueous phase using either LLE or SPE. The overall PFCs in the aqueous phase can be extracted using LLE; but this technique is limited to long-chain PFCs (i.e. hydrophobic PFCs). In contrast, SPE is best suited to PFCs with less than 10 carbon atoms, including the most important contaminants, PFOS and PFOA. Gonzales-Barreiro *et al.* (2006) compared many studies using LLE and SPE in the extraction of PFCs from water and concluded that detection limits using LLE ranged from 0.26 to 0.62 ng/L for enrichment of 900-mL water samples and recovery of PFCs with a carbon chain longer than C7 was excellent (80–93%). With SPE, carboxylates with carbon chains less than C10 could be extracted efficiently (70–98%) under acidic conditions, and PFOS could be extracted efficiently (i.e. 81%) under basic conditions, resulting in a MDL at around 0.25 ng/L (González-Barreiro *et al.* 2006). The commonly applied LLE and SPE methods were also adopted from the review article of Gonzales-Barreiro *et al.* (2006), and those are briefly discussed below.

The typical method of LLE involved addition of NaCl to water samples to give a final concentration of 50 g/L, and the pH was adjusted to 4 with sulfuric acid. After this step, the target analytes were extracted once (10 min) with methyl *tert*-butyl ether (MTBE) (~10:1 v/v) and twice (2 × 10 min) with MTBE (v:v ~20:1) directly in the sampling containers. The solvent was removed by rotary vacuum concentration (75 mbar, 35°C) to a final volume of 1 mL.

The typical method of SPE also started with dissolving NaCl in 500 mL water samples to give a final concentration of 4 g/L prior to SPE. The samples were adjusted to pH 4 with sulfuric acid. SPE cartridges, such as Isolute 1 g, C18 end-capped cartridges and others listed in Table 3.5, were then employed to extract PFCs from water samples. The cartridges were conditioned with 2 × 1.5 mL methanol (MeOH)–ethyl acetate (1:1 v/v) mixture, 2 × 1.5 mL MeOH, and 2 × 1.5 mL H_2O adjusted to pH 4 with sulfuric acid. The sample was loaded at a flow rate of 7 mL/min and the cartridges were rinsed with 1.5 mL H_2O–MeOH (95:5 v/v) solution, and dried completely under a nitrogen stream for 1 h. The analytes were eluted with 5 mL MeOH–ethyl acetate (1:1 v/v) mixture. Finally, the solvents were reduced in volume by vacuum (75 mbar, 35°C) and adjusted to an exact volume of 1 mL with MeOH.

A number of new methods to extract and concentrate PFCs in water included the use of nanoparticles. Zhang *et al.* (2010) synthesized chitosan-coated octadecyl-functionalized magnetite nanoparticles ($Fe_3O_4 - C_{18}$ – chitosan MNPs), as an adsorbent to extract trace PFCs from environmental water samples (Zhang *et al.* 2010). The concentration factor of 1000 was achieved with a 500-mL water sample. The detection limits ranged from 0.075 to 0.24 ng/L. Surfactant-coated Fe_3O_4 magnetic nanoparticles were also developed as an efficient adsorbent for PFCs in water with a concentration factor of 1600 (Zhao *et al.* 2011). Besides nanoparticles, a new passive sampler, developed by modifying a polar organic chemical integrative sampler (POCIS) with a weak anion exchange sorbent (Strata XAW), was created by Kaserzon *et al.* (2012) to quantify perfluoroalkylated carboxylic acids (PFCAs) and perfluoroalkylated sulfonates (PFSAs) in water.

Table 3.5 Quantification methods of PFCs.

Target	Matrix	Extraction	Instrumental Analysis	MDL	References
PFBA, PFPeA, PFHxA, PFHpA, PFOA, PFNA, PFDA, PFUnDA, PFDoDA	River water, STP effluents	SPE (Oasis WAX SPE cartridges)	LC-MS/MS	Recovery: 53–126%	(Ye *et al.* 2014)
PFCAs, PFSAs, PFASAs	Surface seawater	SPE (Oasis WAX SPE cartridges)	UPLC-MS/MS	0.01–10.25 pg/L	(González-Gaya *et al.* 2014)
21 PFASs,	River water	SPE	LC-MS/MS	55–94%	(Campo *et al.* 2015)
PFOS, PFOA	Sewage	SPE (Oasis HLB)	UPLC-MS/MS	91–97% for wastewater	(Castiglioni *et al.* 2015)

Note: PFUnDA – Perfluoroundecanoic Acid; PFDoDA – Perfluorododecanoic Acid.

3.3.1.2 Instrumental determination of perfluorinated compounds

The PFCs, after being extracted and preconcentrated from water samples, are commonly detected with LC/MS in negative-ionization mode. For example, Gonzales-Barreiro *et al.* (2006) applied a Luna C18 column to separate individual PFCs from each other and from background constituents. The mobile phase was prepared from water containing 10 mM ammonium acetate and MeOH. A gradient ramping from 10% MeOH to 98% MeOH was used to elute PFCs. Castiglioni *et al.* (2015) used an XTerra MS C18 and a chromatographic gradient of 10 mM ammonium acetate and acetonitrile. After elution on LC, a triple quadrupole mass spectrometer provides excellent sensitivity for quantification of PFCs. Table 3.5 shows selected recent studies which quantified multiple PFCs in various environmental aqueous phases. For example, Castiglioni *et al.* (2015) applied Ultra Performance Liquid Chromatography (UPLC)-MS/MS to quantify PFOS and PFOA in wastewater, which achieved 91–97% recovery with 0.9–1.6 ng/L PFCs. The MS parameters of this study are shown in Table 3.6.

Table 3.6 Tandem mass detector parameters for the detection of PFOS and PFOA.

Analytes	Precursor ion (Da)	Product ion (Da)	Dwell time (msec)	Declustering Potential (V)	Focusing Potential (V)	Collision Energy (eV)
PFOS	499	80	100	–40	–180	-86
	499	99	100			–68
PFOA	413	169	100	–20	–100	-26
	413	369	100			–16

Source: Adapted from Castiglioni *et al.* (2015).

Alternatively, GC-MS may be employed for the detection of PFCs, although literature is scarce applying GC-based techniques. The ionic PFCAs and PFSAs, after derivatization, may be analyzed with GC. The derivatization agents may include a mixture of 2,4-difluoroaniline and *N,N*-dicyclohexylcarbodiimide or isopropanol (Langlois *et al.* 2007). Tetrabutylammonium hydroxide (TBAH) was used for derivatization via an in situ pyrolytic alkylation reaction that occurred in the GC injector and generated butyl PFOS isomer derivatives. The non-ionic PFCs are amenable to GC without derivatization due to their higher volatility. Langlois *et al.* (2007) developed a high-resolution GC (HRGC) combined with mass spectrometry (MS) method for both ionic and non-ionic PFCs. In this context, the separation of PFOS isomers was performed using a capillary GC-column coated with 5%-phenyl-95%-methylpolysiloxane. Chu and Letcher (2009) developed a method that employed in-port derivatization-gas chromatography-mass spectrometry (GC-MS) to identify and quantitatively determine linear PFOS (L-PFOS) and branched (monotrifluoromethyl and bistrifluoromethyl) isomers in a PFOS technical product and in environmentally relevant biological samples.

In recent years, new methods for quantification of PFCs have been developed. For example, Deng *et al.* (2014) modified a wooden tip probe on ambient mass spectrometry for rapid analysis of PFCs in complex environmental and biological samples. The detection limits ranged from 0.06 to 0.59 ng/L. Boone *et al.* coupled isotope dilution with SPE-LC-MS/MS analysis, which achieved a detection limit as low as 0.034 ng/L for surface water and 0.044 ng/L for drinking water (Boone *et al.* 2014). Sensor technology has also been developed for the detection of PFCs. Gong *et al.* invented a new molecularly imprinted photoelectrochemical sensor to detect PFOA in real water samples with a detection limit of 0.01 μg/L (Gong *et al.* 2015).

3.3.2 Polybrominated diphenylethers

Polybrominated diphenylethers (PBDEs), are organobromine compounds that are the most common brominated flame retardants. They have been widely applied into many commercial products, such as sleepwear, cushions in chairs and sofas, computers, plastics, and electronics. PBDEs have been increasingly manufactured since the polybrominated biphenyls were phased out after the 1980s. PBDEs work as flame retardants by releasing bromine free radicals when heated, which reduce both the rate of combustion and dispersion of fire. The application of flame retardants has decreased the casualties and property damage related to fire incidents. However, PBDEs are persistent in the environment and also frequently detected in human blood. The PBDE intakes from ingestion, inhalation or dermal contact, can act as endocrine system disruptors and also tend to deposit in human adipose tissues. In this section, the analytical methods for quantification of PBDEs are summarized.

3.3.2.1 Sample preparation for polybrominated diphenylethers

Glassware is preferably used for water sampling of PBDEs because PBDEs are lipophilic and tend to adsorb to plastic surfaces. Acidification and low temperature are also recommended to preserve samples prior to analysis. PBDEs are most frequently extracted by LLE and SPE. In addition, hollow fiber–liquid phase microextraction (HFLPME) and cloud point extraction ultrasound-assisted back-extraction (CPE-UABE) are also applied as extraction techniques for PBDEs from the aqueous phase.

Liquid-liquid extraction: Novak *et al.* developed an extraction method by using a Tris-citrate solution and iso-octane (Novak *et al.* 2014). The method is briefly summarized as follows: a 300-mL water sample was acidified with nitric acid to pH 2 and transferred to a 1-L glass bottle. A 300-mL buffer solution of Tris-citrate (pH 6) was then mixed with the water sample, followed by the addition of 3,3′,4,4′-tetrabromodiphenyl ether (BDE 77) as an internal standard. Then, 2 mL of iso-octane was added as an extracting agent. After shaking for 6 h, the organic phase was collected and blown with nitrogen to reduce volume to approximately 25 μL. Finally, the organic phase was transferred into a 2-mL amber GC vial.

Solid phase extraction: Labadie *et al.* developed an SPE method for extracting PBDEs from river water using customized SPE cartridges (Labadie *et al.* 2010). The SPE cartridges were prepared by filling an Oasis MAX cartridge (150 mg) with a homogenous mixture of C18 and SAX sorbents (500 mg each). The sorbent bed was covered by a polyethylene frit. The two-layer cartridge was then conditioned by passing successively 10 mL of hexane/acetone (1:1, v/v), 5 mL of acetone, 10 mL of methanol, and 10 mL of ultra-pure water. Aqueous samples were loaded at a flow rate of approximately 10 mL/min using a vacuum manifold. Cartridges were vacuum-dried for 1 h and PBDEs were then eluted with 3 mL of acetone followed by 12 mL of hexane/acetone (1:1, v/v) and 13 mL of dichloromethane/methanol/formic acid (95:4.5:0.5, v/v). An anhydrous sodium sulfate column (2 g) was then used to remove water content from extracts. Finally, the volume of extracts was reduced to approximately 100 μL by nitrogen gas.

3.3.2.2 Instrumental determination of polybrominated diphenylethers

After extraction, the final analytes of PBDEs are commonly quantified by mass spectrometry. GC-negative chemical ionization (NCI)-MS, GC-high resolution-MS, and GC-ion trap-MS are the most commonly used techniques. Fast tandem-GC-MS and LC-MS have been also applied for the analysis of higher molecular weight PBDEs. This section introduces two examples for the instrumental determination of PBDEs, which used GC-MS and LC-MS, respectively.

GC-MS: Extracts (in organic solvents) were analyzed for PBDEs using high-resolution GC with mass spectrometric detection operated in both negative chemical ionization (GC-NCI-MS) and electron impact (GC-EI-MS) modes (Stapleton *et al.* 2006). In this context, multiple laboratories participated and applied different GC-MS configurations. Analyses were performed on an Agilent (Palo Alto, CA) GC-MS system using GC model 6890 and MS model 5973. With GC-NCI-MS method, bromine atoms (m/z = 79 and 81) were the only target ions. As for GC-EI-MS method, the molecular ion peaks were monitored as well as their corresponding major fragments. The detail GC set up is briefly introduced in Table 3.7.

Table 3.7 GC-MS methods for PBDE quantification.

	Method 1	**Method 2**	**Method 3**
GC column	0.25 mm × 15 m fused silica capillary column coated with a 5% (mole fraction) phenyl methylpolysiloxane column (DB-5MS; J&W Scientific Inc., Folson, CA)	0.25 mm × 60 m fused silica capillary column coated with a 5% phenyl methylpolysiloxane (DB-5MS; 0.25 μm film thickness)	0.25 mm × 15 m fused silica capillary column coated with a 5% phenyl methylpolysiloxane (DB-5MS)
Injection	On-column injection	On-column injection	Split injection
GC-MS temperature programing	0 min 80°C 2 min 80°C 7 min 140°C 35 min 280°C 55 min 280°C	0 min 60°C 1 min 60°C 5 min 150°C 70 min 200°C 120 min 300°C 140 min 300°C	0 min 90°C 1 min 90°C 5 min 160°C 29 min 280°C 49 min 280°C 55 min 300°C 65 min 300°C
	The transfer line was maintained at 280°C and the ion source and quadrupole were maintained at 200°C and 100°C, respectively. Helium was used as the carrier gas with a constant flow rate of 1.5 mL/min.	Helium was used as the carrier gas with a constant flow of 1.2 mL/min. The transfer line was maintained at 300°C and the ion source and quadrupole were maintained at 250°C and 150°C, respectively.	The injector and transfer lines were maintained at 250°C and 290°C, respectively, and the ion source and quadrupole were maintained at 250°C and 105°C, respectively.

Source: Adapted from Stapleton et al. (2006).

LC-MS: An example using the LC-MS technique to detect PBDEs is adopted from Lagalante and Oswald (2008). A Shimadzu Prominence HPLC system (Shimadzu, Colombia, MD, USA) equipped with a Phenomenex Gemini C_{18} column (250 × 3 mm, 5 μm particle) was used. An isocratic mobile phase of 92:8 methanol/water (v/v) at 1.0 mL/min was applied. The detection of PBDEs using multiple reaction monitoring (MRM) was performed on an Applied BioSystems 3200 Q-TRAP equipped with an Applied Biosystems PhotoSpray source (Framingham, MA, USA). For each PBDE studied, the optimized compound declustering potential (DP) and collision energy (CE) are shown in Table 3.8. The collision cell exit potential (CXP) was maintained at −4 V and the entrance potential (EP) was maintained at −10 V for all compounds. The resolution of Q1 for precursor ion selection was fixed at a low resolution and the resolution of Q3 was fixed at open resolution to allow isotope transmission through the quadrupoles and to maximize sensitivity. The dwell time for each MRM transition was 1000 ms.

Table 3.8 MRM conditions for PBDE detection.

PBDE Congener	Precursor (m/z)	Fragment (m/z)	DP (V)	CE (V)
BDE-28	343	79	−47	−55
BDE-47	422	79	−17	−45
BDE-99	501	421	−54	−32
BDE-100	501	266	−52	−38
BDE-153	580	499	−55	−33
BDE-154	580	266	−53	−43
BDE-183	659	499	−57	−40
BDE-209	896	736	−35	−36

Source: Adapted from Lagalante and Oswald (2008).

3.3.3 Polychlorinated compounds

Polychlorinated compounds (PCCs), including polychlorinated biphenyls (PCBs) and polychlorinated alkanes (PCAs), are organochlorine compounds that are mainly used as coolants and in heat transfer fluid. They have adverse environmental and human health effects, such as their endocrine-disrupting properties, and have been classified as persistent organic pollutants (POPs). As a result, the usage of PCCs has been abandoned or greatly reduced (McGovern, 2006). This section discusses quantification of PCCs in a biological system media.

3.3.3.1 Sample preparation for polychlorinated compounds

Extracting PCBs and PCAs from environmental liquid samples applies approaches including SPME, SPE, hollow fiber microextraction (HFME), stir-bar sorptive extraction (SBSE), and liquid-phase microextraction (LPME). Figure 3.4, illustrating essential steps for SPME, SPE, SBSE, and LPME, is adapted from a review article by Grześkowiak *et al.* (2016).

Solid-phase microextraction: PCCs samples are frequently prepared using SPME, including headspace SPME, for preconcentration purposes. In terms of sorbents, organic SPME fibers, such as polydimethylsiloxane (PDMS) fiber, cross-linked polymeric ionic liquid (PIL) SPME fiber, carbowax DVB fiber and polyaniline SPME fibers have been successfully applied in SPME for the detection of PCCs (Castells *et al.* 2003, 2004; Wang *et al.* 2009; Song *et al.* 2012; Kowalski *et al.* 2013; Joshi *et al.* 2014). In addition, the metal–organic framework (MOF) was developed for SPME for PCC extraction, which improved their extraction efficiency (Wu *et al.* 2014).

Solid phase extraction: PCCs are also commonly extracted from aqueous samples by SPE techniques. C_{18} sorbent and hydrophilic polymeric sorbents such as Bond Elut-C_{18}, Strata-X, and Oasis HLB are normally applied for SPE procedures. For example, Wolska *et al.* applied SPE cartridges (C_{18}) to extract seven PCCs from river water (Wolska *et al.* 1999). The cartridges were washed with 2 mL dichloromethane, followed by activation with methanol (2 × 3 mL) and water (3 mL). A sample of 0.5 L was passed through the SPE cartridge with an average sample flow rate of 2.8 mL/min. Analytes trapped by the SPE cartridges were extracted with dichloromethane (2 × 2 mL) and the extract was gently evaporated under nitrogen to 0.3 mL. Besides commercial SPE sorbents, some synthesized materials, such as MOFs and zerovalent metal-based materials, were also applied for the extraction of PCCs (Chen *et al.* 2013; Karamani *et al.* 2013).

Hollow fiber microextraction: Although the hollow fiber (HF) technique is usually used with liquid inside the fiber, Zang *et al.* (2013) and Li *et al.* (2008) used MOF as a micro-SPE sorbent inside a closed HF. The modification is interesting but inherent disadvantages such as long-lasting sample preparation and substantial effort to prepare the extraction device (i.e. pack the fiber with the MOF sorbent) cannot be ignored.

Stir-bar sorptive extraction: PCCs can be isolated from the water samples applying stir bars with a PDMS coating layer. Instead of liquid desorption, Perez-Carrera *et al.* (2007) reported a method with thermal desorption of analytes from the stir bar and their cryofocusing before chromatographic separation. This method allowed for the analysis of all PCBs and many other analytes adsorbed on the stir bar.

Liquid-phase microextraction: Unlike the above methods which use solid sorbent for extraction, LPME methods isolate PCCs using solvent. Chloroform was proved to have proper emulsion and the highest recovery among several tested solvents for ultrasound-assisted emulsification-microextraction by Ozcan *et al.* (2009). Only 10 minutes were needed for extraction and 5 minutes for centrifugation of this procedure. Dispersive liquid-liquid microextraction (DLLME) developed by Rezaei *et al.* can be even faster (Rezaei *et al.* 2008). For the extraction of PCBs, extracting solvent (10 μL of chlorobenzene) and dispersing solvent (500 μL of acetone) were injected into a water sample (5 mL). After the injection, the dispersion was formed

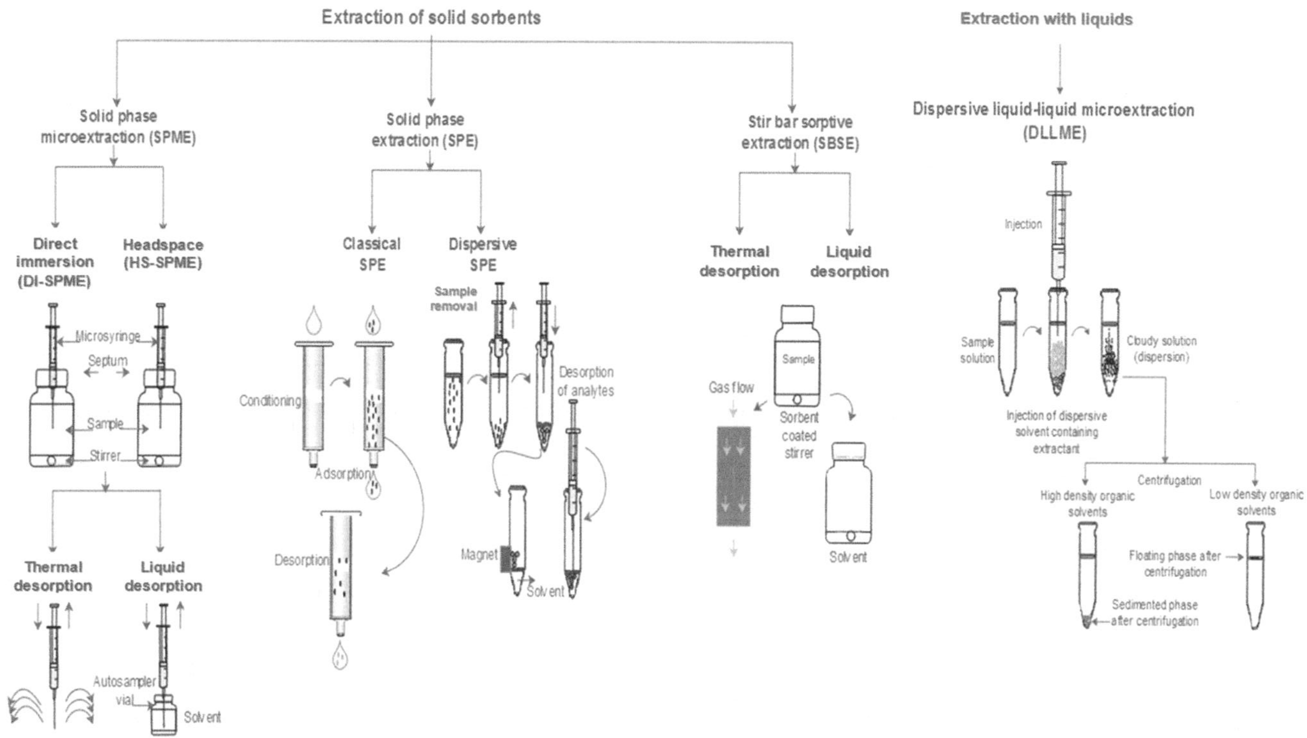

Figure 3.4 Scheme of different extraction techniques using solid sorbents and solvents (adapted from Grześkowiak *et al.* (2016)).

and resulted in a large contact area between the phases and a rapid establishment of the equilibrium state. The extract was centrifuged for 2 min and then subjected to the chromatographic analysis.

3.3.3.2 Instrumental determination of polychlorinated compounds

Gas chromatography with electron capture detection (GC-ECD) and GC-MS are the most commonly applied instruments for the quantification of PCCs. Table 3.9 is adopted from a review article of Grzeskowiak *et al.* (2016), which reviewed approaches in sample preparation for trace analysis of polychlorinated biphenyls, alkyl phenols, and parabens. Joshi *et al.* separated and detected 21 PCB congeners using GC-ECD equipped with a HP-5 capillary column (30 m × 0.25 mm I.D., 0.25 mm film thickness) (Joshi *et al.* 2014). The GC was programed as follows: initial temperature was set at 60°C and ramped to 130°C at 25°C/min. The temperature was then increased to 300°C at 8°C/min and held for 2 min. The ECD detector temperature was set to 300°C and the argon/methane make-up flow was set to 60 mL/min.

Table 3.9 Determination of PCCs (including PCBs and PCAs) in aqueous phase.

Target	Matrix	Extraction	Instrumental Analysis	MDL	References
PCBs	seawater	HS-SPME (PDMS)	GC-MS	0.3–7.5 ng/L	(Song *et al.* 2012)
PCBs	river water	SPE (C18)	GC-MS	N.A.	(Wolska *et al.* 1999)
PCBs	river water	HS-SPME (PANI)	GC-ECD	0.05–0.1 ng/L	(Wang *et al.* 2009)
PCBs	surface water & wastewater	dSPE	GC-MS/MS	1.07–1.57 ng/L	(Chen *et al.* 2013)
PCBs	river water	HF-microSPE	GC-MS/MS	0.16–0.63 ng/L	(Zang *et al.* 2013)
PCBs	seawater	SBSE (PDMS)	GC-MS	0.4–6 ng/L	(Pérez-Carrera *et al.* 2007)
PCBs	well & river water & seawater	DLLME	GC-ECD	1–2 ng/L	(Rezaei *et al.* 2008)
PCBs	lake & tap water	dSPE	GC-MS	N.A.	(Karamani *et al.* 2013)
PCAs	river water	SPE (C_{18})	GC-NCI-MS	N.A.	(Nicholls *et al.* 2001)
PCAs	river water	SPME	GC-ECD	0.3 μg/L	(Castells *et al.* 2003)
PCAs	tap & river water	SPE/SPME	GC-NCI-MS	5–20 ng/L	(Castells *et al.* 2004)

Analysis of PCAs in environmental samples, including short-, medium- and long-chain PCAs, is mainly performed by GC-ECD or GC coupled with electron capture negative-ionization (ECNI) (or negative chemical ionization, NCI) mass spectrometry due to the high selectivity and sensitivity of GC-NCI (Feo *et al.* 2010). In this method, quantification relies on monitoring [M-Cl] ions for each of the groups of congeners with the same carbon-chain length and number of chlorine atoms. The results strongly depend on the degree of chlorination of the applied standard.

3.3.4 Alkyl phenols

The long-chain alkyl phenols (APs) are used extensively as precursors to detergents, as additives for fuels and lubricants, polymers, and as components in phenolic resins. The release of APs to the environment is posing threats to ecological and human health because of their endocrine-disrupting properties. This section summarizes the quantification of APs, which include alkylphenolethoxylates (APEOs), octylphenol (OP), nonylphenol (NP) and bisphenol-A (BPA).

3.3.4.1 Sample preparation for alkyl phenols

To preserve APs in aqueous samples, acidification using strong acid is a common approach to adjust sample pH to below 3. Particularly, formaldehyde solutions have been used to prevent biological degradation of APEOs (Vega Morales *et al.* 2009). To date, SPE is the most commonly applied extraction technique to separate and preconcentrate APs from aqueous phase, along with LLE and other techniques which have been occasionally used.

Solid-phase extraction: Multiple cartridges have been successfully applied in SPE for the extraction of APs. For example, the Oasis HLB cartridge was used in rainwater and wastewater (Jahnke *et al.* 2004; Zhang *et al.* 2012); the PS-DVB cartridge was used for rainwater, wastewater and lake water (Snyder *et al.* 1999; Zgoła-Grześkowiak & Grześkowiak, 2009); and the C_{18} cartridge for water samples from wastewater treatment plants and natural water bodies (Vega Morales *et al.* 2009; Ciofi *et al.* 2014). After cartridges were loaded, organic solvents, including methanol, acetonitrile, dichloromethane, and ethyl acetate have been applied to elute target APs. The volume of extract phase was further reduced by N_2 gas purge or vacuum.

Solid-phase microextraction: SPME has also been applied to extract APs from aqueous samples. Especially, researchers have developed new materials as a highly-selective extraction phase for separating APs from the background matrix. Examples include benzyl functionalized polymeric ionic liquids (López-Darias *et al.* 2010) and zeolitic imidazolate functionalized silicafibers (Yu & Yan, 2013). Commercially available SPME fibers have also been tested for the extraction of APs, in which carbowax/templated resin (CW/TPR) coating fiber was found to be the best one (Cai *et al.* 2004). In addition, headspace SPME has been used to

reproductive dysfunction to male fish (Jobling *et al.* 2005; Liney *et al.* 2005). The group of estrogens, including E1, αE2, βE2, E2, and EE2, has been identified to have high endocrine-disrupting potential. As a result, many studies have made efforts in developing analytical methods to quantify estrogens in aqueous compartments.

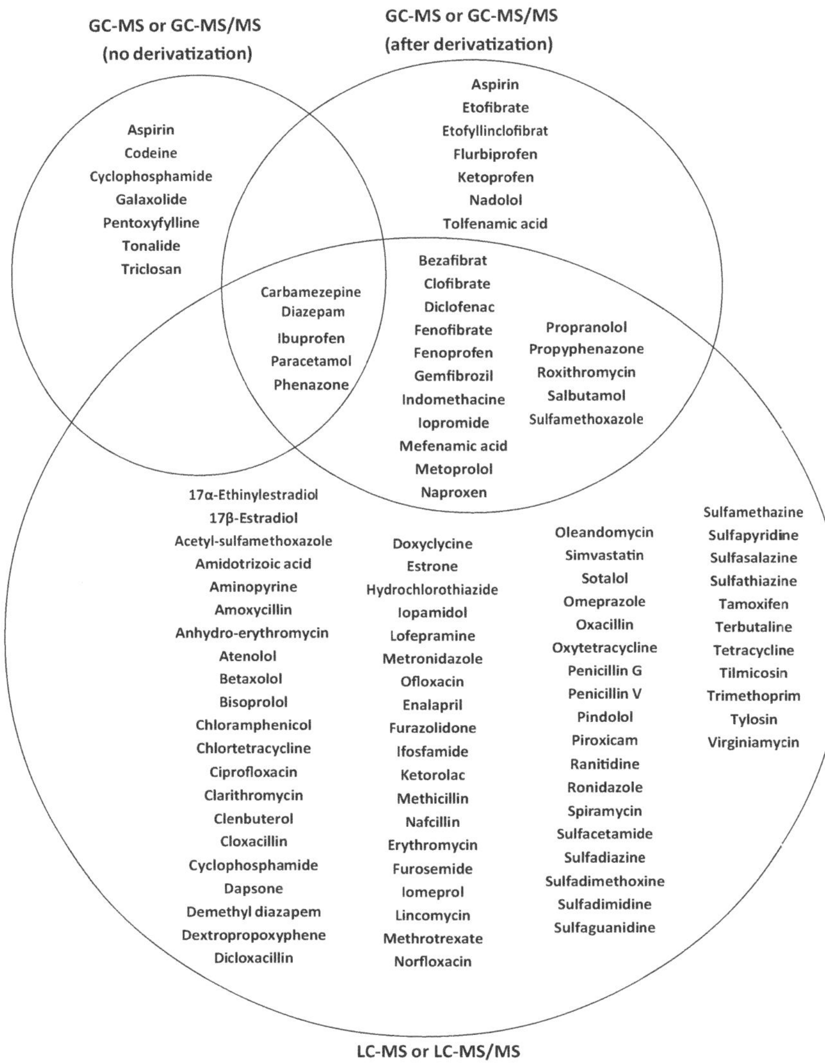

Figure 3.5 Analytical methods applied for the most commonly detected pharmaceuticals in water and wastewater (adapted from Fatta *et al.* (2007)).

3.3.6.1 Sample preparation for estrogens

Sample pretreatment, such as filtration and acidification, is commonly applied to remove particulates from the aqueous phase and preserve samples. As estrogens are hydrophobic, precautions should be taken in the selection of sample containers and filters to prevent adsorption. SPE is frequently used to extract estrogens from biological system media while LLE is also applied sporadically.

Solid-phase extraction: Similar to the extraction of other organic micropollutants from the aqueous phase, SPE is the most commonly applied method to extract and preconcentrate estrogens in surface water and wastewater matrices. To date, Oasis HLB is the most used SPE cartridge (Streck, 2009; Wille *et al.* 2012). Other options may include Strata-X and Strata C18E (Vulliet *et al.* 2008). Extraction disks (e.g. C18XF Speedisk), which have the benefits of enhanced filtration capacity and resistance to clogging, have also been used for estrogen extraction from the aqueous phase (Streck, 2009). A typical SPE protocol for estrogen extraction is described as follows: after the cartridges are conditioned with methanol and nano-pure water sequentially, water samples are loaded through the cartridges at a flow rate of approximately 10 mL/min. Then, the cartridges are dried under vacuum or with nitrogen gas. For dirty samples, such as wastewater and rainfall runoffs, a sequential wash with 10 mL nano-pure water, 10 mL methanol/water mixture (1:1 v/v), and 10 mL acetone/water (1:2 v/v) is applied to remove background compounds before drying the cartridges. Finally, estrogens on the cartridges are eluted with organic solvents (e.g. methanol).

Besides conventional SPE, many studies have used automatic on-line SPE workstations to process large sample quantities. Sorbents that are commonly used are still Oasis HLB. Before analysis on LCMS or GCMS, samples sometimes need further clean-up steps, which may include using silica gel, Florisil or NH_2 cartridges (Wille *et al.* 2012).

Liquid-liquid extraction: Dichloromethane (DCM) is frequently used to extract estrogens in LLE. Below is an example study, which can simultaneously determine 30 estrogenic compounds combining LLE with GC-MS detection (Ikonomou *et al.* 2008). Aqueous samples were extracted with DCM by shaking for 2 min and sonicating for 1 min. The extraction process was repeated three times. The volume of the extracts was further blown down to 1 to 2 mL by nitrogen gas. To remove water residue, anhydrous sodium sulfate was mixed with the extracts. Finally, Florisil was applied to further clean up the matrix. After passing through Florisil, estrogens were eluted with 20 mL DCM and 20 mL 3:7 acetone:DCM, and the volume of the clean extracts was reduced under nitrogen gas. In order to be detected on GC-MS, derivatization was conducted using bis(trimethylsilyl) trifluoroacetamide (BSTFA) containing anhydrous pyridine and trimethylchlorosilane.

3.3.6.2 Instrumental determination of estrogens

Both LC-MS and GC-MS have been applied to detect estrogens. GC-MS offers high separation power and good identification capabilities. However, derivatization of estrogens is commonly required prior to GC analysis, in order to provide better separation and volatility. The most common derivatization reaction is silylation using *N*-methyl-*N*-(trimethylsilyl) trifloroacetamide (MSTFA) or (BSTFA) (Streck, 2009). Besides silylation methods, pentafluoropropionic acid anhydride (PFPA) has also been used for derivatizing estrogens (Díaz-Cruz *et al.* 2003).

Below an example is presented of detection of estrogens using GC-MS (Yu *et al.* 2007). The extracts from SPE were derivatized by mixing with 200 μL MTBSTFA and 1% TBDMSC. The derivatization was conducted in capped vials at 60°C for 90 min. After that, they were cooled down quickly and extracts were transferred into GC vials which were then analyzed by GC-MS. The GC-MS system was installed with a fused-silica column (DB 1701, 30 m × 0.25 mm, 0.25 μm) with helium as the carrier gas (at 1.2 mL/min). Injection of a 4 μL sample was performed with a split/splitless injector at a temperature of 220°C and held splitless for 60 s. The GC oven was programed as follows: 3 min at 45°C, first ramp 20°C/min to 200°C, 5 min at 200°C, second ramp 10°C/min to 250°C, 5 min at 250°C, third ramp 5°C/min to 300°C, 5 min at 300°C. The total analysis time was approximately 41 min. The GC-MS interface temperature was set at 300°C. The mass detector was set at selected ion mode (SIM) by using electron impact ionization (70 eV). For higher sensitivity and lower detection limits, GC-MS/MS methods are also used for analysis of estrogens (Ternes *et al.* 2002).

LC-MS has also been used for the quantification of estrogens because the labor-intensive derivatization is not required. LC columns can efficiently separate different types of estrogens. For example, Vulliet *et al.* applied LC-MS for quantification of 12 selected estrogens with the detection limits from 0.9 to 4.3 ng/L (Vulliet *et al.* 2008). The LC-MS system was equipped with a Zorbax Eclipse XDB C_{18} (100 mm × 2.1 mm, 3.5 μm) and a mass detector with electrospray ionization. The mobile phase was a mixture of formic acid/water (pH 3.4) and acetonitrile with gradient elution from 20 to 66% acetonitrile in 9 min (held for 3.5 min), then from 66 to 100% acetonitrile in 22 min. The flow rate was 0.3 mL/min. The MS detector was set at the following conditions: nebulizer pressure 60 psi; drying gas (N_2) temperature 350°C; drying gas flow 13 L/min; capillary voltage 5,000 V; fragmentor voltage 75 V. Besides the Zorbax Eclipse XDB C_{18} column, other columns have also been shown to be suitable for the separation of estrogens, including Synergi Max-RP C_{12} and Zorbax SB-Phenyl columns (Matějíček & Kubáň, 2007). In recent years, more advanced analytical techniques have been increasingly applied for the detection of estrogens, such as UPLC-MS, LC-QqQ-MS, UPLC-TOF-MS and UPLC-qTOF-MS/MS (Stavrakakis *et al.* 2008; Streck, 2009; Kozlík *et al.* 2011). UPLC offers a shorter analysis time while more advanced MS detectors improve either selectivity or confirmation ability. However, due to the extremely low concentrations of estrogens found in biological systems, advance MS techniques (e.g. MS/MS, TOFMS, HRMS) are preferred.

3.3.7 Disinfection by-products

Disinfection by-products (DBPs) refer to compounds formed through reactions of organic matter with disinfectants. Traditional DBPs include trihalomethanes (THMs) and haloacetic acids (HAAs) while emerging DBPs include nitrogen-containing DBPs which are thought to be highly toxic and carcinogenic (Padhye *et al.* 2009). Although many water and wastewater treatment utilities are moving away from chlorine disinfection, alternative disinfectants, such as monochloramine, chlorine dioxide, or ozone, have been found to be equally culpable in forming DBPs (Padhye *et al.* 2011a). This section introduces quantification of DBPs in a biological system media.

3.3.7.1 Sample preparation for disinfection by-products

This subsection summarizes basic and more recent advances in sample preparation techniques for analysis of DBPs.

Liquid-liquid extraction: Poly-halogenated DBPs, such as THMs and haloacetonitriles, are hydrophobic small organic molecules. Therefore, LLE is suitable for extracting them from the aqueous phase. The commonly applied extraction solvents include MTBE, pentane and hexane. For example, for the analysis of THMs, the U.S. EPA Method 551.1 suggests using 3 mL MTBE for a 50-mL water sample. In order to improve separation of organic solvent and water sample, sodium sulfate is commonly used. For carboxylic DBPs (i.e. HAAs), derivatization is required to transform carboxylic groups to hydrophobic moieties. The commonly applied derivatization agents may include methanol (under acidic conditions) (U.S. EPA Method 552.2).

Head space extraction: Most identified DBPs are also volatile compounds, which can be purged from the aqueous phase into head space where suitable sorbents can be applied to concentrate DBPs. The U.S. EPA Method 524.2 applies head space extraction coupled with GC-MS for the determination of more than 80 small organic compounds in the aqueous phase. Kristinana *et al.* developed a method using headspace solid-phase microextraction to extract the halonitriles from water, which are then analyzed using gas chromatography–mass spectrometry (HS SPME/GC–MS) (Kristiana *et al.* 2012).

Solid-phase extraction: In the context of DBP analysis, SPE is commonly applied to extract relatively hydrophobic DBPs from the aqueous phase. The use of loose activated carbon particles for SPE of nitrosamines was a popular choice but has been shown to generate analytical artifacts for nitrosamine measurements, particularly in wastewater samples (Padhye *et al.* 2010; Padhye *et al.* 2011b). The U.S. EPA Method 521 uses cartridges packed with coconut charcoal to extract *N*-nitrosodimethylamine (NDMA) and six other nitrosamines (*N*-nitrosomethylethylamine (NMEA), *N*-nitrosodiethylamine (NDEA), *N*-nitroso-di-*n*-propylamine (NDPA), *N*-nitroso-di-*n*-butylamine (NDBA), *N*-nitrosopyrrolidine (NPYR), and *N*-nitrosopiperidine (NPIP)) in drinking water. Qin *et al.* employed Waters Oasis HLB cartridges to extract halobenzoquinones from drinking water samples (Qin *et al.* 2010). Fernández-Molina and Silva developed a method based on the continuous in situ derivatization/extraction of aldehydes on a Telos™ ENV -SPE column impregnated with 2,4-dinitrophenylhydrazine for detection of six low-molecular-mass substituted benzaldehydes in treated water (Fernández-Molina & Silva, 2013).

Table 3.10 Determination of DBPs in water samples.

DBP Name	Sample Preparation	Instrumental Analysis	References
Trihalomethanes (THMs)	LLE: A 50-mL sample aliquot is extracted with 3 mL of MTBE or 5 mL of pentane. Head space extraction	GC-ECD GC-MS	U.S. EPA Method 551.1 U.S. EPA Method 524.2
Haloacetic acids (HAAs)	Derivatization-LLE	GC-ECD GC-MS	U.S. EPA Method 552.2
Haloacetonitriles	LLE: A 50-mL sample aliquot is extracted with 3 mL of MTBE or 5 mL of pentane. Head space extraction	GC-ECD GC-MS	U.S. EPA Method 551.1 (Kristiana *et al.* 2012)
Nitrosamines	SPE LLE	GC/CI-MS/MS GC/EI-MS GC/EI-MS	U.S. EPA Method 521 (Padhye *et al.* 2009; Padhye *et al.* 2011a; Padhye *et al.* 2013)
Haloquinone	SPE	LC-MS/MS	(Qin *et al.* 2010; Huang *et al.* 2013; Wang *et al.* 2014; Zhu *et al.* 2016)
Aldehydes	SPE	LC-MS/MS	(Fernández-Molina & Silva, 2013)

3.3.7.2 Instrumental determination of disinfection by-products

GC-MS is the major technique for quantification of DBPs (Table 3.10). Mass spectral libraries are available for rapid identification of DBPs by matching MS spectra with documented data. The NIST and Wiley databases have already included GC-MS spectra of more than 200,000 compounds. As for

GC columns, the U.S. EPA Method 551.1 recommends two types of columns. Column A (a 0.25 mm ID × 30 m fused silica capillary with chemically bonded methyl polysiloxane phase (J&W, DB-1, 1.0 m film thickness or equivalent)) is recommended as the primary analytical column unless routinely occurring analytes are not adequately resolved. Column B (a 0.25 mm ID × 30 m with chemically bonded 6% cyanopropylphenyl/94% dimethyl polysiloxane phase (Restek, Rtx-1301, 1.0 μm film thickness or equivalent)) is recommended for use as a confirmatory column when GC-MS confirmation is not sensitive enough or unavailable.

GC-electron capture detection (GC-ECD) is also commonly applied to measure DBPs, especially for determination of halogenated DBPs, because halogen atoms have strong signals on the ECD. For example, the U.S. EPA Methods 552.2 and 552.3 are commonly used to measure haloacetic acids in drinking water with detection limits as low as 0.012 mg/L.

LC-based analytical techniques are increasingly used for DBP analysis, such as LC-MS/MS, UPLC-MS/MS and LC-qTOF-MS, due to the increasing number of polar DBPs and high molecular weight DBPs that have been identified in recent years. A cross-linked polystyrene resin column with 3% acetic acid in acetonitrile:water (20:80 v/v) has also been used to separate HAAs for analysis with negative ion ESI-MS (Hashimoto & Otsuki, 1998). To separate multiple benzoaldehydes, an Ascentis® Express C18 column (150 mm × 2.1 mm; 2.7 m) was applied with 0.1% formic acid (A) and ACN:methanol (75:25 v/v) (B) solutions as the mobile phase (Fernández-Molina & Silva, 2013).

3.3.8 Nanomaterials

Nanomaterials refer to materials with one dimension of 1–100 nm. The small size of nanomaterials imparts many unique properties, including high strength, thermal stability, high conductivity, and high surface activity. Nanomaterials have been increasingly applied in many products, including cosmetics, sunscreens, clothing, automobiles, and electronics. They have also been promising in applications in medication and cancer treatment. However, the release of nanomaterials to the environment is of significant concern because their unique properties may pose threats to the ecological systems and human health.

3.3.8.1 Sample preparation for nanomaterials

There are multiple preparation methods for the detection of nanoparticles, such as ultra-filtration, dialysis, ultra-centrifugation, ultrasonication, solid phase extraction, and liquid phase extraction. The selection of methods should be based on environmental matrices as well as target nanomaterials. The detailed discussion on the separation techniques of metal nanoparticles (MNPs) and their advantages and disadvantages was provided in a review article by Leopold *et al.* (2016). For example, ultra-filtration may provide the capabilities including size cut-off, purification and preconcentration, whereas size distribution cannot be achieved by using ultra-filtration. Analytical ultra-centrifugation can separate MNPs on the basis of size, whereas its disadvantages may include agglomeration of MNPs. SPE has been frequently applied to separate MNPs from aqueous samples due to its high preconcentration factors. For example, Hadioui *et al.* used an SPE-inductively coupled plasma mass spectrometry (ICP-MS) that was coupled to a column containing a strong metal binding resin (Chelex 100) to study ZnO nanoparticles in natural water and wastewater (Hadioui *et al.* 2015). In addition to the detection of ZnO nanoparticles and the determination of a size distribution in natural waters, this study was possible to investigate the dissolved Zn among free and/or labile and strongly bound Zn fractions.

3.3.8.2 Instrumental determination of nanomaterials

The commonly carbon-based nanoparticles (CNPs) of concern may include fullerenes, carbon nanotubes (CNTs) and graphene. However, to date, only quantitative procedures have been developed for the detection of fullerenes in environmental samples. For the quantification of fullerenes, instruments including HPLC (Fortner *et al.* 2005), (time of flight)-mass spectrometry (TOF)-MS (Kozlovski *et al.* 2004; Deye *et al.* 2008) and tandem TOF-MS (Ilchenko & Cotter, 2007), have been successfully applied. Fortner *et al.* used HPLC for the analysis of fullerenes in toluene extracts from the aqueous phase (Fortner *et al.* 2005). The separation of fullerenes was performed with a 4.6 × 250 mm Cosmosil PYE column at 1 mL/min toluene mobile phase. The characteristic C_{60} absorption peaks at 336, 407, 540, and 595 nm were monitored. Kozlovski *et al.* developed a novel electrospray setup for direct analysis of multiple fullerenes by electrospray time-of-flight mass spectrometry (ES-TOF-MS) (Kozlovski *et al.* 2004). Deye *et al.* developed a HPLC-MS method, which used atmospheric pressure chemical ionization (APCI) at negative ionization (Deye *et al.* 2008). The HPLC mobile phase consisted of an isocratic (60%) hexane and (40%) isopropanol mobile phase at a flow rate of 1.5 mL/min with a YMC ODS-A column (150 mm × 6 mm, 5 m). The corona pin on the MS was operated at 15 μA and the probe was heated at 550°C and the source operated at 120°C with nitrogen gas used for desolvation.

For the quantification of metal NPs, different instruments are available (e.g. ICP-MS, electrothermal atomic absorption spectrometry (ETAAS) (Hartmann *et al.* 2013) and total reflection X-ray fluorescence spectrometry (TXRF) (Menzel & Fittschen, 2014)). For example, Scheffer *et al.* reported that using ICP-MS for the quantification of Au nanoparticles can achieve a detection limit as low as 1 pg/mL (Scheffer *et al.* 2008). ICP-MS has also been widely applied for the quantification of other metal nanomaterials including Ag, TiO_2, CeO_2, and ZnO (Geranio *et al.*

2009; Fabricius *et al.* 2014). Although using ETAAS is less common for the quantification of metal nanoparticles, one study successfully detected Ag NPs with a detection limit of 0.7 ng/L (Hartmann *et al.* 2013). The application of TXRF has been more common for solid samples than for aqueous ones because it allows quantitative direct analysis of solid samples without previous chemical treatment (Fernández-Ruiz, 2009).

3.3.9 Metals

Heavy metal pollution has been shown to pose adverse effects on humans and the ecosystem for more than a century. Conventional industrial activities involve elements such as Cd, Pb, Mn, Cu, Cr, Hg, and As, some of which end up in wastes and are further released to the environment. Historically, numerous incidents of heavy metal contamination from industrial activities have been reported. With the development of information technology, emerging amounts of new toxic metals, such as rare earth elements, have brought more and more concerns. Therefore, the analysis of heavy metals in environmental matrices and wastewater is important. In this section, the most commonly used techniques to determine heavy metals, such as flame atomic absorption spectroscopy (FAAS), inductively coupled plasma-optical emission spectrometry (ICP-OES), and ICP-MS, are introduced.

3.3.9.1 Sample preparation for metals

For the detection of trace levels of heavy metals in aqueous samples, the U.S. EPA Method 1669 specifically describes precautions when using various water sampling techniques. Before samples are collected, all sampling equipment and sample containers are cleaned using detergent, mineral acids, and reagent water. After cleaning, sample containers are filled with weak acid solution, individually double-bagged, and shipped to the sampling site. Sometimes, ligands are used to preconcentrate metals. However, precautions should be taken because a ligand appropriate for the analysis of one target metal may not be suitable for the analysis of other metals. For example, the preconcentration using iminodiacetate functionalized chelating resins is appropriate for the analysis of Cd, Cu, Pb, Ni, and Zn, whereas this procedure is not suitable for As or Ag (U.S. EPA Method 1640).

Filtration and digestion are the pretreatment steps often required. If samples only contain dissolved metals and particulates, filtration will be sufficient for pretreatment. If samples contain significant amounts of organic content, digestion will be required to remove interferences of organic content. Nitric acid is the most commonly used solution for digestion procedures. The typical protocol is, as described in the Standard Methods for the Examination of Water and Wastewater, that 100 mL of sample is heated with 5 mL concentrated nitric acid till a light colored, clear solution is obtained (American Public Health Association [APHA], 2005).

3.3.9.2 Instrumental determination of metals

This subsection summarizes some commonly applied technologies, including atomic absorption spectrometry (AAS), inductively coupled plasma-optical (or atomic) emission spectrometry (ICP-OES or ICP-AES), ICP-MS, voltammetry, and laser induced breakdown spectroscopy (LIBS), as determinative techniques for quantification of metals. In biological systems, quantification of metals is particularly challenging as most of the metals are complexed by strong ligands formed with the organic matter present in the matrix.

AAS instruments quantify elements using the absorption of optical radiation (light) by free atoms in the gaseous state. It can quantify over 70 different elements in solution or directly in solid samples. For example, Khan *et al.* applied a graphite furnace atomic absorption spectrophotometer (GF-AAS) for quantification of Cd, Cr, Cu, Ni, Pb, and Zn in radish, maize, green cabbage, spinach, cauliflower, turnip, and lettuce (Khan *et al.* 2008). El-Shahawi *et al.* used AAS to detect Au, Ag, Fe, Ni, and Cu in tap water and wastewater (El-Shahawi *et al.* 2007).

ICP-OES or ICP-AES uses the inductively coupled plasma to produce excited atoms and ions that emit electromagnetic radiation at wavelengths characteristic of a particular element. It is a flame technique with a flame temperature in the range from 6000 to 10000 K. With the ICP technique, 60 elements can be analyzed in single sample run of less than a minute simultaneously, or in a few minutes sequentially (Baysal *et al.* 2013). For example, Zhu *et al.* coupled ICP-OES with multi-walled carbon nanotube extraction methods for analysis of trace-level Cr(III), Fe(III), Pb(II), and Mn(II) ions in wastewater (Zhu *et al.* 2016).

A review by Novaes *et al.* provided the application of multivariate designs and mathematical-statistical techniques to establish optimal conditions of the analytical methods for elemental analysis using ICP-OES as a detection technique (Novaes *et al.* 2016). De Oliveria Sourze *et al.* applied ICP-OES to detect Na, Ca, Mg, Sr, and Fe in crude oil (de Oliveira Souza *et al.* 2015). Dimpe *et al.* tested different sample preparation methods for extraction of metal ions, including As, Al, Cd, Cr, Cu, Fe, Mn, Ni, Pb, and Zn, in wastewater and sludge. The concentrations of metal ions were then quantified by ICP-OES (Dimpe *et al.* 2014). In ICP-MS, the excited atoms produced by inductively coupled plasma are detected by mass spectrometer, instead of optical spectrometer, as in -ICP-OES. The mass spectrometer separates ions according to their mass-to-charge ratio and has excellent detection limits. Li *et al.* determined Pb, Co, Cu, Ni and Zn using liquid-liquid microextraction based on the solidification of floating organic drops combined with ICP-MS (Li *et al.* 2015).

In order to perform speciation analysis, ICP-MS is sometimes coupled with HPLC. For example, Jia *et al.* analyzed the speciation of Cr(III) and Cr(VI) in wastewater by short-column HPLC–ICP-MS after on-line SPE (Jia *et al.* 2016). Laser induced breakdown spectroscopy (LIBS) uses a highly energetic laser pulse to excite metals and measures their atomic emission spectroscopy. Rai and

Rai applied LIBS to determine Cr in various water samples (Rai & Rai, 2008). Kiptoo *et al.* analyzed speciation of Ni and Cr in wastewater with the use of FAAS and differential pulse adsorptive cathodic stripping voltammetry (DPAdCSV) techniques (Kiptoo *et al.* 2004).

3.3.10 Pathogens

Waterborne pathogens have been a great concern to human health. It was reported that, globally, around 80% of diseases are associated with drinking water contaminated by pathogens. Therefore, a proper assessment of the microbial contaminants in water is important for water treatment entities to meet regulations and to provide early warning of waterborne diseases. In the United States, under the National Primary Drinking Water Regulations (NPDWR), five groups of microbial contaminants (or corresponding indicators) are currently regulated, including Cryptosporidium, Giardia lamblia, Legionella, total coliforms and viruses. In addition, due to the fact that many recent outbreaks of waterborne illnesses in the United States and other parts of the world resulted from unregulated microorganisms, analytical methods for detecting emerging microbial contaminants are of urgent necessity. Several microorganisms are included on the new CCL-3 list of the U.S. EPA (https://www.epa.gov/ccl/contaminant-candidate-list-3-ccl-3). This section focuses on the detection methods of regulated microbial contaminants in a biological medium.

3.3.10.1 Sample preparation for pathogens

Filtration is the most commonly used preparation method for pathogen detection due to the diluted state of pathogens in aqueous samples. The collected cells on the filters are then eluted with either buffer solution or culture media. For viral pathogens, charged filters are applied to absorb viruses from aqueous media because size-exclusion is not suitable due to the small size of viruses. For example, the U.S. EPA Method 1615 suggests using an electropositive filter to collect enteroviruses and noroviruses from water samples. To optimize the sensitivity of target pathogens, preparation after filtration sometimes includes magnetization.

For the pathogens to be detected by molecular techniques, such as qPCR, DNA or RNA extraction is required after collecting cells. RNA extraction for hepatitis A virus is shown as an example: Samples (250 μL of lab specimens or 100 μL of water samples) were treated with sodium dodecyl sulfate and proteinase K at final concentrations of 0.5% and 5 mg/mL, respectively, incubated at 55°C for 30 min, and followed by two extractions with phenol-chloroform-isoamyl alcohol (Jothikumar *et al.* 2006). RNA was precipitated by the addition of 0.3 M ammonium acetate and 2.5 volumes of ice cold ethanol, followed by incubation at –70°C for 30 min. The pellet was centrifuged at 11,000 × *g* for 30 min, washed with 75% ethanol, then resuspended in 50 μL Tris-EDTA (TE) buffer (Jothikumar *et al.* 2006). The protocol of DNA extraction has been standardized, and is usually performed with commercially available kits.

3.3.10.2 Instrumental determination of pathogens

After sample preparation, quantification of microbial contaminants is either performed with culture-based methods or molecular-based methods.

Culture-based method: Culture-based methods have been extensively used to detect microbial pathogens in environmental matrices and are capable of detecting most of the pathogens of concern. After filtration, the typical protocol is to place the collected cells on selective media, which are kept in an incubation chamber to let individual cells form observable colonies. For example, the counting of total coliforms is suggested using MI agar, on which blue colonies are formed from coliforms (U.S. EPA Method 1604). As for *E. coli* O157 detection, the filter membrane used to collect *E. coli* O157 cells is directly put on selective growth media such as sorbitol MacConkey (SMAC) agar, or cefiximetellurite sorbitol MacConkey (CT-SMAC) agar. After incubation, the colorless colonies are formed by *E. coli* O157, distinguishing from the pink color of other *E. coli* isolates (Quilliam *et al.* 2011).

For infectious viruses, their detection relies on their inflicted damage to corresponding host cells. For example, norovirus can be detected by Buffalo Green Monkey (BGM) kidney cells (U.S. EPA Method 1615). After filtration, the concentrated sample is inoculated onto flasks of BGM cells to measure infectious viruses. Cultures are examined for the development of cytopathic effects for two weeks and then re-passaged onto fresh cultures for confirmation (U.S. EPA Method 1615).

Molecular-based method: Compared with culture-based methods, the advantages of molecular tools may include the reduction of detection time and higher true positivity rate. In a review article by Aw and Rose (Aw & Rose, 2012), molecular tools used for pathogen detection were systematically reviewed and discussed. To date, qPCR is the most popular molecular tool for pathogen detection. The commonly recognized pathogens, including *E. coli* O157, *Legionella*, *Salmonella*, adenoviruses, noroviruses and hepatitis A, have all been successfully detected by qPCR methods (Aw & Rose, 2012). Primers have been designed for multiple pathogens of concern. Commercial qPCR kits are also available for the detection of pathogens, such as *Legionella* (Merault *et al.* 2011).

3.4 CONCLUSIONS

Trace level environmental analysis is an ever-changing field as new compounds with unknown health effects are released into the environment every day, and environmental regulations are updated frequently. Due to increased water stress in many regions of the world, wastewater reuse is becoming a necessity for many water treatment utilities. Hence, it is critical to understand concentrations of contaminants not only in drinking waters, but also in wastewaters and in surface water bodies. PFCs, PBDEs, PCCs, APs, PPCPs, natural and synthetic hormones, DBPs, nanomaterials, heavy metals, and pathogens are currently a focus of environmental analytical chemistry due to their significant occurrence levels in aquatic environments around the globe.

For environmental analysis of pollutants in biological systems, the most common method of sampling is still grab sampling due to simplicity and ease of collection. However, composite sampling is preferred due to better representativeness of the collected sample in time and space. Passive sampling techniques have been gaining more popularity recently, with further advances in passive samplers in terms of their designs and efficiencies. Due to time-integrated analysis and pre-concentration of analytes, passive sampling will be a preferred sampling technique of the future.

LLE was the most common method of sample extraction in the last century. However, SPE and SPME are replacing LLE as preferred methods of sample extraction due to availability of customized adsorbent materials, reduction in quantity of toxic solvents involved in experiments, pre-concentration of analytes, and simplicity of extraction protocols. Due to rapid improvements in detection techniques, many new contaminants are getting detected every year. The ever-growing availability and increasing affordability of high resolution mass spectrometric instruments, such as QToF-MS and Orbitrap-MS, coupled with GC or LC, has enabled considerable progress in the characterization of hazardous pollutants at laboratory scale and in biological system samples where interferences from matrix components are commonplace. As knowledge of these hazardous pollutants increases and new regulations are implemented, sampling and analytical methods for pollutants will become commonplace; however, detection of new emerging contaminants will mean that environmental analysis will have to continuously adapt to changing needs.

3.5 ACKNOWLEDGEMENTS

Authors would like to acknowledge assistance of Ms Monica Blanca Monterozo for recreating some of the Figures in this Chapter.

3.6 REFERENCES

Alvarez D. and Jones-Lepp T. (2010). Sampling and analysis of emerging pollutants. In: Water Quality Concepts, Sampling, and Chemical Analysis, Li Y. and Migliaccio K. (eds), CRC Press, Boca Raton, FL, pp. 199–226.

APHA (2005). WEF (2005) Standard Methods for the Examination of Water and Wastewater. American Public Health Association, American Water Works Association, and Water Environment Federation.

Aw T. G. and Rose J. B. (2012). Detection of pathogens in water: from phylochips to qPCR to pyrosequencing. *Current Opinion in Biotechnology*, **23**(3), 422–430.

Basheer C. and Lee H. K. (2004). Analysis of endocrine disrupting alkylphenols, chlorophenols and bisphenol-A using hollow fiber-protected liquid-phase microextraction coupled with injection port-derivatization gas chromatography–mass spectrometry. *Journal of Chromatography A*, **1057**(1–2), 163–169.

Baysal A., Ozbek N. and Akman S. (2013). Determination of trace metals in waste water and their removal processes. In: Waste-Water Treatment Technologies and Recent Analytical Developments (ed. Einschlag F.S.G., Carlos L.), pp. 145–171. INTECH Open Access Publisher, Rijeka.

Boone J. S., Guan B., Vigo C., Boone T., Byrne C. and Ferrario J. (2014). A method for the analysis of perfluorinated compounds in environmental and drinking waters and the determination of their lowest concentration minimal reporting levels. *Journal of Chromatography A*, **1345**, 68–77.

Buchberger W. W. (2011). Current approaches to trace analysis of pharmaceuticals and personal care products in the environment. *Journal of Chromatography A*, **1218**(4), 603–618.

Cai Y., Jiang G., Liu J., Liang X., Yao Z., Liu J., Liu J. and Zhou Q. (2004). Solid-phase microextraction coupled with high performance liquid chromatography-fluorimetric detection for the determination of bisphenol A, 4-n-Nonylphenol, and 4-tert-Octylphenol in environmental water samples. *Analytical Letters*, **37**(4), 739–753.

Campo J., Pérez F., Masiá A., Picó Y., la Farré M. and Barceló D. (2015). Perfluoroalkyl substance contamination of the Llobregat River ecosystem (Mediterranean area, NE Spain). *Science of The Total Environment*, **503**, 48–57.

Cantero M., Rubio S. and Pérez-Bendito D. (2006). Determination of alkylphenols and alkylphenol carboxylates in wastewater and river samples by hemimicelle-based extraction and liquid chromatography–ion trap mass spectrometry. *Journal of Chromatography A*, **1120**(1–2), 260–267.

Castells P., Santos F. and Galceran M. (2003). Solid-phase microextraction for the analysis of short-chain chlorinated paraffins in water samples. *Journal of Chromatography A*, **984**(1), 1–8.

Castells P., Santos F. and Galceran M. (2004). Solid-phase extraction versus solid-phase microextraction for the determination of chlorinated paraffins in water using gas chromatography–negative chemical ionisation mass spectrometry. *Journal of Chromatography A*, **1025**(2), 157–162.

Castiglioni S., Valsecchi S., Polesello S., Rusconi M., Melis M., Palmiotto M., Manenti A., Davoli E. and Zuccato E. (2015). Sources and fate of perfluorinated compounds in the aqueous environment and in drinking water of a highly urbanized and industrialized area in Italy. *Journal of Hazardous Materials*, **282**, 51–60.

CFR (2017). Title 40 Code of Federal Regulations. http://www.ecfr.gov/cgi-bin/text-idx?tpl=/ecfrbrowse/Title40/40tab_02.tpl, last accessed January 2017.

Chary N. S. and Fernandez-Alba A. R. (2012). Determination of volatile organic compounds in drinking and environmental waters. *TrAC Trends in Analytical Chemistry*, **32**, 60–75.

Chen X., Ding N., Zang H., Yeung H., Zhao R.-S., Cheng C., Liu J. and Chan T. W. D. (2013). Fe3O4@MOF core–shell magnetic microspheres for magnetic solid-phase extraction of polychlorinated biphenyls from environmental water samples. *Journal of Chromatography A*, **1304**, 241–245.

Chu S. and Letcher R. J. (2009). Linear and branched perfluorooctane sulfonate isomers in technical product and environmental samples by in-port derivatization-gas chromatography-mass spectrometry. *Analytical Chemistry*, **81**(11), 4256–4262.

Ciofi L., Ancillotti C., Chiuminatto U., Fibbi D., Checchini L., Orlandini S. and Del Bubba M. (2014). Liquid chromatographic–tandem mass spectrometric method for the simultaneous determination of alkylphenols polyethoxylates, alkylphenoxy carboxylates and alkylphenols in wastewater and surface-water. *Journal of Chromatography A*, **1362**, 75–88.

Deng J., Yang Y., Fang L., Lin L., Zhou H. and Luan T. (2014). Coupling solid-phase microextraction with ambient mass spectrometry using surface coated wooden-tip probe for rapid analysis of ultra trace perfluorinated compounds in complex samples. *Analytical Chemistry*, **86**(22), 11159–11166.

de Oliveira Souza M., Ribeiro M. A., Carneiro M. T. W. D., Athayde G. P. B., de Castro E. V. R., da Silva F. L. F., Matos W. O. and de Queiroz Ferreira R. (2015). Evaluation and determination of chloride in crude oil based on the counterions Na, Ca, Mg, Sr and Fe, quantified via ICP-OES in the crude oil aqueous extract. *Fuel*, **154**, 181–187.

Deye J. R., Shiveley A. N., Oehrle S. A. and Walters K. A. (2008). Separation of substituted fullerenes using non-aqueous reversed-phase liquid chromatography–mass spectrometry. *Journal of Chromatography A*, **1181**(1), 159–161.

Díaz-Cruz M. S., López de Alda M. J., López R. and Barceló D. (2003). Determination of estrogens and progestogens by mass spectrometric techniques (GC/MS, LC/MS and LC/MS/MS). *Journal of Mass Spectrometry*, **38**(9), 917–923.

Dimpe K. M., Ngila J. C., Mabuba N. and Nomngongo P. N. (2014). Evaluation of sample preparation methods for the detection of total metal content using inductively coupled plasma optical emission spectrometry (ICP-OES) in wastewater and sludge. *Physics and Chemistry of the Earth, Parts A/B/C*, **76–78**, 42–48.

El-Shahawi M. S., Bashammakh A. S. and Bahaffi S. O. (2007). Chemical speciation and recovery of gold(I, III) from wastewater and silver by liquid–liquid extraction with the ion-pair reagent amiloride mono hydrochloride and AAS determination. *Talanta*, **72**(4), 1494–1499.

Fabricius A.-L., Duester L., Meermann B. and Ternes T. A. (2014). ICP-MS-based characterization of inorganic nanoparticles – sample preparation and off-line fractionation strategies. *Analytical and Bioanalytical Chemistry*, **406**(2), 467–479.

Fatta D., Achilleos A., Nikolaou A. and Meric S. (2007). Analytical methods for tracing pharmaceutical residues in water and wastewater. *TrAC Trends in Analytical Chemistry*, **26**(6), 515–533.

Feo M. L., Eljarrat E. and Barcelo D. (2010). A rapid and sensitive analytical method for the determination of 14 pyrethroids in water samples. *Journal Chromatogr A*, **1217**(15), 2248–2253.

Fernández-Molina J. M. and Silva M. (2013). Trace determination of low-molecular-mass substituted benzaldehydes in treated water using micro solid-phase extraction followed by liquid chromatography–mass spectrometric detection. *Journal of Chromatography A*, **1300**, 180–186.

Fernández-Ruiz R. (2009). Three empirical cases of the deposition morphology influence in the analytical quality of direct solid suspension measurements by total-reflection X-ray fluorescence. *Spectrochimica Acta Part B: Atomic Spectroscopy*, **64**(7), 672–678.

Fortner J., Lyon D., Sayes C., Boyd A., Falkner J., Hotze E., Alemany L., Tao Y., Guo W. and Ausman K. (2005). C60 in water: nanocrystal formation and microbial response. *Environmental Science & Technology*, **39**(11), 4307–4316.

Geranio L., Heuberger M. and Nowack B. (2009). The behavior of silver nanotextiles during washing. *Environmental Science & Technology*, **43**(21), 8113–8118.

Gong J., Fang T., Peng D., Li A. and Zhang L. (2015). A highly sensitive photoelectrochemical detection of perfluorooctanic acid with molecularly imprined polymer-functionalized nanoarchitectured hybrid of AgI–BiOI composite. *Biosensors and Bioelectronics*, **73**, 256–263.

González-Barreiro C., Martínez-Carballo E., Sitka A., Scharf S. and Gans O. (2006). Method optimization for determination of selected perfluorinated alkylated substances in water samples. *Analytical and Bioanalytical Chemistry*, **386**(7–8), 2123–2132.

González-Gaya B., Dachs J., Roscales J. L., Caballero G. and Jiménez B. (2014). Perfluoroalkylated substances in the global tropical and subtropical surface oceans. *Environmental Science & Technology*, **48**(22), 13076–13084.

Grześkowiak T., Czarczyńska-Goślińska B. and Zgoła-Grześkowiak A. (2016). Current approaches in sample preparation for trace analysis of selected endocrine-disrupting compounds: focus on polychlorinated biphenyls, alkylphenols, and parabens. *TrAC Trends in Analytical Chemistry*, **75**, 209–226.

Hadioui M., Merdzan V. and Wilkinson K. J. (2015). Detection and characterization of ZnO nanoparticles in surface and waste waters using single particle ICPMS. *Environmental Science & Technology*, **49**(10), 6141–6148.

Hartmann G., Hutterer C. and Schuster M. (2013). Ultra-trace determination of silver nanoparticles in water samples using cloud point extraction and ETAAS. *Journal of Analytical Atomic Spectrometry*, **28**(4), 567–572.

Hashimoto S. and Otsuki A. (1998). Simultaneous determination of haloacetic acids in environmental waters using electrospray ionization liquid chromatography mass spectrometry. *Journal of High Resolution Chromatography*, **21**(1), 55–58.

Henze M., Loosdrecht M. C. M. v., Ekama G. A. and Brdjanovic D. (2008). Biological Wastewater Treatment Principles, Modelling and Design. IWA Publishing, London.

Houde F., DeBlois C. and Berryman D. (2002). Liquid chromatographic–tandem mass spectrometric determination of nonylphenol polyethoxylates and nonylphenol carboxylic acids in surface water. *Journal of Chromatography A*, **961**(2), 245–256.

Huang R., Wang W., Qian Y., Boyd J. M., Zhao Y. and Li X.-F. (2013). Ultra pressure liquid chromatography–negative electrospray ionization mass spectrometry determination of twelve halobenzoquinones at ng/L levels in drinking water. *Analytical Chemistry*, **85**(9), 4520–4529.

Ikonomou M. G., Cai S. S., Fernandez M. P., Blair J. D. and Fischer M. (2008). Ultra-trace analysis of multiple endocrine-disrupting chemicals in municipal and bleached kraft mill effluents using gas chromatography–high-resolution mass spectrometry. *Environmental Toxicology and Chemistry*, **27**(2), 243–251.

Ilchenko S. and Cotter R. J. (2007). Collision energetics in a tandem time-of-flight (TOF/TOF) mass spectrometer with a curved-field reflectron. *International Journal of Mass Spectrometry*, **265**(2), 372–381.

Jahnke A., Gandrass J. and Ruck W. (2004). Simultaneous determination of alkylphenol ethoxylates and their biotransformation products by liquid chromatography/ electrospray ionisation tandem mass spectrometry. *Journal of Chromatography A*, **1035**(1), 115–122.

Jia X., Gong D., Xu B., Chi Q. and Zhang X. (2016). Development of a novel, fast, sensitive method for chromium speciation in wastewater based on an organic polymer as solid phase extraction material combined with HPLC–ICP-MS. *Talanta*, **147**, 155–161.

Jobling S., Williams R., Johnson A., Taylor A., Gross-Sorokin M., Nolan M., Tyler C. R., van Aerle R., Santos E. and Brighty G. (2005). Predicted exposures to steroid estrogens in UK rivers correlate with widespread sexual disruption in wild fish populations. *Environmental Health Perspectives*, **114**, 32–39.

Jonsson G., Cavcic A., Stokke T. U., Beyer J., Sundt R. C. and Brede C. (2008). Solid-phase analytical derivatization of alkylphenols in fish bile for gas chromatography–mass spectrometry analysis. *Journal of Chromatography A*, **1183**(1–2), 6–14.

Joshi M. D., Ho T. D., Cole W. T. S. and Anderson J. L. (2014). Determination of polychlorinated biphenyls in ocean water and bovine milk using crosslinked polymeric ionic liquid sorbent coatings by solid-phase microextraction. *Talanta*, **118**, 172–179.

Jothikumar N., Cromeans T. L., Robertson B. H., Meng X. and Hill V. R. (2006). A broadly reactive one-step real-time RT-PCR assay for rapid and sensitive detection of hepatitis E virus. *Journal of Virological Methods*, **131**(1), 65–71.

Karamani A. A., Douvalis A. P. and Stalikas C. D. (2013). Zero-valent iron/iron oxide-oxyhydroxide/graphene as a magnetic sorbent for the enrichment of polychlorinated biphenyls, polyaromatic hydrocarbons and phthalates prior to gas chromatography–mass spectrometry. *Journal of Chromatography A*, **1271**(1), 1–9.

Kaserzon S. L., Kennedy K., Hawker D. W., Thompson J., Carter S., Roach A. C., Booij K. and Mueller J. F. (2012). Development and calibration of a passive sampler for perfluorinated alkyl carboxylates and sulfonates in water. *Environmental Science & Technology*, **46**(9), 4985–4993.

Keith L. H. (1992). Environmental Sampling and Analysis: a Practical Guide. Lewis Publishers, Chelsea, Mich.

Khan S., Cao Q., Zheng Y., Huang Y. and Zhu Y. (2008). Health risks of heavy metals in contaminated soils and food crops irrigated with wastewater in Beijing, China. *Environmental Pollution*, **152**(3), 686–692.

Kiptoo J. K., Ngila J. C. and Sawula G. M. (2004). Speciation studies of nickel and chromium in wastewater from an electroplating plant. *Talanta*, **64**(1), 54–59.

Kowalski C. H., da Silva G. A., Godoy H. T., Poppi R. J. and Augusto F. (2013). Application of Kohonen neural network for evaluation of the contamination of Brazilian breast milk with polychlorinated biphenyls. *Talanta*, **116**, 315–321.

Kozlík P., Bosáková Z., Tesařová E., Coufal P. and Čabala R. (2011). Development of a solid-phase extraction with capillary liquid chromatography tandem mass spectrometry for analysis of estrogens in environmental water samples. *Journal of Chromatography A*, **1218**(15), 2127–2132.

Kozlovski V., Brusov V., Sulimenkov I., Pikhtelev A. and Dodonov A. (2004). Novel experimental arrangement developed for direct fullerene analysis by electrospray time-of-flight mass spectrometry. *Rapid Communications in Mass Spectrometry*, **18**(7), 780–786.

Kristiana I., Joll C. and Heitz A. (2012). Analysis of halonitriles in drinking water using solid-phase microextraction and gas chromatography–mass spectrometry. *Journal of Chromatography A*, **1225**, 45–54.

Labadie P., Tlili K., Alliot F., Bourges C., Desportes A. and Chevreuil M. (2010). Development of analytical procedures for trace-level determination of polybrominated diphenyl ethers and tetrabromobisphenol A in river water and sediment. *Analytical and Bioanalytical Chemistry*, **396**(2), 865–875.

Lagalante A. F. and Oswald T. D. (2008). Analysis of polybrominated diphenyl ethers (PBDEs) by liquid chromatography with negative-ion atmospheric pressure photoionization tandem mass spectrometry (LC/NI-APPI/MS/MS): application to house dust. *Analytical and Bioanalytical Chemistry*, **391**(6), 2249–2256.

Langlois I., Berger U., Zencak Z. and Oehme M. (2007). Mass spectral studies of perfluorooctane sulfonate derivatives separated by high-resolution gas chromatography. *Rapid Communications in Mass Spectrometry*, **21**(22), 3547–3553.

Leopold K., Philippe A., Wörle K. and Schaumann G. E. (2016). Analytical strategies to the determination of metal-containing nanoparticles in environmental waters. *TrAC Trends in Analytical Chemistry*, **84**, 107–120.

Li, G., Zhang L. and Zhang Z. (2008). Determination of polychlorinated biphenyls in water using dynamic hollow fiber liquid-phase microextraction and gas chromatography–mass spectrometry. *Journal of Chromatography A*, **1204**(1), 119–122.

Li Y., Peng G., He Q., Zhu H. and Al-Hamadani S. M. Z. F. (2015). Dispersive liquid–liquid microextraction based on the solidification of floating organic drop followed by ICP-MS for the simultaneous determination of heavy metals in wastewaters. *Spectrochimica Acta Part A: Molecular and Biomolecular Spectroscopy*, **140**, 156–161.

Liney K. E., Jobling S., Shears J. A., Simpson P. and Tyler C. R. (2005). Assessing the sensitivity of different life stages for sexual disruption in roach (Rutilus rutilus) exposed to effluents from wastewater treatment works. *Environmental Health Perspectives*, **113**(10), 1299–1307.

Loconto P. R. (2006). Trace Environmental Quantitative Analysis Principles, Techniques, and Applications. CRC/Taylor & Francis, Boca Raton, FL.

López-Darias J., Pino V., Meng Y., Anderson J. L. and Afonso A. M. (2010). Utilization of a benzyl functionalized polymeric ionic liquid for the sensitive determination of polycyclic aromatic hydrocarbons; parabens and alkylphenols in waters using solid-phase microextraction coupled to gas chromatography–flame ionization detection. *Journal of Chromatography A*, **1217**(46), 7189–7197.

Matějíček D. and Kubáň V. (2007). High performance liquid chromatography/ion-trap mass spectrometry for separation and simultaneous determination of ethynylestradiol, gestodene, levonorgestrel, cyproterone acetate and desogestrel. *Analytica Chimica Acta*, **588**(2), 304–315.

McGovern V. (2006). PCBs are endocrine disruptors: mixture affects reproductive development in female mice. *Environmental Health Perspectives*, **114**(6), A368.

Menzel M. and Fittschen U. E. A. (2014). Total reflection X-ray fluorescence analysis of airborne silver nanoparticles from fabrics. *Analytical Chemistry*, **86**(6), 3053–3059.

Merault N., Rusniok C., Jarraud S., Gomez-Valero L., Cazalet C., Marin M., Brachet E., Aegerter P., Gaillard J. and Etienne J. (2011). Specific real-time PCR for simultaneous detection and identification of Legionella pneumophila serogroup 1 in water and clinical samples. *Applied and Environmental Microbiology*, **77**(5), 1708–1717.

Nicholls C., Allchin C. and Law R. (2001). Levels of short and medium chain length polychlorinated n-alkanes in environmental samples from selected industrial areas in England and Wales. *Environmental Pollution*, **114**(3), 415–430.

Novaes C. G., Bezerra M. A., da Silva E. G. P., Santos A. M. P. d., Romão I. L. d. S. and Santos Neto J. H. (2016). A review of multivariate designs applied to the optimization of methods based on inductively coupled plasma optical emission spectrometry (ICP OES). *Microchemical Journal*, **128**, 331–346.

Novak P., Zuliani T., Milačič R. and Ščančar J. (2014). Development of an analytical procedure for the determination of polybrominated diphenyl ethers in environmental water samples by GC–ICP-MS. *Analytica Chimica Acta*, **827**, 64–73.

Ozcan S., Tor A. and Aydin M. E. (2009). Determination of selected polychlorinated biphenyls in water samples by ultrasound-assisted emulsification-microextraction and gas chromatography-mass-selective detection. *Analytica Chimica Acta*, **647**(2), 182–188.

Padhye L., Tezel U., Mitch W. A., Pavlostathis S. G. and Huang C.-H. (2009). Occurrence and fate of nitrosamines and their precursors in municipal sludge and anaerobic digestion systems. *Environmental Science & Technology*, **43**(9), 3087–3093.

Padhye L., Wang P., Karanfil T. and Huang C.-H. (2010). Unexpected role of activated carbon in promoting transformation of secondary amines to N-nitrosamines. *Environmental Science & Technology*, **44**(11), 4161–4168.

Padhye L., Luzinova Y., Cho M., Mizaikoff B., Kim J.-H. and Huang C.-H. (2011a). PolyDADMAC and dimethylamine as precursors of N-Nitrosodimethylamine during ozonation: reaction kinetics and mechanisms. *Environmental Science & Technology*, **45**(10), 4353–4359.

Padhye L. P., Hertzberg B., Yushin G. and Huang C.-H. (2011b). N-Nitrosamines formation from secondary amines by nitrogen fixation on the surface of activated carbon. *Environmental Science & Technology*, **45**(19), 8368–8376.

Padhye L. P., Kim J.-H. and Huang C.-H. (2013). Oxidation of dithiocarbamates to yield N-nitrosamines by water disinfection oxidants. *Water Research*, **47**(2), 725–736.

Padhye L. P., Yao H., Kung'u F. T. and Huang C.-H. (2014). Year-long evaluation on the occurrence and fate of pharmaceuticals, personal care products, and endocrine disrupting chemicals in an urban drinking water treatment plant. *Water Research*, **51**(0), 266–276.

Pan Y.-P. and Tsai S.-W. (2008). Solid phase microextraction procedure for the determination of alkylphenols in water by on-fiber derivatization with N-tert-butyl-dimethylsilyl-N-methyltrifluoroacetamide. *Analytica Chimica Acta*, **624**(2), 247–252.

Pérez-Carrera E., León V. M. L., Parra A. G. and González-Mazo E. (2007). Simultaneous determination of pesticides, polycyclic aromatic hydrocarbons and polychlorinated biphenyls in seawater and interstitial marine water samples, using stir bar sorptive extraction–thermal desorption–gas chromatography–mass spectrometry. *Journal of Chromatography A*, **1170**(1–2), 82–90.

Petrović M., Hernando M. D., Díaz-Cruz M. S. and Barceló D. (2005). Liquid chromatography–tandem mass spectrometry for the analysis of pharmaceutical residues in environmental samples: a review. *Journal of Chromatography A*, **1067**(1), 1–14.

discussion is provided on the role of nitrifying sludges in cometabolizing a number of well-known hazardous organic pollutants. In addition, examples are provided on biodegradation rates of specific organic pollutants under different redox conditions. Then, inhibitory effects of hazardous pollutants are discussed using several inhibition models. Another section of the chapter is allocated to the impact and transformation abilities of inorganic hazardous pollutants, a group that is of great interest nowadays in biological treatment. In that regard, special attention is paid to speciation, biotransformation, biosorption and inhibitory effects of metals. Also, a discussion is provided on the behavior and effects of nanometals and hazardous ions.

4.2 IMPORTANCE OF HAZARDOUS POLLUTANTS IN BIOLOGICAL TREATMENT

Hazardous pollutants include a high number of chemicals consisting of the main groups of fragrances, gasoline additives, industrial solvents, engineered nanoparticles, perfluoroalkylated substances, personal care products, pesticides, pharmaceuticals, wood preservatives, algal toxins, anticorrosives, antifoaming agents, antifouling compounds, antioxidants, biocides, complexing agents, detergents, disinfection by-products and flame retardants. As literature indicates, among these pollutants, a special group consisting of pharmaceuticals and personal care products, collectively abbreviated as PPCPs, receives special attention since in everyday life their use has dramatically increased.

Most of the hazardous pollutants discussed in this chapter are encountered nowadays in domestic wastewater and thus are found in wastewater treatment plants (WWTPs). Hazardous pollutants are also found in the treatment of industrial wastewaters, or the like, in which they may reach quite appreciable levels. As a result of the input of industrial wastewaters without or with pretreatment into the sewer system, such pollutants enter domestic WWTPs. Another source of the pollutants found in WWTPs is hospital wastewaters and surface runoff. As discussed in **Chapter 2**, in the water environment, these pollutants can lead to adverse effects. Therefore, nowadays great efforts are spent in understanding their behavior under the different operating conditions of WWTPs, in particular in biological treatment units.

Depending on the type of carbon and energy source, the main biological processes occurring in bioreactors can be listed as follows:

(1) Aerobic heterotrophic (e.g. removal of organics from domestic wastewater under aerated conditions),
(2) Aerobic autotrophic (e.g. nitrification as the initial step in nutrient removal from domestic or other types of wastewaters),
(3) Anoxic heterotrophic (e.g. denitrification of the oxidized nitrogen formed in nitrification of domestic wastewater etc.),

(4) Fermentative (e.g. formation of volatile fatty acids as the first step of anaerobic removal in sludge digestion),
(5) Anaerobic (e.g. digestion of treatment sludge and formation of methane, removal of organics from highly concentrated wastewaters etc.),
(6) Sulfate reduction (e.g. the anaerobic treatment of industrial wastewaters based on microbial sulfate reduction), and
(7) Anaerobic autotrophic (e.g. the Anaerobic Ammonium Oxidation (Anammox) process in which anaerobic bacteria oxidize ammonium nitrogen and reduce nitrite nitrogen to nitrogen gas).

In all biological treatment systems involving one or more of these biological processes, the importance of hazardous pollutants arises due to the following:

(1) *Hazardous pollutants can affect biological processes*: For example, in a municipal WWTP, depending on the type and concentration, many organic and inorganic hazardous pollutants can affect aerobic organic carbon removal, nitrification and denitrification processes. After treatment of wastewater, these pollutants still continue to exert their effects, this time in processes such as anaerobic sludge digestion. Inhibitors may also affect the treatment of high ammonia wastewaters by the Anammox process.
(2) *Biological processes may affect hazardous pollutants*: Biological processes, such as aerobic oxidation, nitrification, denitrification or anaerobic digestion, may have different effects on hazardous pollutants. The type of a biological process determines
    (a) the probability of biodegradation/biotransformation of an organic pollutant or the biotransformation of an inorganic pollutant,
    (b) the extent of removal or biotransformation as well as formation of innocuous or harmful intermediate products, and
    (c) pathways of biodegradation or biotransformation.

Although biological treatment is often the best alternative in removal of most hazardous pollutants, experience indicates that most of these pollutants are either poorly eliminated or not eliminated at all in biological treatment units.

4.3 HAZARDOUS ORGANIC POLLUTANTS: BASIC REMOVAL MECHANISMS

As shown in Figure 4.1, in biological treatment units one possibility for elimination of hazardous organics is biodegradation/biotransformation. In almost any case, the preferred elimination mechanism for a hazardous pollutant is biodegradation, since biodegradation mostly leads to the complete conversion of a pollutant into harmless end products. In many cases abiotic mechanisms also come into play. Abiotic mechanisms can be divided into two, as physical and chemical. Although

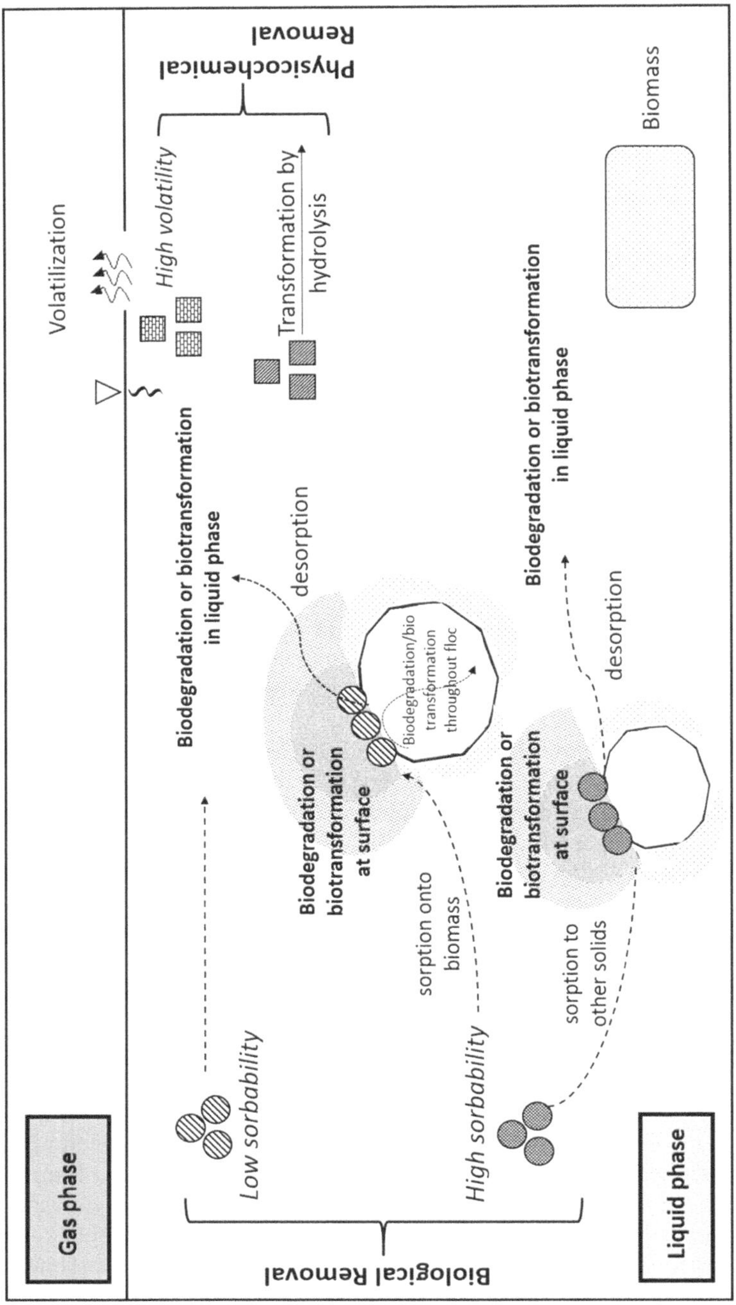

Figure 4.1 Removal mechanisms of hazardous organic pollutants in a biological reactor.

the boundaries between these two are often not clearly defined, biosorption and volatilization are often listed among physical removal mechanisms. Biosorption of a hazardous pollutant is more often encountered than volatilization. In contrast to biodegradation, removal of a pollutant by biosorption or volatilization results in a phase transfer only. Chemical removal mechanisms include hydrolysis, photooxidation, chemooxidation and chemoreduction.

4.3.1 Biodegradation and biotransformation

4.3.1.1 Primary, acceptable and ultimate biodegradation

Primary biodegradation is the effective breakdown of an organic substrate by microorganisms. It is often determined using analytical measurements. Biodegradation takes place in many steps; it may involve a variety of microorganisms such as bacteria and fungi. But, also enzymes alone may be used for biodegradation purposes.

In primary biodegradation the chemical structure of an organic substrate is altered to such an extent that it loses its original characteristics. Therefore, if a test monitors the disappearance of the initial substrate alone, this may lead to a wrong idea about its removal since it may not have completely decayed into end products yet. Primary biodegradation of a substrate may take place by the action of non-specific degraders that can feed on a number of different substrates or alternatively, by specific degraders that have the ability to degrade the substrate of interest. In some cases the metabolites of degradation may even be more problematic than the parent substrate itself.

Acceptable biodegradation means the change of a molecule by microorganisms to such an extent that toxic or undesirable characteristics of the substrate are eliminated. This term has often been used to denote the changes taking place in pollutants such as surfactants that may be rendered into a harmless form if discharged into surface waters.

Ultimate biodegradation shows the transformation potential of a parent substrate by microorganisms into minerals. Under aerobic conditions, the substrate is converted into carbon dioxide, biomass, water and other inorganic substances like ammonia. Similarly, ultimate anaerobic biodegradation is the breakdown of an organic substrate by microorganisms in the absence of oxygen to carbon dioxide, methane, water and mineral salts of any other elements present (mineralization) plus biomass (ISO, 2013). Generally, anaerobic biodegradation takes place at much slower rates than aerobic biodegradation.

While the definitions above indicate the extent of biodegradation, there are also terms describing the "readiness" of biodegradation. Biodegradable organics are roughly divided into two, as readily and slowly biodegradable. In that regard, a readily biodegradable organic pollutant is classified as one which has passed specified screening tests for ultimate biodegradability. Based on these tests, it is assumed that it will rapidly and completely biodegrade in the water environment under aerobic conditions (European Commission, 1999).

Readily biodegradable substances include volatile fatty acids and low molecular weight (LMW) compounds, such as acetic acid and glucose, respectively. On the other hand, most high molecular weight (HMW) or synthetically produced compounds are slowly biodegradable. Nowadays, many compounds found in industrial effluents and in domestic wastewaters, such as pharmaceuticals, hormones and household cleaners fall into the category of slowly biodegradable organics (Grady *et al.* 2011).

As explained in detail in **Chapter 5**, to predict the biodegradability of hazardous pollutants in wastewater treatment, many computational tools may be used. Also, standard tests exist for the assessment of biodegradability of single substances or their mixtures under different redox conditions. Examples of such tests are provided in **Chapter 5**.

4.3.1.2 Effect of substrate properties on biodegradation

The intrinsic biodegradation potential of an organic pollutant depends largely on its molecular structure and physicochemical characteristics. Among molecular characteristics one can list the molecular structure, molecular weight, functional groups and extent of branching. The physicochemical properties of a pollutant such as its solubility, hydrophobicity and volatility are also of importance. Regarding biodegradation, the significance of all factors can be summarized as follows:

(a) *Functional groups and biodegradation*: Literature shows that biodegradation of a compound correlates with its molecular characteristics (Tunkel *et al.* 2000). In that regard, a relationship exists between the functional groups of a compound and its biodegradation. On the basis of a large set of organic chemicals, it was found that compounds including esters, nitriles and aromatic alcohols have functional groups that may increase biodegradability, while aromatic amines, iodide, nitro and azo groups increase the persistence.

(b) *Functional groups and redox conditions*: There is also a relationship between functional groups and redox conditions. For example, a study examining the relationship between the specific molecular features of 40 trace organic contaminants and their removal in a laboratory-scale membrane bioreactor (MBR) showed that an apparent correlation existed between chemical structures and removal (Tadkaew *et al.* 2011).

In that respect, occurrence and types of electron withdrawing or donating functional groups appear to be important factors. Biodegradation of a pollutant is closely related to electron transfer. When an organic pollutant is degraded, this results in a release of electrons. If there are electron withdrawing functional groups in a compound, this generates an electron deficiency and thus renders the compound less susceptible to oxidative catabolism. As reported in literature, all hydrophilic and moderately hydrophobic compounds possessing strong electron withdrawing functional groups showed relatively low removal efficiencies. On the other hand, most compounds bearing electron donating functional groups, such as hydroxyl and primary amine groups, exhibited high removal efficiencies (Tadkaew *et al.* 2011).

Similarly, high biodegradation of some pharmaceuticals such as ibuprofen (IBF), salicylic acid, primidone, and naproxen (NPX) was attributed to the presence of strong electron donating functional groups. On the other hand, some pharmaceuticals, such as the antiepileptic drug carbamazepine (CBZ), containing electron withdrawing groups in the structure were shown to be highly resistant to biodegradation (Luo *et al.* 2014).

Another study reported that the pharmaceutical atenolol (ATN) containing one electron donating (alcohol) and one electron withdrawing group (amide) should be attacked more readily by a reductive or nucleophilic rather than by an oxidative or electrophilic mechanism. Therefore, this compound was thought to be removed by facultative anaerobes although it was classified as recalcitrant under aerobic conditions (Rattier *et al.* 2014). This finding is worthy of consideration since it contradicts most studies which have shown that hazardous pollutants are more amenable to aerobic biodegradation.

(c) *Molecular structure and biodegradation*: As reported in a review (Jones *et al.* 2005), long and highly branched side chains render a compound more persistent. Also, complicated aromatic ring structures and the presence of halogens in the structure are known to increase the persistence of a compound. In general, biodegradability of a compound decreases upon chlorination. For example, molecules such as diclofenac including a chlorine group in the structure are resistant to biodegradation. Also branching usually decreases biodegradability. As an example, in domestic wastewater treatment systems, the branched-chain alkyl benzene sulfonates (ABS) and linear alkylbenzene sulfonates (LAS) were observed to have different biodegradation capacities. Therefore, years ago there was a switch to LAS detergents.

(d) *Molecular weight and biodegradation*: The biodegradation potential of a pollutant also depends on its molecular weight. For example, HMW humic acids often found in drinking water or HMW hydrocarbons are largely nonbiodegradable.

(e) *Solubility and biodegradation*: Another physicochemical factor determining the biodegradability of a pollutant is its solubility. High solubility increases the accessibility of a pollutant for microorganisms, a phenomenon referred to as “bioavailability”. Usually, a pollutant is considered bioavailable if its concentration is high enough in the soluble phase. This consideration is based on the assumption that biodegradation takes place in the soluble phase only.

(f) *Hydrophobicity and biodegradation*: In a biological treatment unit, hydrophobic pollutants having a high capacity for sorption to either biological sludge and/or to the inorganic particulates, may become less bioavailable for microorganisms. For example, for pharmaceuticals displaying a low hydrophobicity, biodegradation was observed to be the major pathway of removal rather than sorption (Luo *et al.* 2014). However, as discussed in Section 4.3.2.4, sorption onto sludge cannot be regarded as a retarding factor for biodegradation in every case.

(g) *Volatility and biodegradation*: Volatilization occurs by stripping and surface volatilization. However, for volatilization, the threshold of the Henry's law constant is not clearly defined (Pomies *et al.* 2013). Volatilization leads to a decrease in the soluble-phase concentration of a pollutant. Therefore, the pollutant will become unavailable for microorganisms even if it is amenable to biodegradation. Volatilization further leads to the emission of the pollutant into the air phase.

4.3.1.3 Hazardous organics: Elimination as primary, secondary or cometabolic substrates

As shown in Figure 4.2, hazardous organic pollutants can be consumed by biomass as primary, secondary and cometabolic substrates.

Direct metabolism: Use of a pollutant as a primary substrate

The removal of background or bulk organic matter from domestic wastewater, industrial wastewater or landfill leachates, as defined by the sum organic parameters Biochemical Oxygen Demand (BOD_5), Chemical Oxygen Demand (COD) or Total Organic Carbon (TOC), occurs by a direct mechanism, meaning that the organic matter, usually composed of a large number of unidentifiable compounds, acts as a primary substrate which supplies the necessary energy and carbon for biomass. This type of removal is also called growth-associated removal.

Also, some hazardous pollutants found at mg/L or μg/L levels in domestic wastewater, industrial wastewater or the like, may undergo direct metabolism, meaning that they can be used as primary substrates providing the necessary carbon and energy for bacteria. For example, under aerobic conditions cationic surfactants such as Quaternary Ammonium Compounds (QACs) can be used as primary substrates.

Use of a pollutant as a secondary substrate

In addition to direct metabolism, also other bioremoval mechanisms may be relevant for hazardous organic pollutants, as shown in Figure 4.2. One of them is the use of the pollutant as a secondary rather than primary substrate. A secondary substrate is one that can be removed in the presence of other substrates that are found at concentrations to support biomass growth.

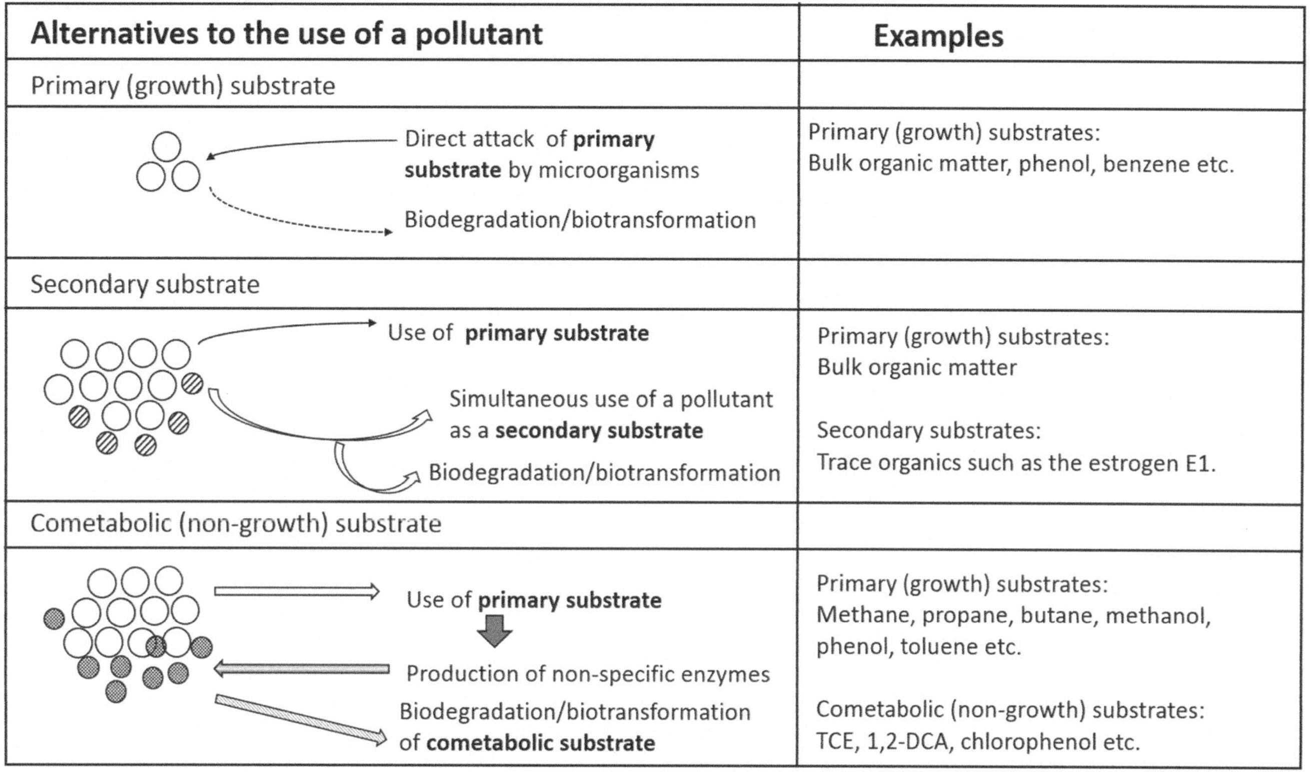

Figure 4.2 Biological removal of hazardous pollutants as primary, secondary and cometabolic substrates.

(a) *Importance of Solids Retention Time (SRT)*: SRT is an operating parameter which is of great importance in biodegradation of both non-hazardous and hazardous pollutants. As shown theoretically for biodegradable pollutants, the effluent concentration depends on the SRT (Rittmann & McCarty, 2001):

$$S = K_m \frac{1 + b \cdot SRT}{Y \cdot q_m SRT - (1 + b \cdot SRT)} \tag{4.1}$$

where:
S : the effluent substrate concentration (M_s/L^3).
K_m : half-velocity constant (M_s/L^3).
b : decay rate ($1/T$).
Y : yield coefficient (M_x/M_s).
q_m : maximum specific substrate removal rate ($M_s/M_x \cdot T$).

In the above notation, the subscripts s and x represent substrate and biomass, respectively.

SRT has two types effects on pollutant removal. First, as shown in Equation 4.1, long SRT enables the decrease of effluent concentration (S) to low levels. The second effect of long SRT is that the sludge becomes more diversified in terms of bacteria. This occurs because a long SRT favors growth of slowly-growing bacteria that would otherwise not be retained in the sludge. In aerobic systems, the most important slowly-growing bacterial group is the nitrifiers that have growth rates of about one tenth of heterotrophs. As discussed in detail in Section 4.6, with regard to biodegradation of most hazardous organic pollutants, growth of nitrifying bacteria is important since this enables the production of a monooxygenase that brings about cometabolic removal of many hazardous organics.

(b) *Presence of very reduced substrate concentrations*: According to Equation 4.2, at very high SRT values a minimum concentration (S_{min}) is reached below which the substrate cannot be further decreased since the growth rate of biomass will be lower than its decay rate under these conditions (Rittmann & McCarty, 2001):

$$S_{min} = K_m \frac{b}{Yq_m - b} \tag{4.2}$$

Thus, biodegradation of a pollutant can only occur if its initial concentration is higher than the corresponding S_{min} value. Even if the pollutant were biodegradable, it would not be prone to biodegradation when its concentration is lower than the S_{min} value. This is a key factor in understanding the behavior of most hazardous organic pollutants nowadays.

In most biological treatment systems, biodegradation of hazardous organic pollutants is often complicated, not due to their high concentrations but, on the contrary, due to extremely reduced concentrations, ranging from a few μg/L to ng/L. For example, a hormone present at 20 ng/L level in a domestic wastewater would not be removed since it does not support biomass growth.

However, it is often observed that many pollutants can still be biologically eliminated despite their low concentrations. In biological wastewater treatment of domestic wastewater, industrial wastewaters or landfill leachates, usually the background organic matter is high enough. This organic matter acts as a primary substrate supplying the necessary energy and carbon for biomass. But, also in drinking water supplies, if enough biodegradable organic matter (BOM) is available as a primary substrate, for micropollutants such as petroleum hydrocarbons, chlorinated hydrocarbons, and taste and odor compounds, conditions can be provided for them to be degraded as secondary substrates (Rittmann, 1995a). Besides being defined as "secondary substrate removal", sometimes, this type of utilization in conjunction with other substrates is also termed as "multiple substrate utilization".

Use of a pollutant as a cometabolic substrate

Cometabolism is the biological transformation of a non-growth substrate (also called the cometabolic substrate or the cometabolite) by microorganisms through enzymes which are induced in the presence of a growth-substrate (primary substrate) providing reducing power and energy for cell growth and maintenance. Several types of organic compounds can serve as primary substrates while in their presence many apparently nonbiodegradable compounds can be removed as cometabolites. In Figure 4.2 cometabolism of some organic compounds is illustrated.

In biological treatment units, microorganisms such as nitrifiers, methanotrophs and methanogens can lead to cometabolic transformation of anthropogenic compounds. If aerobic conditions prevail, several bacterial strains feeding on organic and inorganic growth-substrates, such as phenols, methane or ammonia, respectively, produce mono- or dioxygenase enzymes. These enzymes are non-specific, meaning that they can also lead to oxidation of others. During oxidation of the primary substrate, simultaneously oxygen is inserted into another substrate, called the cometabolic or non-growth substrate, leading to an initial enzymatic transformation. In some cases, such a transformation may even render the cometabolic pollutant biodegradable. This would then allow the pollutant to be further consumed as an energy and carbon source. The oxygen incorporated in mono- or dioxygenase reactions is called the cosubstrate. The kinetics of pollutant degradation is directly determined by concentrations of cosubstrates in some instances (Rittmann, 1995). In such reactions, for generation of energy and for regeneration of the reductive force in the form of NAD(P)H, a primary substrate is always needed. On the other hand, the non-growth substrate (the cometabolite) consumes the NAD(P)H that is generated in the presence of the primary substrate.

Cometabolism receives wide attention in the treatment of domestic wastewaters that contain a wide range of hazardous pollutants. Also in the treatment of industrial wastewaters, landfill leachates, and in preparation of surface water for drinking purposes, as well as in bioremediation of groundwater, cometabolism is often encountered.

Cometabolism can occur under aerobic, anoxic or anaerobic conditions while pathways of removal and by-product formation differ in each case. For example, the pollutant, trichloroethylene (TCE), often found in groundwater, can be biodegraded under anaerobic conditions. On the other hand, TCE cannot be used as a carbon and energy source under aerobic conditions, but it can be eliminated by aerobic cometabolism. While anaerobic degradation of TCE may sometimes lead to accumulation of a known carcinogenic product, the vinyl chloride (VC), aerobic cometabolism of TCE does usually not generate hazardous by-products.

Removal by direct metabolism or cometabolism?

In some cases it is uncertain whether pollutants are removed by direct metabolism or cometabolism in biological systems. Such uncertainties arise often in the treatment of chlorinated compounds. In recent decades, in bioremediation of polluted groundwater, removal of aliphatic and aromatic chlorinated pollutants has received much attention. Less chlorinated compounds such as VC and dichloroethylene (DCE) are amenable to direct metabolism, meaning that they can be used as primary substrates acting as electron donors. Yet, for many others cometabolism is the main mechanism of removal.

Under aerobic conditions oxidative dehalogenation of the pollutant is of interest whereby the pollutant releases its halogens into the medium. For example, in biological treatment of groundwater the conversion of TCE to DCE by cometabolism leads to the release of a chloride ion into the medium.

Under anaerobic conditions the main process leading to biological removal of hazardous chlorinated pollutants is reductive dehalogenation. In reductive dehalogenation, the hazardous pollutant is used as an electron acceptor while organic matter, or more often H_2 gas, acts as an electron donor. In this type of reaction, electrons are added to the pollutant to reduce its charge, while also removing a halogen molecule. An example to reductive dehalogenation would be the breakdown of perchloroethylene (PCE) to TCE in groundwater treatment. This type of halogen loss often occurs by cometabolism. Yet, in most cases, it is difficult to determine whether such a reaction is energy-yielding or cometabolic. If microorganisms derive energy out of reductive dehalogenation, the process is termed as dehalorespiration. Dehalorespiration has been confirmed for chlorinated ethenes.

In wastewater treatment, aerobic cometabolism is often observed in activated sludge or other aerobic treatment systems. On the other hand, reductive dehalogenation occurs under the anaerobic conditions of a sludge digester. In such cases, usually anaerobic cometabolism is encountered too, as discussed in Section 4.3.2.6.

Comparison of removal as a secondary and cometabolic substrate

Nowadays a number of pollutants found in biological treatment units are believed not to be removed by direct metabolism, but as secondary or cometabolic substrates. Yet, there is a great confusion about these two terms in literature. Misleadingly, "secondary substrate utilization" and "cometabolism" are often equated with each other, although the removal mechanisms are entirely different. Unlike a secondary substrate, a cometabolic substrate is not biodegradable at all under the indicated redox condition. This means that the molecular or physicochemical properties of this pollutant would not allow its degradation under the present redox condition, even if it were present at higher concentrations. For example, TCE is a pollutant that is practically nonbiodegradable under aerobic conditions as a primary substrate. However, it can be degraded as a cometabolite in the presence of a growth-substrate such as ammonia. In the case of ammonia oxidation, the enzyme ammonia monooxygenase (AMO) produced also attacks TCE, as discussed in detail in Section 4.6.

In the case of hazardous pollutants, it is often difficult to distinguish whether removal occurs by secondary substrate utilization or by cometabolism. However, distinguishing between these two mechanisms is important because each may have different implications for process optimization. For example, a study showed that the removal of the estrogen estrone (E1), found at very low concentrations in domestic wastewater, occurred rather by multiple substrate utilization (secondary substrate utilization). If cometabolism were the key E1 degradation mechanism, prior exposure of sludge to estrogens would not affect the removal rates (Tan *et al.* 2015).

Besides multiple substrate utilization, another possibility is that such pollutants are utilized by specialist bacteria (specialists). However, also in that case the presence of natural substrates is essential for growth of such specialists since micropollutants found at extremely low concentrations cannot support biomass growth. As exemplified in the case of bisphenol A (BPA), often present in domestic wastewaters at very low concentrations, specialists degrading BPA grew using natural substrates (Lindblom *et al.* 2006). There is always a delicate balance between bulk organic matter and micropollutants. In the case of E1, the point was to generate a sufficient quantity of biomass while maintaining a selective pressure for E1 degraders at the same time. If the pollutant is degraded by specialists, the presence of high amounts of bulk organic carbon would hinder E1 removal since E1 specialists would not be selected in the sludge then (Tan *et al.* 2015). If no bulk organic carbon is present, however, also the E1 specialists would not grow.

4.3.2 Removal of hazardous organic pollutants by biosorption

Usually, when the fate of hazardous pollutants is monitored in laboratory- or pilot-scale biological reactors, their apparent removal is attributed to "biodegradation" which is a simplistic approach that is often not valid. While in a biological system biodegradation is the main removal mechanism, the second most important one is biosorption which can be defined as the sorption of organics or other species onto biological sludge. Depending on the type and concentration of a hazardous organic or inorganic pollutant, biological activity may be adversely affected upon sorption.

4.3.2.1 Relative rates of biosorption and biodegradation

In a bioreactor, the relative magnitudes of sorption, desorption and biodegradation rates need to be known. In a continuous-flow stirred tank reactor (CSTR), the partitioning of a pollutant between the solid (biomass or other solid material) and liquid phases depends on the retention time of the reactor as well as the rates of sorption, desorption and biodegradation (Falas *et al.* 2013). In such reactors, the retention time of a pollutant in the reactor is higher than other timescales. The typical hydraulic retention time (HRT) of the wastewater in a biological tank is about a few hours while sorption takes place more rapidly. For example, a study reports that solid-water equilibrium is reached within 30 minutes for many pharmaceuticals, including diclofenac, ibuprofen and clofibric acid (Ternes *et al.* 2004). Therefore, practically a direct equilibrium is assumed between the solid and liquid phases of a bioreactor in most cases. Also, sorption of a pollutant is assumed to occur at much faster rates than its biodegradation (Pomies *et al.* 2015b). Therefore, the kinetics of sorption can usually be disregarded when compared to biodegradation.

4.3.2.2 Expression of biosorption

The sorption potential of a pollutant onto sludge may be predicted based on several parameters:

(1) *Octanol-water partition coefficient, K_{ow} (P)*: K_{ow}, also denoted as the P value, is a physicochemical parameter that stands for the hydrophobicity of a species. It is based on the partitioning of a neutral species between octanol (organic phase) and water phases. Thus, the K_{ow} value is a measure for the partitioning of species that do not ionize. Micropollutants with a low K_{ow} are less soluble in water and would have a high affinity for the solid phase. In general, low, medium and high sorption potentials are indicated by $\log K_{ow} < 2.5$, $2.5 < \log K_{ow} < 4$ and $\log K_{ow} > 4$, respectively (van Beelen, 2007).

(2) *Distribution coefficient, D*: The distribution coefficient D is similar to K_{ow}. But, it is an expression based on the distribution of a species between octanol and water phases in ionized plus unionized forms. For non-ionizable

species, K_{ow} and D values are identical. But, for ionizable species, the D value changes with the pH of water, since ionization depends on pH.

In biological systems, some researchers also take into consideration the effective hydrophobicity of a pollutant that is expressed as log D at pH 8. For example, in activated sludge systems, very hydrophobic organic pollutants (log D > 3.2) were observed to be removed by sorption to sludge. However, as the log D value of pollutants decreased below 3.2, their removal was rather attributed to intrinsic biodegradability (Tadkaew *et al.* 2011).

(3) *The solid-water partition coefficient (K_d)*: If sufficient time is allowed in a bioreactor, a pollutant partitions between the solid (biomass or other solid material) and liquid phases. At equilibrium conditions, sorption of a pollutant that is found at low concentrations is usually described by a linear model as given in Equation 4.3 (Siegrist *et al.* 2005):

$$C_{sorb} = K_d \cdot SS \cdot C_{diss} \tag{4.3}$$

where C_{sorb} is the concentration of sorbed pollutant (M_s/L^3), K_d is the solid-water partition (or sorption) coefficient (L^3/M_{SS}), SS is the suspended solids concentration (M_{SS}/L^3) and C_{diss} is the concentration of the same pollutant in dissolved phase (M_s/L^3).

The K_d parameter in Equation 4.3 is often reported in units of L/g SS or L/kg SS. By measuring the distribution of a pollutant between the dissolved and sorbed phases, the K_d parameter can be experimentally assessed. But, there is not a unified procedure for determination of K_d yet.

In contrast to the K_{ow} parameter, which is based on the physicochemical properties of a pollutant, the K_d parameter is used in characterizing the uptake potential of the pollutant onto biological sludge. Thus, this parameter also considers the presence of a specific solid phase, namely the sludge, which has its own characteristics. Therefore, this parameter may be a better indicator of sorption than K_{ow}.

Although no distinct limits have been set for the sorption potential, the removal by sorption onto sludge and particles in a domestic WWTP is considered negligible for $K_d < 0.500$ L/g SS (Ternes *et al.* 2004) or for $K_d < 0.300$ L/g SS (Joss *et al.* 2005).

(4) *Organic carbon-water partition coefficient, (K_{oc})*: This coefficient shows the sorption potential of organic pollutants to an organic sorbent like the organic matter in biological sludge or soil. The coefficient is expressed in units of mg/g which denotes the concentration of organic pollutant per the organic carbon of the respective sorbent. The K_{oc} parameter can be estimated using K_{ow} value and solubility.

Using the K_d parameter as the criterion, Table 4.1 exemplifies the sorbability of some hazardous organic pollutants to the sludge phase. Some micropollutants, such as the synthetic musks galaxolide and tonalide, are highly sorbed to biological

Table 4.1 Biosorption of selected hazardous organic pollutants.

	K_d (L/g SS)	Remarks	Reference
Examples of highly sorbing (K_d > 1.0 L/g SS) pollutants			
Galaxolide (HHCB)	2.428 ± 1.297	Sorption to NAS	(Fernandez-Fontaina *et al.* 2012)
	5.2	Sorption to CAS sludge	(Joss *et al.* 2005)
	7.9	Sorption to MBR sludge	
Tonalide (AHTN)	3.347 ±1.900	Sorption to NAS	(Ternes *et al.* 2004)
	10.8	Sorption to CAS sludge	(Joss *et al.* 2005)
	16.0	Sorption to MBR sludge	
Fluoxetine (FLX)	1.603 ± 0.905	Sorption to NAS sludge	(Fernandez-Fontaina *et al.* 2012)
	372 (948)	Sorption under aerobic (anoxic) conditions (sorption was enhanced by anoxic conditions)	(Pomies *et al.* 2015a)
17 α-ethinylestradiol (EE2)	0.316± 0.037	Sorption onto secondary sludge taken from the nitrification compartment	(Ternes *et al.* 2004)
Examples of poorly sorbing (K_d < 0.100 L/g SS) pollutants			
Atenolol (ATN)	0.002	Extremely low sorption under aerobic and anoxic conditions.	(Pomies *et al.* 2015b)
Ibuprofen (IBP)	0.023	Sorption coefficient was determined under anoxic conditions. Under aerobic conditions biodegradation was very fast and this coefficient could not be determined.	(Pomies *et al.* 2015b)
	0.0071 ± 0.002	The test was conducted by minimizing biodegradation.	(Ternes *et al.* 2004)
Diclofenac (DCF)	0.019, 0.09	Aerobic sorption parameters were taken parameter from literature and full-scale data. Since diclofenac ionizes, K_d selection depends on pH.	(Plosz *et al.* 2012)
	0.019	Anoxic sorption	(Plosz *et al.* 2012)
	0.016 ± 0.003	The test was conducted by minimizing biodegradation.	(Ternes *et al.* 2004)
Carbamazepine (CBZ)	0.0012 ± 0.0005	The test was conducted by minimizing biodegradation. The same value is assumed in aerobic and anoxic sorption (Plosz *et al.* 2012)	(Ternes *et al.* 2004)

NAS: nitrifying activated sludge, CAS: conventional activated sludge, MBR: Membrane bioreactor.

sludge. Also, for some estrogens, such as the 17 α-ethinylestradiol (EE2), a very potent endocrine disrupting compound (EDC) often encountered in domestic wastewater treatment systems, sorption onto sludge is an important removal mechanism. Yet, for others, such as some pharmaceuticals, sorption is negligibly small whereas biological transformation is the main elimination mechanism. Compounds such as CBZ that neither sorb to sludge (Table 4.1) nor biodegrade would pass through the WWTPs and be found in the effluent (Ternes *et al.* 2004). As shown in a modeling work with three antibiotics, sorption of xenobiotic organic micropollutants also exhibits different patterns under aerobic and anoxic conditions (Plósz *et al.* 2010).

4.3.2.3 Importance of speciation in sorption of hazardous organics

Sorption of many organic pollutants to solids depends also on their speciation. Therefore, in sorption of organics, an important factor to be taken into account is the acid dissociation constant (pK_a). In particular, pharmaceutical pollutants may possess a charge state(s) that leads to more complex ionic, ion pairing or complexation mechanisms (Kümmerer, 2008). Certain acidic trace organics were removed more effectively when MBRs were operated under acidic rather than neutral conditions. This was explained by the formation of more hydrophobic species at pH values lower than the pK_a (Tadkaew *et al.* 2011). Moreover, some studies take into consideration the molecular charge at pH 7 since this indicates the potential of a pollutant to create electrostatic interactions with the surface of biomass that is usually negatively charged (Verlicchi *et al.* 2012).

As shown in Section 4.3.2.2, sorption onto biological sludge may be expressed by several parameters, each having a different implication. A study comparing the sorption of several pharmaceuticals showed that for compounds without hydrophilic carboxyl functional groups such as estrone (El), estradiol (E2), 17 α-ethinylestradiol (EE2), bisphenolA (BPA), benzylpiperazine (BZP), propyphenazone (PPZ) and carbamazepine (CBZ), a linear relationship was valid between log K_d and log K_{ow} values. However, no such relationship was observed for acidic pharmaceutical compounds like clofibric acid (CA), gemfibrozil (GFZ), ibuprofen (IBP), fenoprofen (FEP), ketoprofen (KEP), naproxen (NPX), diclofenac (DCF) and indomethacin (IDM) since these existed in ionic form under neutral pH conditions (Urase & Kikuta, 2005). A recent study showed that IBF remained mostly in the soluble phase although it had a high log K_{ow} value of 3.97. The low sorption of this compound was attributed to its ionization resulting in a log D value of 0.83 (Pomies *et al.* 2015a).

Under the real conditions of a WWTP, pH values do not often remain constant. Therefore, for modeling purposes, partitioning of such organics to the sludge phase cannot be accounted for by a single sorption coefficient. The assumption of a single K_d value can be only valid for neutral chemicals such as CBZ (Plosz *et al.* 2012).

4.3.2.4 Impact of sorption on biodegradation

An important point to consider in biodegradation studies is whether biodegradation of the pollutant takes place in the bulk phase or following its sorption to the solid phase. Most studies only measure the liquid-phase concentration of a pollutant and then conclude that the pollutant disappeared due to biodegradation taking place in the bulk phase. However, it is very likely that a hydrophobic hazardous organic pollutant sorbing to biological sludge is biodegraded. This occurs because sorption allows a long retention of the pollutant on the biomass surface, which is then accompanied by biodegradation or biotransformation on the floc surface. In addition, the pollutant may also infiltrate through the floc and then biodegrade as shown in Figure 4.1.

The positive effect of biosorption on biodegradation may gain particular importance in the case of slowly biodegradable or apparently nonbiodegradable pollutants. For example, a study revealed that pharmaceuticals sorbing to biomass underwent biodegradation whereas those that did not sorb were not biodegraded, suggesting the existence of a relationship between sorption and biodegradation (Khunjar & Love, 2011).

As shown in Figure 4.1, in a biological reactor, eventually some of the sorbed pollutants may also be desorbed from the surface of biomass. In many cases, the desorption is closely related with the biodegradation taking place in the bulk phase. When biodegradation occurs, the bulk-phase concentration of a pollutant often drops to very low levels while that in the solid phase remains high (Çeçen & Aktaş, 2011). This phenomenon, called the reversal of concentration gradient, is mostly the underlying reason for desorption from the biomass surface.

4.3.2.5 Sorption of hazardous organics to different solids

The following factors appear to be of interest in sorption of hazardous organics to a solid phase in a bioreactor:

Presence of different solid phases: Organic (micro)pollutants usually have a high tendency to sorb to the solid phase in a biological reactor. In that respect, the main solid phase is the biological phase, namely the biomass. However, what is named as biomass is actually not a unique entity, but consists of different fractions. Pollutants may be sorbed to Extracellular Polymeric Substances (EPS) found outside of cells, to cell walls or membranes. Sorption onto EPS is usually regarded as extracellular uptake.

Sorption abilities of different sludges: Different types of biomass may have different sorption abilities. However, to date few studies have focused on this issue. For example, a group has shown that for most PPCPs, sorption to Conventional Activated Sludge (CAS) and nitrifying activated sludge (NAS) was similar (Fernandez-Fontaina *et al.* 2012). Yet, contradictory results were also reported. A study was conducted on sorption of the pharmaceuticals,

CBZ, iopromide (IOP), trimethoprim (TMP) and EE2 onto the four types of biomass, namely pure ammonia oxidizing bacteria (AOB), two heterotrophic enrichment cultures with varying levels of oxygenase activity, and a full-scale nitrifying activated sludge culture (Khunjar & Love, 2011). The EPS character of each sludge was different (Khunjar & Love, 2011). Experiments with AOB indicated that sorption of CBZ, IOP and EE2 was not significant at 95% confidence level. CBZ and IOP did not sorb to any type of biomass. Conversely, EE2 and TMP sorbed to biomass.

Also, when inorganics were sorbed onto sludge, such as Ag^+ and silver nanoparticle (AgNP), the same observation was made. Different fractions in EPS were shown to vary in their ability to sorb these two silver species (Geyik & Çeçen, 2016). Generally, it can be concluded that sorption of pollutants is very much dependent on the EPS production of sludge. In that respect, both composition and fractionation of EPS are of importance.

Existence of nonbiological solid phases: Mostly, the solid phase in a bioreactor is assumed to consist of a biological phase (biomass) only. However, hazardous pollutants can also be removed by sorption onto a nonbiological phase. As another approach, a study has considered that an organic micropollutant might be present in freely dissolved form, or be sorbed to particles and dissolved and colloidal matter (DCM) (Barret *et al.* 2010a). However, in contrast to biomass, nonbiological matter can act as an adsorbent material only and has no biodegradation potential. Despite this fact, the long retention on a sludge surface and eventual desorption may affect the biodegradation potential of a pollutant.

A nonbiological sorbent of utmost interest is the activated carbon which is deliberately added to biological treatment units, in particular when the removal of hazardous pollutants is of interest. In contrast to non-adsorbing media such as suspended inorganic matter, addition of activated carbon to a biological system in the form of Powdered Activated Carbon (PAC) or Granular Activated Carbon (GAC) has tremendous effects. As discussed in detail in literature (Çeçen & Aktaş, 2011), if a pollutant is slowly biodegradable or is found at very reduced concentrations, adsorption onto activated carbon enhances its biodegradation.

4.3.2.6 Fate of hazardous organics in anaerobic sludge digestion

Most hydrophobic pollutants found in domestic or other types of wastewater tend to accumulate in the sludge phase. In that respect, their fate deserves special attention in anaerobic sludge digestion.

A review shows that in biological sludge the concentration of emerging pollutants ranges from few μg/kg (estrogens, some pharmaceuticals, polyfluorinated compounds (PFCs)) to some g/kg (LAS) levels. The fate of these pollutants in anaerobic digestion is often affected by the SRT, temperature, sludge characteristics, adaptation of biomass and bioavailability. Nonylphenol ethoxylates (NPE), some pharmaceuticals and most phthalates are biotransformed to some extent during anaerobic digestion of sludge. However, contradictory results exist for LAS, nonylphenol (NP), synthetic musks, estrogens and bis(2-ethylhexyl)phthalate (DEHP). No significant removal has been observed for triclosan (TCS), triclocarban (TCC), fluoroquinolones, CBZ, tributyltin (TBT) and PFCs (Stasinakis, 2012).

As calculated by researchers, despite the low sorption of the statin drugs atorvastatin and simvastatin, their concentrations in sludge could increase up to 32 and 26 g/kg, respectively (Ottmar *et al.* 2012). Thus, even in the case of small sorption the concentration of pollutants in sludge cannot be neglected.

Others have studied the fate and removal of 48 pharmaceuticals and PPCPs during anaerobic digestion of treatment sludge in full-scale domestic WWTPs. During anaerobic digestion, changes occurred in the distribution of PPCPs between the liquid and solid phases, probably due to the increase in pH. Further, sulfamethoxazole (SMX) and TMP were almost completely degraded (>90%); triclosan, triclocarban, and ofloxacin were moderately degraded (around 30–50%); but CBZ was not eliminated. Such residues might then pose problems for land application of sludge (Narumiya *et al.* 2013).

In anaerobic digestion, pollutants that are sorbed to sludge may also be removed by cometabolism. For example, a study has shown that biodegradation of polycyclic aromatic hydrocarbons (PAHs) is linked to cometabolism in anaerobic sludge digestion (Delgadillo-Mirquez *et al.* 2011). Other researchers monitoring the removal of 13 PAHs, 7 polychlorobiphenyls (PCBs) and NP during the continuous anaerobic digestion of five different sludge samples concluded that both micropollutant bioavailability and cometabolism were important factors for removal of these compounds (Barret *et al.* 2010).

4.3.3 Other abiotic mechanisms leading to removal of hazardous organics

There are also physical and chemical removal mechanisms such as volatilization, hydrolysis, photodegradation, chemooxidation and chemoreduction.

Photodegradation is the transformation or degradation of a substance with light energy. Photodegradation may be not very relevant in most wastewater treatment systems where the penetration of light into reactor systems is minimal. On the other hand, in constructed wetlands (CWs) photodegradation was shown to be an important removal mechanism (Zhang *et al.* 2014). Photodegradation was suggested as the most plausible mechanism for diclofenac (DCF) and KEP removal from a secondary effluent of a domestic WWTP in a full-scale free-water-surface constructed wetland (FWS CW) (Llorens *et al.* 2009). Also in water treatment studies, micropollutants such as SMX, sulfamethazine, caffeine, DCF, IBF, and acetaminophen were removed by photodegradation (Nam *et al.* 2014).

4.4 IMPACT OF PROCESS CONFIGURATION ON REMOVAL OF HAZARDOUS ORGANICS

4.4.1 Biomass configuration: suspended- versus attached-growth (biofilm) operation

Figure 4.3 compares the two different configurations, namely suspended- and attached-growth (biofilm) reactors:

Activated sludge (AS), sequencing batch reactors (SBRs) and membrane bioreactors (MBRs) are examples of suspended-growth reactors. Examples of attached-growth (biofilm) reactors are trickling filters (TFs), rotating biological contactors (RBCs), submerged biofilters, moving bed biofilm reactors (MBBRs) and fluidized-bed (FB) reactors. A literature survey shows that most of the information gathered about the removal and behavior of hazardous pollutants stems from suspended-growth reactors although biofilm reactors are equally important. Clear differences exist in the kinetics and extent of micropollutant removal between suspended- and biofilm systems. The latter usually exhibit considerably higher removal (Falas *et al.* 2013).

In order to understand the differences in the performance of suspended- and biofilm reactors, the following factors need to be addressed:

Retention of biomass: As shown in Figure 4.3, in a suspended-growth reactor, like activated sludge, the flocs are dispersed throughout the reactor. However, in a biofilm reactor, such as a submerged biofilter, biomass is retained on the surface of some packing medium such as a plastic ring. This retention and the high sludge age in biofilm systems allow growth of slowly-growing microorganisms. In the case of hazardous organic pollutants this provides an advantage, since specific bacteria utilizing such pollutants usually belong to the group of slowly-growing microorganisms. In comparison to a suspended-growth reactor, in a biofilm reactor microorganisms are also not easily lost due to the tight attachment. In some activated sludge systems, for retention of slowly-growing microorganisms such as nitrifiers inside the reactor, carriers are even added to the reactor to lead to biofilm formation, a configuration known as hybrid suspended/attached growth.

Sorption and biodegradation: As shown in Section 4.3.2.4, sorption of a pollutant onto biomass may also facilitate its biodegradation. This is even more pronounced in biofilm reactors which are operated at high SRTs. Particularly in biofilm systems, after sorption hazardous organic pollutants are retained in the bioreactor for a long period of time which allows a higher biotransformation and biomineralization. In the case of toxic pollutants, the long retention may even lead to their detoxification. In that regard, EPS may also play a crucial role. Many organic pollutants have to cross through EPS in order to reach cells where they will be eventually metabolized (Flemming *et al.* 2000).

Stratification of biomass: In biofilm reactors, microorganisms attaching to a packing medium are often found in layers (Figure 4.3). This layering is

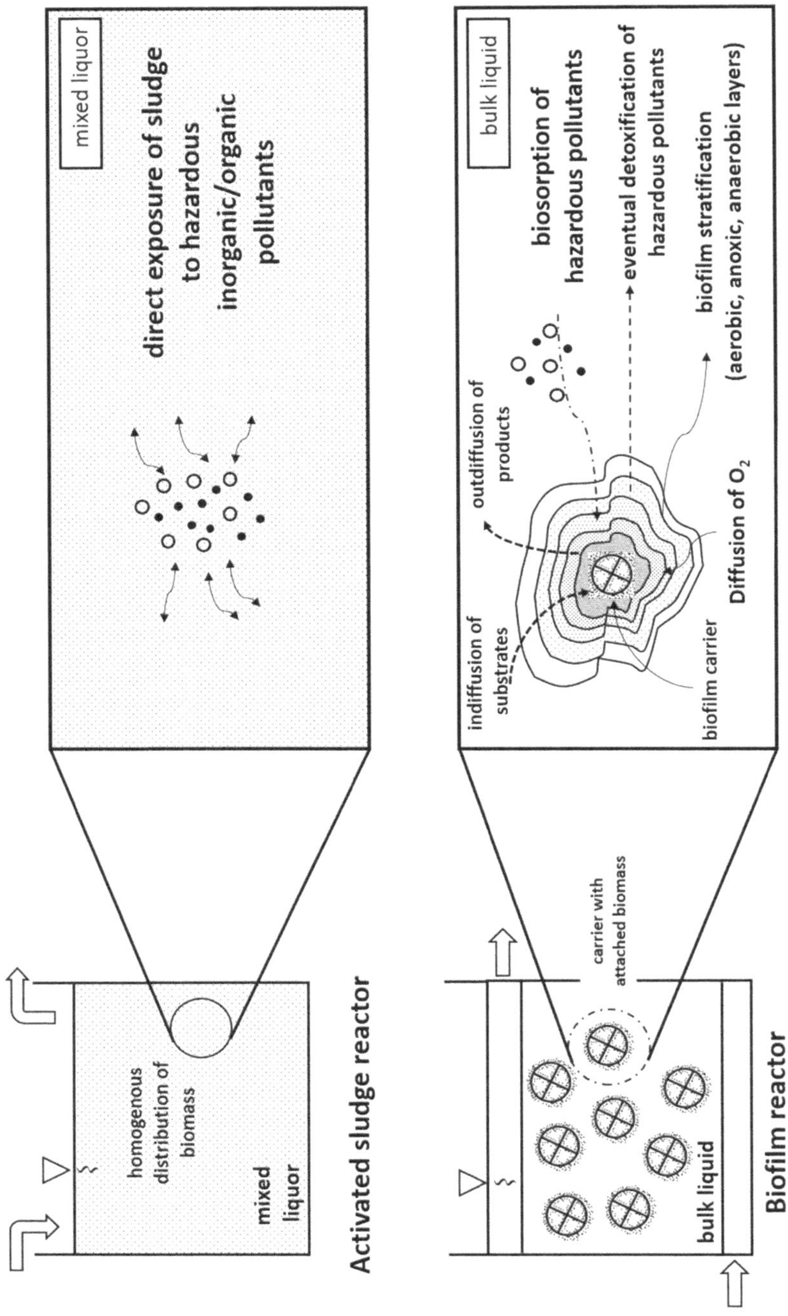

Figure 4.3 Behavior of hazardous compounds in suspended- and attached-growth (biofilm) systems.

usually dictated by the redox conditions inside the biofilm. The simultaneous existence of aerobic, anoxic, sulfate-reducing and anaerobic conditions in a biofilm, referred to as biofilm stratification, may facilitate the removal of a wide range of pollutants.

Operation at low substrate concentrations: In comparison to suspended-growth systems, biofilm systems can function even if operated at very low substrate concentrations. Even if the concentration of the substrate in the feed is less than the S_{min} value, as calculated from Equation 4.2, the substrate can be decreased to very low values. This occurs because the biofilm thickness at any time is not dependent on the availability of substrate at that time, but prior development of biofilm is rather effective in biodegradation (Rittmann, 1994). On the other hand, in suspended-growth reactors, the biomass growth and retention in bulk liquor depend on the availability of substrate in the bulk liquor.

Response to shock conditions: If a biofilm reactor receives inhibitory matter or shock loadings, this does not necessarily affect the whole biomass, but rather the fraction occupying the uppermost layer. In that regard, EPS surrounding the cells may also serve an important function. Production and composition of EPS may differ between biofilm and suspended-growth reactors. In biofilm reactors, production of EPS is usually higher and hazardous pollutants are mostly retained in EPS, leading to the protection of biomass from toxic/inhibitory effects. Similarly, EPS can also sorb inorganic pollutants.

While EPS have the ability to retain a number of organic and inorganic pollutants, in some cases also the presence of pollutants can affect EPS. At the concentrations normally observed in typical domestic wastewater, it is unlikely that pollutants at microgram or nanogram per litre levels have an effect on substrate removal; yet they may affect the production of EPS. For example, a study has shown that many pharmaceuticals and household products had no effect on COD or ammonium removal, but they induced effects on EPS production (Pasquini *et al.* 2013).

4.4.2 Importance of hydraulic regime in bioreactors

The hydraulic regime in a bioreactor is decisive in removal of pollutants. If the substrate removal rate follows zero-order kinetics, which happens in the case when substrate concentration is much higher than the half-velocity constant ($S \gg K_m$), a CSTR and a plug-flow reactor (PFR) will have the same efficiency in the same reactor volume. On the other hand, for reaction rates below zero-order, plug-flow operation is more effective. Degradation of most micropollutants follows the first-order reaction shown in Equation 4.3, since they are found at very low concentrations. Therefore, a reactor cascade or plug-flow reactor is significantly more efficient than a single-reactor system (Andersen *et al.* 2003), if other factors are not taken into consideration.

In addition to the hydraulic regime in a reactor, the properties of the substrate need to be considered in each case, as discussed in Section 4.3.1.2. For example, research has shown that the effectiveness of micropollutant removal depended on the type of reactor, such as CAS and MBR, as well as on hydraulic regime and sorption of micropollutants to sludge. If the sorption of a micropollutant onto sludge was negligibly small, biodegradation was of main interest. In such a case, in both CAS and MBR operation, due to the first-order kinetics, the PFR was more effective than the CSTR configuration in decreasing the pollutant to low levels. In contrast to this, with increasing sorption ability, the differences between PFR and CSTR vanished (Joss *et al.* 2006). As discussed in Section 4.3.2.4, if a micropollutant is highly sorbed this may favor its biodegradation on the surface of biomass rather than in the bulk phase. This happens because the concentration of the pollutant rises to much higher levels on the surface of biomass. Then, the biodegradation rate of a micropollutant will no longer depend on its bulk concentration.

4.4.3 Impact of different treatment units

The type of WWTP and the sequence of treatment units are also of importance in removal of hazardous pollutants. This was shown in a study comparing the performances of facultative and aerated lagoons, chemically-enhanced primary treatment, secondary activated sludge, and advanced biological nutrient removal. In the study, various PPCPs (62 antibiotic, analgesic/anti-inflammatory, and antifungal compounds) in the liquid and biosolid samples from six WWTPs were monitored. Removal efficiencies varied highly with the type of compound, WWTP, and season. Mass balances showed that in biological treatment units analgesic/anti-inflammatory compounds were predominantly removed by biodegradation, while antibiotics and antifungal compounds more likely sorbed to sludge. However, some PPCPs remained soluble and were detected in effluent samples (Guerra *et al.* 2014).

Another study investigated the level and removal of the estrogens E1, E2 and EE2 and NP in different treatment processes including primary and chemical treatment only, submerged aerated filter, oxidation ditch, activated sludge and combined trickling filter with activated sludge. Compared to primary treatment, secondary treatment was shown to improve the removal of estrogens. In general, the study provided some evidence that increased HRT and SRT helped E1 removal (and probably also for other biodegradable steroids) within domestic WWTPs (Johnson *et al.* 2005).

4.4.4 Conventional biological treatment and Biological Nutrient Removal (BNR)

Domestic WWTPs were initially designed with the aim to remove bulk organic matter, often expressed in terms of the parameter BOD_5. Later, eutrophication of lakes became a great issue and to avoid the input of phosphates into such water bodies, phosphate removal was introduced in domestic wastewater treatment in the 1960s. The process used for this purpose was chemical precipitation. Nitrogen removal by nitrification began at the end of the 1970s whereas denitrification was introduced in the 1980s. Biological phosphorus removal began in the 1990s (Siegrist *et al.* 2005). This new biological treatment configuration involving simultaneous removal of nitrogen and phosphorus, leading at the same time to the removal of organic carbon, is known as Biological Nutrient Removal (BNR). Today, most of the WWTPs are designed according to this principle.

The activated sludge process is the most widely used one in biological treatment of domestic as well as other types of wastewater. Also, the BNR configuration is derived from the activated sludge process. A typical BNR configuration consists of a primary clarifier, anaerobic, anoxic, aerobic reactors and a secondary settler. Activated sludge systems aiming at BNR are operated at higher SRTs than conventional systems. It is likely that conventional treatment systems have a different effectiveness in removing hazardous substances than BNR. Yet, so far relatively few studies have compared the difference between conventional and BNR systems.

The positive effect of SRT in BNR systems was evidenced in many studies. Increased SRT has two types of positive effects:

(a) High SRT decreases biodegradable organic substrates to very low values. In that respect, biodegradable hazardous organics can also be reduced to low levels at high SRT, as shown in Equation 4.1 in Section 4.3.1.3.
(b) High SRT also has an indirect effect. It allows the slowly-growing nitrifying bacteria to be retained in the sludge, leading to improved nitrification. Nitrification is the first and most important step of BNR. Although nitrifiers themselves do not consume organic matter as an energy source, the nitrification process itself has a great effect on biological elimination of several hazardous organics. This is attributed to the occurrence of cometabolism in nitrifying sludges, as discussed in detail in Section 4.6. Similarly, BNR can also be achieved in biofilm systems. Since biofilm systems are usually operated at higher sludge ages and thus achieve high nitrification rates, they may be even more advantageous for the removal of micropollutants by cometabolism. High SRT also allows the maintenance of other slowly-growing specific degraders in sludge that act on hazardous substances.

However, in many cases it is not clear whether the improvement observed in removal of hazardous organics occurs as a result of (a) high SRT which leads to a decrease in substrate, or rather due to (b) the activity of slowly-growing microorganisms.

The importance of SRT can best be exemplified by examining the discrepancies between MBR and CAS operation with regard to removal of pollutants. As concluded in a literature review presenting a large amount of data on a high number of pharmaceuticals in WWTPs, pharmaceuticals were more consistently and effectively removed in an MBR than a CAS (Verlicchi *et al.* 2012). The main distinguishing feature here is the SRT. Commonly, SRT is about 15–80 d in an MBR whereas in a CAS it varies in the range of 7–20 d (Verlicchi *et al.* 2012). This improved performance in the MBR can be attributed to the lowering of effluent substrate concentration at higher SRTs. At the same time, a long SRT in MBRs allows more time for the growth of slowly-growing strains that would feed on specific organic compounds. In contrast to that, CAS systems operating at lower SRT have rather the ability to remove organic carbon of natural origin. They are not very effective in removing specific organics.

Yet, some studies also report contradictory results about the relative removal of specific organics in CAS and MBR. For example, for the compounds BPA, IBP and bezafibrate (BZF), effluent concentrations and removal rates were comparable between MBR and CAS (Clara *et al.* 2005b). In another study the differences between MBR and CAS were visible only in the case of the poorly biodegradable micropollutants DCF, mecoprop and sulfophenylcarboxylates, while for the micropollutants ethylenediamine tetraacetic acid (EDTA) and CBZ no differences could be found since these were not removed at all by biological means (Bernhard *et al.* 2006).

Another study reported that an increase in SRT led to a higher removal rate for some compounds like fluoxetine (FLX), while for others like ibuprofen it had no effect (Suarez *et al.* 2010). Other researchers concluded that the effluent concentrations of EDCs and pharmaceuticals went down to very low values at SRTs greater than 10 days. This high SRT condition is satisfied in WWTPs incorporating nutrient removal by nitrification and denitrification (Clara *et al.* 2005b). Supporting this, in another study EE2 removal was observed to improve significantly after upgrading of a full-scale plant to include nutrient removal with an SRT of 11–13 days while the old plant operating at an SRT < 4 days was ineffective (Andersen *et al.* 2003). Although it is often difficult to distinguish which factor is more effective, consumption of hazardous organics as cometabolic substrates seems to be more probable when there is a high nitrifying activity in a biological sludge.

4.4.5 Combination of biological treatment with advanced physicochemical treatment

4.4.5.1 Combination of biological treatment with activated carbon adsorption

As seen from the previous discussion, in conventional biological treatment systems, organic micropollutants like hormones, PPCPs, and household products, are often inadequately removed. Therefore, beginning from the 1970s more elaborate wastewater treatment methods were sought to increase the elimination efficiency. In that regard, the use of activated carbon in conjunction with biological treatment became established. Besides adsorbing a number of organic compounds, activated carbon can also facilitate biodegradation of hazardous compounds (Çeçen & Aktaş, 2011).

Integration of activated carbon adsorption with biological treatment can be realized using both PAC and GAC. To suspended-growth biological reactors, such as activated sludge and MBR, activated carbon is added in the form of PAC. Such configurations are then referred to as Powdered Activated Carbon Treatment (PACT) and PAC-MBR, respectively. Activated carbon can also be employed in biofilm reactors. In that case, activated carbon in the form of GAC serves as a medium on which a biofilm layer develops. This configuration is known as the Biological Activated Carbon (BAC) process.

4.4.5.2 Combination of biological treatment with oxidative treatment

To bring about a more efficient removal of micropollutants, also oxidative methods may be combined with biological treatment. In that regard, ozonation and Advanced Oxidation Processes (AOPs) are widely tested in removal of (micro)pollutants. The main aim of these processes is to oxidize specific organics in such a way that they are converted into less harmful and/or more biodegradable compounds. Treatment of a wastewater by such methods generally leads to detoxification and an increase of biodegradability.

Ozonation and AOP may be applied as pre-oxidation methods before biological treatment. Since these oxidation techniques are non-specific, they usually also oxidize other compounds than the targeted ones. Therefore, they are usually preferred in the tertiary treatment step to remove residual nonbiodegradable and toxic compounds that are found at trace levels in secondary effluents.

4.5 HAZARDOUS POLLUTANTS AND THEIR INHIBITORY EFFECTS

4.5.1 Brief review of inhibition

Inhibition is defined as a decrease in substrate removal or biomass growth rate. Mostly, in environmental science and engineering studies, the terms "inhibition" and "toxicity" are used interchangeably. Ideally, the term "toxicity" should stand for irreversible inhibition.

In aerobic, anoxic, and anaerobic processes, removal of bulk organic matter can be inhibited due to the presence of a vast number of hazardous organic and inorganic pollutants. Similarly, in biological nitrogen removal processes (nitrification, denitrification) removal of nitrogen may be inhibited.

Hazardous organic pollutants do not only biodegrade or biotransform by the mechanisms listed in Section 4.2. Depending on their type and concentration, they may also cause inhibitory effects on the removal of other substances as well as their own removal. Under normal cases, in domestic wastewater, hazardous organics are often found at extremely low concentrations so that it is unlikely that they inhibit organic carbon or nutrient removal. However, high concentrations of hazardous organic pollutants may be encountered in industrial wastewaters such as phenolic effluents, or in the case of industrial inputs to WWTPs, and in landfill leachate treatment. In some cases, the concentrations of hazardous organics may reach a few mg/L (Dionisi, 2014), leading often to inhibitory effects. Also, in bioremediation of groundwater, some pollutants such as chlorinated aliphatic compounds (e.g. TCE) might be present at quite high concentrations leading to inhibition.

4.5.2 Inhibition models

Since literature provides much information on inhibition mechanisms and models, only a brief review is made in this chapter. To understand inhibition models, it is useful to consider the Michaelis-Menten expression first:

$$q = q_m \frac{S}{K_m + S} \tag{4.4}$$

where

q: specific substrate removal rate, $(M_s/M_x \cdot T)$.
q_m: maximum specific substrate removal rate, $(M_s/M_x \cdot T)$.
S: substrate concentration, (M_s/L^3).
K_m: half-velocity constant (affinity of the substrate for the enzyme, also called the Michaelis-Menten constant), (M_s/L^3).

Mostly, to give a general overview of inhibition patterns, the Michaelis-Menten equation is modified to the following form:

$$q = q_m^{app} \frac{S}{K_m^{app} + S} \tag{4.5}$$

According to this expression, in the presence of an inhibitor both the maximum specific substrate removal rate and half-velocity constant may take new values termed as the apparent (*app*) ones.

Figure 4.4 illustrates the general reaction scheme for inhibition and shows the alternatives for binding of a substrate (S) and inhibitor (I) to an enzyme (E).

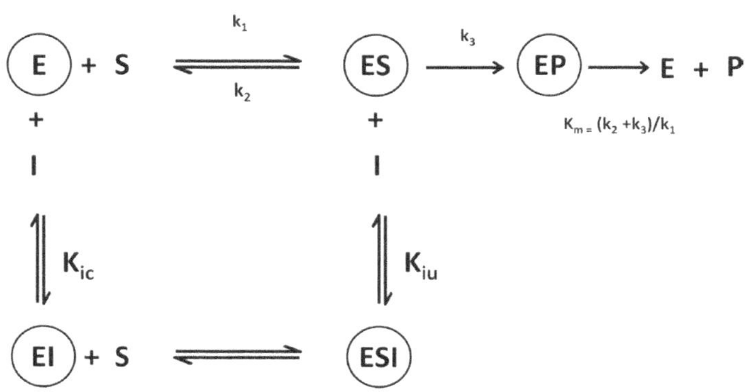

Figure 4.4 General reaction scheme for inhibition.

4.5.2.1 Competitive inhibition

In biological treatment of domestic wastewater, industrial wastewater or landfill leachates, competitive inhibition is often encountered between two different substrates. In competitive inhibition, the inhibiting substrate (I), which is the (micro)pollutant in this case, has structural similarities to the main substrate (S). In a biological wastewater or water treatment system, the main substrate is often bulk organic matter, but it may also be an inorganic substrate such as the ammonia in nitrification. As the name implies, the inhibiting substrate competes with the main substrate for the same active site in the enzyme (E). To model this type of inhibition, the following equation is used:

$$q = q_m \frac{S}{K_m(1 + I/K_{ic}) + S} \tag{4.6}$$

where

K_{ic}: competitive inhibition constant demonstrating the affinity of the free enzyme (E) for the inhibitor, (M_I/L^3).

I: concentration of inhibitor (M_I/L^3).

Thus, the net effect of a competitive inhibitor is to increase the K_m to a new value, namely the K_m^{app}. However, q_m remains constant in competitive inhibition. If enough S is present in the system, the effects of the inhibitor I can be overcome.

Most hazardous organic pollutants found in domestic or other types of wastewater act as competitive inhibitors. Competitive inhibition may also be observed when one of the substrates is consumed as a cometabolite. For example, a study focusing on the removal of TCE in a nitrification system showed that ammonia (the primary substrate) and TCE (the cometabolic substrate) acted competitively (Alpaslan Kocamemi & Çeçen, 2005).

The readily biodegradable substrate in domestic wastewater often serves as a primary (growth) substrate in cometabolic degradation of various micropollutants. However, in some cases the readily biodegradable substrate itself may appear as a competitive inhibitor, since its bulk concentration is incomparably higher than the micropollutant (Table 4.2). This happens because the microorganisms preferentially feed on the readily biodegradable substrate that is abundant and easy to consume. As an example, a study has shown that the readily biodegradable substrate in an activated sludge system may compete with specific antibiotic micropollutants (Plósz *et al.* 2010). In another study comparing the performance of a conventional activated sludge plant with that of a fixed-bed reactor in the same WWT, the researchers hypothesized that degradation of the natural estrogens E1 and E2 was competitively inhibited by the substrate contained in the primary effluent (Joss *et al.* 2004). More commonly, however, it is the cometabolite that inhibits the removal of a primary substrate serving as an energy and carbon source (Table 4.2).

Table 4.2 Examples of cometabolic degradation of hazardous organic pollutants.

Organic Pollutant	Type of Biodegradation	Primary (Growth) Substrate	Inhibition Type	Inhibited Substrate	Inhibitory Substrate	Other Information	Reference
1,2-dichloroethane (1,2-DCA)	Cometabolism	Ammonia-N	Mixed inhibition	Ammonia-N	1,2-DCA	The results were obtained in a nitrifying packed-bed biofilm reactor. Adverse effect of 1,2-DCA on nitrification was reversible due to the ability of cells to recover from inhibition.	(Alpaslan Kocamemi & Çeçen, 2007a)
Trichloroethylene (TCE)	Cometabolism	Ammonia-N	Competitive	Ammonia-N	TCE	Inhibitory effect of TCE was studied in an activated sludge enriched for nitrifiers.	(Alpaslan Kocamemi & Çeçen, 2005)
Atenolol (ATN)	Cometabolism	Ammonia-N	Competitive	Ammonia-N	Atenolol (at conc. >100μg/L)	AOB made up 5% of the total biomass, but they contributed to 7 – 17% of ATN biodegradation.	(Sathyamoorthy *et al.* 2013)
Ibuprofen (IBF)	Cometabolism	Biodegradable organic carbon	–	–	–	Aerobic conditions were present; no inhibition was reported.	(Pomies *et al.* 2015b)

It is often overlooked that the inhibitor itself may be removed in a biological system. If the inhibitor is amenable to biodegradation, its concentration may be reduced by biodegradation or biotransformation. It may also be eliminated by abiotic processes such as biosorption. Thus, the "I" value representing the inhibitor concentration in Equation 4.6 may not necessarily remain constant, but may change during the whole biological process.

4.5.2.2 Uncompetitive inhibition

The uncompetitive inhibition is described as shown in Equation 4.7.

$$q = \frac{q_m \left[1/(1 + I/K_{iu})\right] S}{K_m \left[1/(1 + I/K_{iu})\right] + S} \tag{4.7}$$

where

K_{iu}: uncompetitive inhibition constant demonstrating the affinity of the enzyme-substrate complex (ES) for the inhibitor.

In this type of inhibition, the inhibitor can bind only to the ES complex (ES). When substrate concentration increases, the effect of inhibition also increases because more substrate causes more enzyme-substrate complex (ES) formation and the inhibitor can bind to this complex. As seen in Equation 4.7, uncompetitive inhibition decreases both K_m and q_m at the same time and at the same rate; thus the new q_m/K_m ratio stays constant. A study examining the acute impact of erythromycin on activated sludge at high concentrations revealed that the major impact of this antibiotic was partial obstruction/binding of available substrate for microbial utilization, suggesting uncompetitive inhibition (Pala-Ozkok *et al.* 2014).

4.5.2.3 Mixed inhibition

$$q = \frac{q_m \left[1/(1 + I/K_{iu})\right] S}{K_m \left[(1 + I/K_{ic})/(1 + I/K_{iu})\right] + S} \tag{4.8}$$

The inhibitor can bind either the enzyme (E) or the enzyme-substrate (ES) complex to form EI and EIS, respectively. If the inhibitor binds the enzyme, the dissociation constant of the EI complex is K_{ic}. In the case of binding to ES, the dissociation constant of the EIS complex is expressed as K_{iu}. As shown in Equation 4.8, q_m will be reduced to a new value; but also, an increase will be observed in K_m. Thus, depending on the level of K_{ic} and K_{iu}, both competitive and uncompetitive effects will be observed in mixed inhibition. For example, a study examining the cometabolic degradation of 1,2-dichloroethane (1,2-DCA) revealed that this compound had mainly mixed inhibitor characteristics on nitrification, but it acted rather as a competitive inhibitor at concentrations lower than 25,000 μg/L (Alpaslan Kocamemi & Çeçen, 2010b).

4.5.2.4 Non-competitive inhibition

In non-competitive inhibition, the inhibitor binds the enzyme (E) or enzyme-substrate (ES) complex randomly, producing an inactive enzyme-substrate-inhibitor (ESI) which does not decay to give a product. As the name implies, the inhibitor does not compete with the substrate since it is dissimilar to the substrate and binds to another site of the enzyme than the substrate. If the inhibitor has the same affinity for the EI and ES complexes which are shown in Figure 4.4, K_{iu} and K_{ic} are equal to each other. As shown in Equation 4.9, in such cases the maximum specific substrate removal rate (q_m) is lowered whereas the half-velocity constant (K_m) remains unchanged.

$$q = \frac{q_m \left[1/(1 + I/K_{iu})\right] S}{K_m + S} \tag{4.9}$$

Mostly, metals in ionic form or anions at high concentrations act as non-competitive inhibitors in biological reactions. A study evaluating the removal of fluorinated sulfonates, a type of surfactant difficult to biodegrade, indicated that difluoromethanesulfonate (DFMS) utilization was inhibited by some forms of sulfur and non-competitive inhibition kinetics were observed (Key *et al.* 1998).

An analysis of competitive, uncompetitive and non-competitive inhibition shows that they are in fact limiting cases of mixed inhibition. In competitive inhibition, the right side in Figure 4.4, namely the binding of I to the ES complex, is missing. In uncompetitive inhibition, the left side in Figure 4.4, namely the binding of I directly to the enzyme (E), is missing. In non-competitive inhibition, I has the same affinity for E and ES. Binding of I does not alter apparent affinity of the enzyme for the substrate. Therefore the K_m value does not change.

4.5.2.5 Substrate inhibition

Another type of inhibition is substrate inhibition that is also known as self-inhibition. In this type of inhibition, the removal rate of the substrate is slowed down due to its high concentration. The self-inhibition kinetics is often represented by the Haldane equation shown below where the K_I value is the self-inhibition constant:

$$q = q_m \frac{S}{K_m + S + (S^2/K_I)} \tag{4.10}$$

Substrate inhibition often occurs in nitrification if the concentration of ammonia nitrogen is high and pH is raised. For example, such an inhibition is often encountered in the treatment of landfill leachates that have a high level of Total Kjeldahl Nitrogen (TKN). In such a case, the high concentration of free ammonia acts as an inhibitor in oxidation of ammonia (nitrification) itself. Another well-known case of substrate inhibition is the inhibitory effect of phenol on its own oxidation under aerobic conditions, as observed in effluents from coke production. Substrate inhibition may also be observed in anaerobic processes; for example, in the case of free ammonia accumulation in anaerobic digestion.

4.5.2.6 Product inhibition

There are also other inhibition mechanisms such as product inhibition. In this type of inhibition a product accumulates during a biological reaction and inhibits this reaction itself. An example of this is the inhibition of fermentation of waste activated sludge by excess production of volatile fatty acids (VFAs).

The inhibitory characteristics of hazardous organics on their own removal (self- or substrate inhibition) or on the removal of organic carbon and nitrogen in a domestic or any other wastewater can be determined experimentally. In aerobic systems, often the growth or respiration rates of sludges are monitored or the consumption of substrate is measured. Likewise, in anaerobic systems growth rates or specific methane formation rates may be measured. A detailed discussion about the experimental assessment of inhibition can be found in **Chapter 5**.

4.6 IMPACT OF NITRIFICATION ON REMOVAL OF HAZARDOUS ORGANICS BY COMETABOLISM

4.6.1 Role of nitrifiers in cometabolism

In the biological treatment of wastewater or water, the cometabolic removal of many hazardous pollutants relies on the activity of Ammonia Oxidizing Bacteria (AOB) that are responsible for the first step of the nitrification process, namely the ammonia oxidation. These bacteria derive their metabolic energy from oxidation of ammonia whereas the enzyme AMO oxidizes ammonia to nitrite, as shown in Figure 4.5. Among nitrifying autotrophs, *Nitrosomonas europaea* (*N.europaea*) is a ubiquitous bacterium producing AMO by using ammonia as a growth-substrate and leading to its oxidation in soils, natural waters and nitrifying activated sludge.

It has often been demonstrated that the enzyme AMO is capable of cooxidizing many organic compounds that would be otherwise not biodegraded. This enzyme can, for example, cooxidize halogenated hydrocarbons, aromatics, ethers, thioethers as well as hydrocarbons such as alkanes and alkenes. However, fortuitous oxidation of such organic compounds by AMO does not result in energy generation. In fact, such compounds or their intermediate products may even reduce the rate of nitrification, by acting often as competitive inhibitors on ammonia oxidation, as described in Equation 4.6.

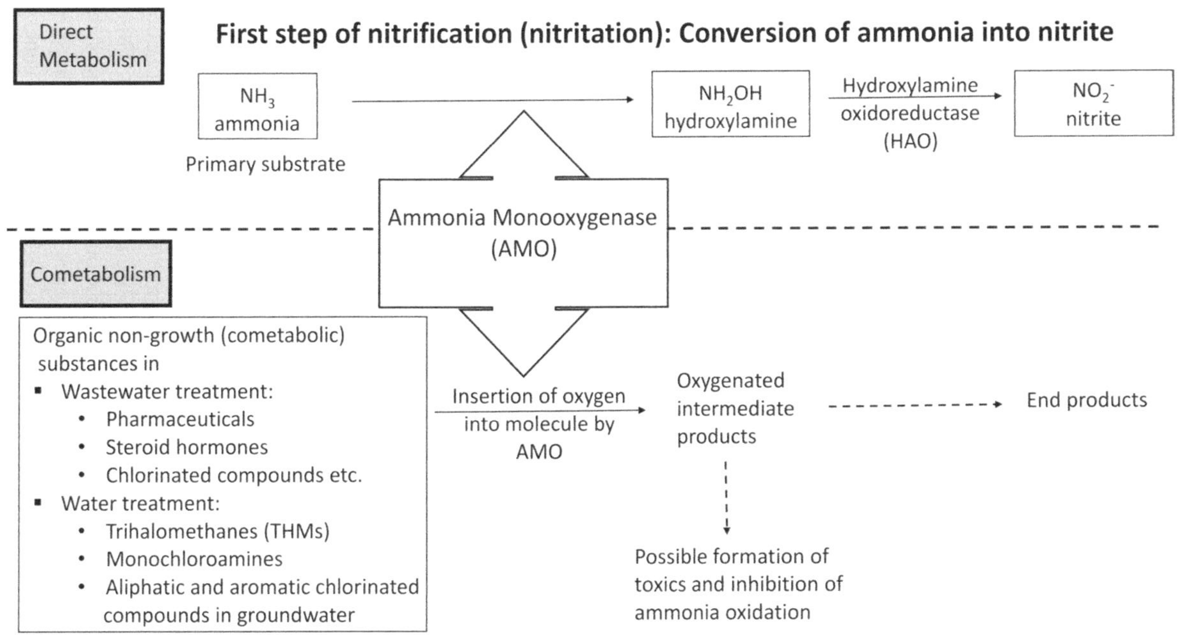

Figure 4.5 Cometabolic removal of hazardous organics in water/wastewater treatment by the AMO enzyme produced by nitrifiers.

4.6.2 Occurrence of cometabolism in nitrifying sludges

Nitrification can take place in different types of aerobic sludges. In testing the effectiveness of nitrification in cometabolic degradation of (micro)pollutants, mainly the following types of sludge are employed in laboratory-scale studies:

(a) *Nitrifying activated sludge (NAS)*: This type of sludge is derived from an activated sludge sample by feeding it with ammonium salts and other minerals in the absence of organic carbon. The idea here is to enrich the sludge in terms of nitrifiers. Examples of this procedure can be found in literature (Alpaslan Kocamemi & Çeçen, 2007b; Fernandez-Fontaina *et al.* 2012). The enrichment of sludge in terms of AOB and nitrite-oxidizing bacteria (NOB) can be shown by molecular microbiology techniques.

(b) *Pure nitrifier cultures*: Commercially available pure cultures such as *N. europaea* are often used to test the intrinsic capacity of single species to degrade a pollutant, by excluding any interference from other microorganisms.

(c) *Activated sludge samples*: An activated sludge sample may be taken from a single-sludge WWTP that includes the nitrification step. In almost every case, in all aerobic sludges there is a certain level of nitrifying activity. Both heterotrophic and autotrophic biomass can be active in cometabolism of many (micro)pollutants from domestic or other types of wastewater. However, as stated in a recent review (Pomies *et al.* 2013) most cometabolic removal seems to be carried out by nitrifiers. Therefore, for a more accurate process characterization, the role of heterotrophs and nitrifiers needs to be assessed separately. To achieve this, comparative experiments have to be carried out to differentiate the potential of nitrifiers from that of heterotrophs in cometabolizing hazardous organic pollutants.

In most experimental studies, the first two types of biomass ((a) and (b)) are employed. Literature studies point to effective degradation of some estrogens by both types of sludges. For example, a study has shown that both types of sludges ((a) NAS and (b) *N. europaea*)) degraded estrogens, yet differences were observed in the rates of estrogen removal and formation of degradation products. NAS degraded estrogens and their degradation intermediates, while *N. europaea* only degraded estrogens with no further degradation of their intermediates (Shi *et al.* 2004). Thus, NAS, composed of different nitrifiers, proved to be more effective than a pure culture.

Another study compared the biodegradation potential of triclosan, bisphenol A, and ibuprofen by *N. europaea* and mixed AOB in nitrifying activated sludge. *N. europaea* could degrade triclosan and bisphenol A, but not ibuprofen. The enzyme AMO produced by nitrifiers was thought to be responsible for triclosan and bisphenol A degradation. Competitive inhibition of ammonia oxidation was observed when triclosan or bisphenol A were present. Additionally, inactivation of *N. europaea* was observed after degrading a mixture of triclosan and bisphenol A. On the other hand, ibuprofen was more likely degraded by the heterotrophic microorganisms found in activated sludge (Roh *et al.* 2009).

4.6.3 Factors affecting cometabolic removal of hazardous organics in nitrification

So far, few studies focused solely on the role of nitrifiers in cometabolizing hazardous organics in activated sludge and biofilm systems. Some studies clearly delineated the cometabolic removal of the chlorinated xenobiotics TCE and 1,2-DCA in nitrifying activated sludges and packed-bed nitrifying biofilm reactors (Alpaslan Kocamemi & Çeçen, 2005, 2007a, b, 2009, 2010a, b).

From all studies conducted so far in nitrifying systems, it can be deduced that the following points need to be mainly considered in cometabolic removal of organic pollutants:

(a) *Feeding of the primary (growth) substrate (ammonia)*: Cometabolic removal of a hazardous substrate depends largely on the concentration of the primary substrate, namely the ammonia. In addition, cometabolism depends on the constant supply of ammonia into a bioreactor as a primary substrate. However, other results are also reported. As shown in a study conducted on an aerated submerged fixed bed bioreactor (Forrez *et al.* 2009), removal of EE2 was sustained even after 2 months of operation when ammonia was no longer present in the feed. This long-lasting ability of microorganisms is termed "resting cell condition", indicating that cells have accumulated enough monooxygenases to function for a while even in the absence of ammonia.

(b) *Nitrification rate*: One of the objectives in cometabolism studies is to clarify the relationship between the rate of primary substrate removal (nitrification rate) and that of cometabolite.

While the effect of nitrifying bacteria on other organic compounds such as LMW chlorinated aliphatics is known, relatively little information exists about other pollutants. Today, there is a great concern about the estrogens E1, E2 and EE2 since they significantly contribute to the estrogenic activity in WWTP effluents. Vader *et al.* showed that a nitrifying sludge had the capacity to convert EE2 by cometabolism where ammonium served as the sole energy source. In the case of insignificant nitrifying capacity no such degradation was detected (Vader *et al.* 2000).

In the case of the micropollutant EE2, commonly encountered in domestic wastewater, a linear relationship was shown between nitrification and EE2 removal, implying that EE2 cometabolism could be enhanced if the sludge in a bioreactor was enriched for nitrifiers (Yi & Harper, 2007). Another study confirmed that an activated sludge that was directly taken from a WWTP without acclimation removed EE2 only through sorption. However, it was able to remove EE2 completely when it became enriched in nitrifiers (Clouzot *et al.* 2010). This points clearly to the role of nitrification in removal of such estrogens.

Also, for PPCPs such as galaxolide (HHCB), tonalide (AHTN), IBP, naproxen (NPX), erythromycin (ERY), trimethoprim (TMP), roxithromycin (ROX) and FLX, the rates of micropollutant degradation were seen to correlate with specific nitrification rate. High nitrogen loading rates led to high nitrifying activities. As a consequence, AMO production became higher leading to an effective removal of different emerging micropollutants (Fernandez-Fontaina *et al.* 2012).

The same type of behavior was also observed in the case of chlorinated aliphatics. For example, a study on cometabolic degradation of 1,2-DCA in a suspended-growth nitrifying system revealed that an increase in NH_4^+-N utilization favored cometabolic degradation of this compound (Alpaslan Kocamemi & Çeçen, 2010b). An important factor in cometabolism is also the transformation yield which shows the degradation potential of biomass for the cometabolic substrate per amount of growth substrate consumed. In a nitrifying biofilm reactor where the primary and cometabolic substrates were NH_4^+-N and 1,2-DCA, respectively, the removal rates of both were dependent on their respective loading rates to the column (Alpaslan Kocamemi & Çeçen, 2009). Another study revealed that the transformation yield of TCE strongly correlated with the initial level of NH_4^+-N and TCE (Alpaslan Kocamemi & Çeçen, 2010a).

Few studies tackle the issue of why micropollutants are biotransformed to varying extents among different WWTPs (Helbling *et al.* 2012). Researchers have shown that each biomass taken from 10 different WWTPs had different biotransformation capabilities. However, neither biomass concentration nor the SRT parameter alone could explain the variability of results. On the other hand, biotransformation ability of micropollutants was dependent on effluent ammonia concentration and effectiveness of ammonia removal. This observation emphasized the role of the enzyme AMO in cometabolic removal of several micropollutants. Yet, it was also highlighted that it was not necessarily bacteria alone which produced this enzyme and contributed to the removal of micropollutants, but also ammonia oxidizing archaea (AOA) found in sludge. There was a strong association between the expressed archaeal *amoA* content and oxidative micropollutant biotransformation. The presence of AOA in ammonia-removing activated sludge bioreactors is well documented in literature (Park *et al.* 2006).

(c) *Inhibitory effects of cometabolic substrate on oxidation of primary substrate (ammonia)*: Although in domestic or other types of wastewater, micropollutants are present at much lower levels than the primary substrate, they may often compete with the primary substrate. This observation can be attributed to their high affinity for the relevant enzyme. For example, the half-saturation constant (K_N) of ammonia nitrogen, a measure of its affinity for the relevant enzyme, was about 458–558 µM while the inhibition constant (K_{ic}) of TCE was found in the range of 5.0–6.1 µM. This indicated that TCE had a much higher affinity for AMO than ammonium nitrogen (Alpaslan Kocamemi & Çeçen, 2007b).

(d) *Solids Retention Time (SRT)*: For an adequate nitrification to take place, a high SRT should be maintained in a bioreactor. This is particularly important at low temperatures since nitrifier growth rates significantly decrease at low temperatures. A study pointed to the importance of SRT in cometabolic removal of estrogens (Vader *et al.* 2000). Since EE2 removal correlated with nitrification, disturbance of nitrification would simultaneously mean a lower removal efficiency.

(e) *Organic carbon loading*: High organic loading rates (OLRs) increase the competition between heterotrophic and nitrifying biomass. At high organic loading rates, nitrification is not favored. The effects of bacterial activity on the degradation of E1, E2, estriol (E3) and EE2 in nitrifying activated sludge were studied using different substrates, at different OLRs and low temperature conditions. At high OLRs, a lower nitrification efficiency was achieved. Consequently, this reduced the degradation of E1, E2 and EE2, indicating that cometabolic removal took place mainly due to the presence of AOB. On the other hand, E3 was removed largely by heterotrophic activity (Ren *et al.* 2007).

The removal of micropollutants was also studied in tertiary filtration of effluents with a relatively low level of organic carbon. Compared to secondary treatment, activity of nitrifiers was much favored in such cases. For example, in anthracite filters spiked with a broad range of micropollutants, nitrifying bacteria clearly had a role in degradation of several micropollutants (Rattier *et al.* 2014).

(f) *Impact of different cultures*: As mentioned in Section 4.6.2, cometabolism can take place in all types of nitrifying cultures. To investigate the effect of different cultures, the biological fate of the estrogen EE2 (500 ng/L to 1 mg/L) and TMP (1 µg/L to 1 mg/L) was examined using three different cultures:

- An AOB culture
- Two enriched heterotrophic cultures devoid of nitrifier activity
- Nitrifying activated sludge (NAS) cultures

The AOB culture biotransformed EE2 but not TMP, whereas heterotrophs mineralized EE2, biotransformed TMP, and mineralized EE2-derived metabolites generated by AOBs. Kinetic bioassays showed that AOBs biotransformed EE2 five times faster than heterotrophs. The importance of AOBs in removing EE2 and TMP was evaluated further by performing NAS experiments at lower feed concentrations (500–1000 ng/L). EE2 removal slowed markedly after AOBs were inhibited, while TMP removal was not affected by AOB inhibition. Two key EE2 metabolites formed by AOB and heterotrophic laboratory-scale chemostats were also found in independent laboratory-scale mixed culture bioreactors; one of these, sulfo-EE2, was largely resistant to further biodegradation. AOBs and heterotrophs may cooperatively enhance the reliability of treatment systems where efficient removal of EE2 is desired (Khunjar *et al.* 2011).

The effectiveness of nitrifiers has also been tested and modelled in the case of several pharmaceuticals (Sathyamoorthy *et al.* 2013). Among the beta blockers studied, metoprolol and sotalol were not biodegraded by the enriched nitrifying culture. Biodegradation of ATN was linked to the activity of AOB and heterotrophs, but not nitrite-oxidizing bacteria (NOB). The research suggested that even when AOB made up 5% of the total biomass, their contribution to biodegradation of ATN was higher (between 7 – 17%). In partial nitrification systems such as Sharon, AOB have a share greater than 10% in biomass. Therefore, this would imply a higher removal of these organics. Moreover, ATN competitively inhibited the growth of AOB, though model simulations suggest inhibition is most relevant at ATN concentrations greater than approximately 200 ng/L. An inhibition of AOB would mean a reduction in the ability to meet stringent effluent nitrogen limits. In addition, biodegradation of pharmaceuticals would be hindered in this case.

4.7 IMPACT OF REDOX CONDITIONS ON BIODEGRADATION OF HAZARDOUS ORGANIC POLLUTANTS

4.7.1 Expression of biodegradation rates

The following equation is used to delineate the substrate removal rate (r_s) in a bioreactor:

$$r_s = q_m \frac{S}{K_m + S} X_{ss} \tag{4.11}$$

where

r_s : substrate removal rate, $(M_s/L^3 \cdot T)$.
q_m : maximum specific substrate removal rate, $(M_s/M_x \cdot T)$.
S : substrate concentration, (M_s/L^3).
K_m : half-velocity constant (indicating the affinity between the substrate and enzyme), (M_s/L^3).
X_{ss} : suspended solids concentration, (M_x/L^3).

This nonlinear rate expression is valid for a certain substrate range only. For example, when bulk organic matter is decreased to 50 mg/L COD in a typical domestic wastewater treatment unit, this expression is valid since the half-velocity constant may be as high as 20 mg/L COD. However, hazardous organic pollutants are often found at very reduced concentrations ("micropollutants"). In such cases, compared to the K_m value, the substrate concentration in the denominator of Equation 4.11 becomes negligibly small. As a result, the equation reduces to a first-order rate expression as shown below:

$$r_s = k_{biol}\, S\, X_{ss} \tag{4.12}$$

where

k_{biol} : first-order biodegradation rate constant, $(L^3/M_s \cdot T)$.
S : substrate concentration, (M_s/L^3).
X_{ss} : suspended solids concentration, (M_x/L^3).

The k_{biol} is a significant parameter that is widely used throughout literature studies in comparing the effectiveness of micropollutant removal.

4.7.2 Biodegradation rates at different redox conditions

Biodegradation rates of hazardous organic pollutants depend also on the redox conditions in a bioreactor. Nowadays, in the treatment of domestic wastewater the main aim is the biological removal of organic carbon and nutrients (nitrogen and phosphorus). For that purpose, often single-sludge systems are used in which the biological sludge is subjected to alternating aerobic, anoxic and anaerobic conditions. This applies to both suspended- and biofilm systems. Also in the treatment of industrial effluents or concentrated wastewaters, such as landfill leachates, microorganisms may be exposed to different redox conditions.

The importance of redox conditions in biodegradability and biodegradation rates of (micro)pollutants can mainly be attributed to the following factors: (i) Redox conditions affect the production of enzymes that are responsible for degradation of pollutants, and (ii) The molecular structure of a pollutant dictates under which specific redox condition it would be best degraded, as discussed in Section 4.3.1.2.

Table 4.3 exemplifies the biodegradation rate constants (k_{biol}) of various pollutants under different redox conditions. It is often difficult to compare the results between different studies since there is no standardization in the conduction of experiments yet (Pomies *et al.* 2015b). The experimental assessment of biodegradability is discussed in detail in **Chapter 5**.

When the data presented in Table 4.3 as well as others are taken into consideration, the following factors come to the fore in evaluation of biodegradation rate constants under aerobic, anoxic (denitrifying) or anaerobic conditions:

(a) *High variability in biodegradation rate constants*: As seen in Table 4.3, biodegradation rate constants (k_{biol}) are usually expressed per amount of suspended solids (SS) or volatile suspended solids (VSS) using the units

L/g SS · d or L/g VSS · d. The classification of biodegradability is rather arbitrary. For example, a group of researchers classify very highly, highly, moderately and hardly biodegradable compounds with $k_{biol} > 5$ L/g SS · d, $1 < k_{biol} < 5$ L/g SS · d, $0.5 < k_{biol} < 1$ L/g SS · d, and $k_{biol} < 0.5$ L/g SS · d, respectively (Suarez *et al.* 2010). In another classification, compounds with $k_{biol} < 0.1$ L/g SS · d are regarded as hardly biodegradable whereby their removal is lower than 20%. Other compounds within the range of $0.1 > k_{biol} > 10$ L/g SS · d and $k_{biol} > 10$ L/g SS · d are considered partially and highly biodegradable, respectively (Joss *et al.* 2006). It is likely that 20–90% and 90% removal is achieved for these compounds, respectively.

Table 4.3 also indicates that many discrepancies exist in removal rates of the same substrate, even under the same redox condition. One of the reasons for this is believed to be the fluctuation of the active biomass in total (as shown in Equation 4.12) or volatile suspended solids (Pomies *et al.* 2015b).

(b) *Occurrence of cometabolism*: Some studies have shown that unexpectedly high biodegradation rates are achieved in the presence of growth-substrates (Plosz *et al.* 2012), suggesting that hazardous pollutants were removed cometabolism. Feeding of sludge with different biodegradable substrates was effective in biodegradation of some hazardous pollutants. For example, it was shown that the biodegradation rate of ibuprofen was clearly improved if the sludge was fed with a biodegradable carbon source, suggesting that ibuprofen was essentially removed by the cometabolic activity of heterotrophs (Pomies *et al.* 2015b). More often, however, it is the nitrifying ability of sludge that brings about the cometabolic removal of most hazardous organics.

(c) *High rates under aerobic conditions*: Most studies show that biodegradation rates can vary significantly with redox conditions. Yet, biodegradation rate constants have often been reported for aerobic conditions, for example for polycyclic aromatic hydrocarbons (PAHs), surfactants and a few pharmaceutical compounds (Joss *et al.* 2004; Urase & Kikuta, 2005; Clouzot *et al.* 2013). Few researchers studied the effect of different redox conditions on the transformation rates or elimination efficiency of trace or industrial chemicals (Suarez *et al.* 2010; Plosz *et al.* 2012; Ogunlaja & Parker, 2015; Pomies *et al.* 2015b). Thus, compared to data under aerobic conditions, there is still a lack of information under anoxic and anaerobic conditions.

As exemplified in the case of pharmaceuticals, often a higher biodegradation rate is observed under aerobic conditions (Table 4.3). An example of this is the degradation of the estrogen E1. While degradation of E1 took place under all redox conditions, in the case of transition from anaerobic to anoxic conditions as well as transition from anoxic to aerobic conditions, biodegradation rates became increasingly higher (Joss *et al.* 2004). Also, another study revealed that E2 was readily converted to E1, faster under nitrifying than anoxic (denitrifying) conditions (Dytczak *et al.* 2008). Natural estrogens (E1 + E2) also exhibited a much higher rate of biodegradation under aerobic conditions than anoxic conditions. Compared to this, as shown in Table 4.3, EE2 was persistent under anoxic conditions, whereas it could be degraded during the active nitrification phase of treatment (Dytczak *et al.* 2008). As mentioned in former sections, aerobic conditions lead to generation of mono- or dioxygenase enzymes which catalyze cometabolic removal of most organic pollutants. Under the anoxic conditions of a bioreactor the persistence of compounds such as EE2 and the like is attributed to this lack of oxygenases which can only be induced under aerobic conditions. As seen in Table 4.3, generally a significant nitrifying activity in sludge seems to be the key factor in increasing biodegradation rates.

The effect of aerobic conditions on biodegradation rates is also observed in the case of other compounds such as FLX (Table 4.3), musk fragrances HHCB, AHTN and celestolide (ADBI). They were all transformed to a larger extent under aerobic conditions than anoxic conditions. Some compounds such as naproxen, roxithromycin (ROX) and erythromycin (ERY) were only significantly transformed under aerobic conditions. Others such as the anti-depressant citalopram (CTL) (Table 4.3) were moderately biotransformed under both aerobic and anoxic conditions. Some compounds, such as CBZ (Table 4.3), diazepam (DZP), SMX and trimethoprim (TMP) had a high resistance to biological transformation (Suarez *et al.* 2010).

(d) *Effect of SRT*: As shown in Equation 4.1 in Section 4.3.1.3, generally a high SRT favors biodegradation. Supporting this, as seen from Table 4.3, biodegradation rate constants belonging to some micropollutants mostly increased at higher SRTs (Clara *et al.* 2005a). In addition, as reported in a former study, for the natural estrogens E1 and E2, degradation activity was seen to be higher in MBR sludge than in CAS. This was attributed mostly to the higher SRT in the MBR (30 d as compared to 11 d in the conventional system) (Joss *et al.* 2004). But, a recent study also indicated that there was a limit to the positive effect of SRT. Micropollutant degradation rates, as expressed by k_{biol}, were not strongly affected if SRT increased to extremely high values, namely from 25 to 80 days (Falas *et al.* 2016).

(e) *Acclimation of biomass to hazardous pollutants*: An important factor in biodegradation of hazardous organic substrates is acclimation or adaptation of biomass. If a microbial community is previously exposed to a pollutant, changes such as gene transfer, mutation, enzyme induction and population shifts may occur that increase the rate of biodegradation (van Leeuwen & Vermeire, 2007). The changes taking place in a microbial community with exposure to hazardous pollutants are discussed in **Chapter 7.**

However, even upon acclimation some substrates are often incompletely mineralized, meaning that they are transformed into organic metabolites only. The effect of acclimation on biological removal of hazardous compounds has been demonstrated to date for few compounds. For example, studies showed

Table 4.3 First-order biodegradation rate constants (k_{biol}) of hazardous organic pollutants under different redox conditions.

Organic Micropollutant	Redox Condition	First-Order Biodegradation Rate Constant k_{biol} (L/solids* · d)	Conditions	Reference
Natural estrogens (E1 + E2)	Aerobic	170 L/g SS · d	High degradation rate in the nitrifying activated sludge is attributed to cometabolic removal by nitrifiers.	(Suarez *et al.* 2010)
	Anoxic	3 L/g SS · d	In the anoxic reactor, a slight increase in transformation was observed with increasing SRT.	
Estrone +17β-estradiole + estriol (E1 + E2 + E3)	Aerobic + anoxic	0.9, 12.0, 4.6, 7.6 L/g VSS · d	Single-sludge taken from a WWTP; data taken from laboratory-scale reactors operated at 10°C and at SRTs of 2, 10, 34, 68 days, respectively.	(Clara *et al.* 2005a)
17 α-ethinylestradiol (EE2)	Aerobic Aerobic	9 L/g SS · d 20 L/g SS · d	Aerobic sludge with heterotrophic activity. Aerobic sludge after development of nitrifying activity.	(Suarez *et al.* 2010)
	Anoxic	0 L/g SS · d	No removal under anoxic conditions.	(Dytczak *et al.* 2008)
Estrone (E1)	Aerobic Aerobic	~96–120 L/g VSS · d ~36–72 L/g VSS · d	In the presence of low organic carbon loads. In the presence of high organic carbon loads.	(Tan *et al.* 2015)
Ibuprofen (IBP)	Aerobic	2.13 L/g SS · d 0.34 L/g SS · d	The sludge was supplied with biodegradable organic carbon and nitrogen. The sludge did not receive biodegradable organic carbon.	(Pomies *et al.* 2015b)
	Aerobic	20 L/g SS · d	Results were obtained in a nitrifying activated sludge. Aerobic transformation of IBP was between 93 and 96%.	(Suarez *et al.* 2010)
	Anoxic	~1.5 L/g SS · d	In anoxic sludge, efficient transformation of ibuprofen (75%) was observed after an adaptation period of 340 d.	

(Continued)

Table 4.3 First-order biodegradation rate constants (k_{biol}) of hazardous organic pollutants under different redox conditions (*Continued*).

Organic Micropollutant	Redox Condition	First-Order Biodegradation Rate Constant k_{biol} (L/solids* · d)	Conditions	Reference
	Aerobic+ anoxic	1.0, 37.9, 47.5, 44.6 L/g VSS · d	Single-sludge taken from a WWTP; data taken from laboratory-scale reactors operated at 10°C and at SRTs of 2, 10, 34, 68 days, respectively.	(Clara *et al.* 2005a)
Ketoprofen (KEP)	Aerobic	≤0.34–0.57 L/g SS · d	Oxic sludge (batch experiments at 16°C).	(Falas *et al.* 2013)
	Anoxic	≤0.2 L/g SS · d	Anoxic sludge (batch experiments at 16°C).	
Fluoxetine (FLX)	Aerobic	<0.09 L/g SS · d	The sludge was supplied with biodegradable organic carbon and nitrogen.	(Pomies *et al.* 2015b)
	Anoxic	<0.11 L/g SS · d	The sludge was supplied with biodegradable organic carbon and nitrogen.	
	Aerobic	9 L/g SS · d	Nearly complete removal in nitrifying activated sludge (>90%).	(Suarez *et al.* 2010)
	Anoxic	~5 L/g SS · d	Significant transformation in the anoxic treatment (79–89%).	
Carbamazepine (CBZ)	Aerobic + anoxic	<0.01 L/g SS · d	Single-sludge from a nutrient-removing domestic WWTP.	(Joss *et al.* 2006)
	Aerobic	2 L/g SS · d	Aerobic experiments in activated sludge with readily biodegradable substrate (unexpectedly high rate).	(Plosz *et al.* 2012)
	Anoxic	1.2 L/g SS · d	Anoxic experiments in activated sludge with readily biodegradable substrate (unexpectedly high rate).	
	Aerobic or anoxic	0.01 L/g SS · d	Low removal rates under both aerobic and anoxic conditions with readily biodegradable substrate becoming limiting.	

Organic Micropollutant	Redox Condition	First-Order Biodegradation Rate Constant k_{biol} (L/solids* · d)	Conditions	Reference
	Aerobic Anoxic	<0.06 L/g SS · d <0.03 L/g SS · d	High resistance to removal in aerobic (nitrifying) and anoxic sludges.	(Suarez *et al.* 2010)
		0.1 L/g VSS · d	Single-sludge taken from a WWTP, data taken from laboratory-scale reactors operated at 10°C and at SRTs of 10 days.	(Clara *et al.* 2005a)
Diclofenac (DCF)	Aerobic	1.6 L/g SS · d	Aerobic experiments in activated sludge with readily biodegradable substrate.	(Plosz *et al.* 2012)
		0.14 L/g SS · d	Aerobic experiments in activated sludge with readily biodegradable substrate becoming limiting.	
	Anoxic	0.96 L/g SS · d	Anoxic experiments in activated sludge with readily biodegradable substrate.	
		0.10 L/g SS · d	Anoxic experiments in activated sludge with readily biodegradable substrate becoming limiting.	
	Aerobic Anoxic	1.2 L/g SS · d <0.04 L/g SS · d	Nitrifying sludge fed with inorganics only. Denitrifying sludge.	(Suarez *et al.* 2010)
	Aerobic+ anoxic	0.4, 0.8 L/g VSS · d	Single-sludge taken from a WWTP; data taken form laboratory-scale reactors operated at 10°C and at SRTs of 2 and 10 days, respectively	(Clara *et al.* 2005a)
Citalopram (CTL)	Aerobic Anoxic	3 L/g SS · d 0.5 L/g SS · d	Nitrifying activated sludge (NAS). Denitrifying sludge.	(Suarez *et al.* 2010)

that N,N-diethyl-m-toluamide (DEET), found in liquids, lotions, sprays, and impregnated materials, could be degraded upon acclimation and at influent concentrations above the threshold value of 1 μg/L (Knepper, 2004). However, as emphasized in another work, there is still little knowledge about the evolution or adaptation of a biological sludge when exposed to hazardous pollutants (Clouzot *et al.* 2012a).

Acclimation of microbial communities to micropollutants in WWTPs is a debated issue because very low concentrations are believed not to promote population shifts (Fischer & Majewsky, 2014). Researchers argue that rather organic loading and the promotion of non-specific enzymes, such as AMO, would govern the degradation of pharmaceuticals, and not acclimation. Also, a recent study supports this view. It was shown that long-term exposure of activated sludge to organic micropollutants was not a precondition for biodegradation (Falas *et al.* 2016).

4.8 BRIEF LOOK AT MODELING OF HAZARDOUS ORGANICS REMOVAL

Modeling the removal of biogenic as well as hazardous compounds is an immense field. Regarding this, the aim of this section is only to provide a very brief description of the modeling approaches used in removal of micropollutants.

A literature review investigating 18 literature models describing micropollutant removal in activated sludge processes showed that investigated micropollutants were mainly volatile organic compounds (VOCs), metals, surfactants, pesticides and pharmaceutical compounds (Pomies *et al.* 2013). Further, the models examined were differentiated into two, as static and dynamic, while the latter were more adapted to account for temporal variations in the influent of a WWTP.

4.8.1 Basic mass balance describing removal of pollutants

In biological treatment systems, model equations take into account various processes such as biodegradation, sorption to solid phases, volatilization from surface and stripping, and eventually other processes. Mass balances can be constructed for (micro)pollutants considering both the dissolved and solid phases of a bioreactor. For example, if a mass balance is made in the dissolved phase of a suspended-growth CSTR, the accumulation rate of a (micro)pollutant can be expressed as follows:

$$V\frac{dS_{MP}}{dt} = F_{MP,in}(t) - F_{MP,out}(t) + V\left(\frac{dS_{MP}}{dt}\right)_{stripping} + V\left(\frac{dS_{MP}}{dt}\right)_{surface\ vol.} + V\left(\frac{dS_{MP}}{dt}\right)_{sorption} + V\left(\frac{dS_{MP}}{dt}\right)_{bio\deg radation} \quad (4.13)$$

where

$V \cdot dS_{MP}/dt$: accumulation rate of (micro)pollutant inside the reactor, (M_s/T).

S_{MP} : concentration of (micro)pollutant inside the reactor, (M_s/L^3).

V : reactor volume, (L^3).

$F_{MP,in}$ (t) and $F_{MP,out}$ (t) : mass flow rate of (micro)pollutant into and out of the reactor, (M_s/T).

$V\,(dS_{MP}/dt)_X$: removal rate of (micro)pollutant as a result of different mechanisms, (M_s/T).
(x: stripping, surface volatilization, sorption, biodegradation).

Each process shown in Equation 4.13 can then be further expressed using detailed expressions (Pomies *et al.* 2013). Depending on pollutant and operational characteristics, in each treatment system one or more of these processes may be relevant. For example, volatilization may become relevant in wastewater treatment units with large surface areas or high aeration rates (Clouzot *et al.* 2013). In models dealing with VOCs, acetone and phenol, volatilization is also considered. However, literature studies indicate that most micropollutants of emerging interest have a low volatility and thus modeling of volatility is of limited value (Plosz *et al.* 2013). For example, for pollutants like pharmaceuticals, hormones and metals, volatilization is disregarded.

In addition, occurrence or relevance of each removal process depends on the extent of another one. For example, if pollutants are nonbiodegradable, volatilization by stripping is of relevance at relatively high Henry's law constants. In contrast to this, a highly biodegradable pollutant may be rapidly removed by biomass. In such a case, volatilization by stripping may lose its importance because biodegradation will proceed much more rapidly.

4.8.2 Extension of biodegradation models to include hazardous organics

Several models can be used in describing biodegradation of (micro)pollutants. Relatively recently, activated sludge models (ASM), initially proposed by the Task Group of IWA, have been extended to include also xenobiotic trace chemicals. In these models incorporating xenobiotics (ASM-X), values of k_{biol} are estimated under both aerobic and anoxic conditions. The "X" in this expression is attributed to Xenobiotic Organic Compounds (XOCs), a group name for compounds that are alien to the biota.

Although the removal of a pollutant as a primary substrate received attention, to date, in modeling studies there was not enough focus on cometabolism. Relatively recently, models took into consideration also the possibility of cometabolic degradation in the presence of growth-substrates. For example, a model described the fate of EE2 using an unadapted biomass. In modeling the cometabolic degradation of EE2 by nitrifiers, researchers have transformed the

ASM1 equation for aerobic growth of nitrifiers to include a kinetic limitation term for EE2. The total model proposed by the researchers also took into consideration sorption. This synthetic hormone was sorbed quickly, it was then degraded by cometabolism after an adaptation period. It was also shown that SRT was very effective. At very low SRTs like 2 days, no nitrification took place and therefore no cometabolic removal took place. The substance was then merely removed by sorption (Clouzot *et al.* 2012). In activated sludge models some researchers take into consideration also the conjugation of pharmaceuticals as well as their retransformation possibility into parent compounds (Plosz *et al.* 2012).

In modeling the removal of hazardous organics, it is also important to know the role of biomass. As shown in Equation 4.12 in Section 4.7.1, the X_{ss} term used in most rate expressions indicates a non-specific biomass. However, this type of approach is often inadequate since it can neither explain the removal mechanism nor can be extrapolated to practice. The importance of specifying the biomass can be best exemplified in a nitrifying system. In a nitrifying reactor, often the whole mixed liquor volatile suspended solids (MLVSS) concentration is considered in the calculation of ammonium (ammonia) or nitrite degradation rates whereas actually only a tiny fraction of total biomass consists of nitrifiers. Therefore, there is often a wide discrepancy in the reported nitrification rates of different sludges. In recent years, attempts were made to use more specific terms for "biomass" in modeling studies. Thus, a differentiation was made between heterotrophic (X_{HET}) and ammonia oxidizing bacteria (X_{AOB}). In the enriched nitrifying community, quantitative polymerase chain reaction (qPCR) results suggested that AOB represented 75–85% of the nitrifying population (i.e. AOB + NOB). The researchers used model equations to link cometabolic biodegradation rates of pharmaceuticals to specific fractions in biomass like X_{AOB} (Sathyamoorthy *et al.* 2013).

While in most cases biodegradation rate is of first-order in terms of micropollutant concentration, as shown in Equation 4.12 in Section 4.7.1, recent studies consider also the concentration of other substrates, namely the availabilities of dissolved oxygen, biodegradable organic carbon and ammonium nitrogen in aerobic degradation. Similarly, if the micropollutant is degraded under anoxic conditions, the availability of oxidized nitrogen and biodegradable organic carbon is taken into account. In that case the presence of dissolved oxygen would hinder the reaction rate (Pomies *et al.* 2015a).

As a novel approach, some wastewater treatment studies also delineate the inhibitory effect of micropollutants. For example, a study modelled the effect of the pharmaceutical ATN on AOB growth in a nitrification system according to the competitive inhibition model for ammonia oxidation (Sathyamoorthy *et al.* 2013). As mentioned in Section 4.5.2, also the readily biodegradable substrate in a wastewater may inhibit the removal of a micropollutant. The competitive inhibition of cometabolic substrates by the readily biodegradable substrate was accounted for in modeling of antibiotic micropollutants SMX, tetracycline (TCY) and ciprofloxacin (CIP) (Plósz *et al.* 2010).

Modeling of micropollutants still remains a difficult task since in most cases not enough information is provided in experimental studies about operational parameters, such as SRT and HRT (Clouzot *et al.* 2013). In addition, the absence of a unified experimental protocol may lead to biased modeling results. Therefore, efforts have been spent in recent years to develop a laboratory-scale experimental strategy (Pomies *et al.* 2015b). A thorough examination of experimental strategies is presented in **Chapter 5**.

4.9 BEHAVIOR OF HAZARDOUS INORGANIC POLLUTANTS

In a wastewater or water undergoing biological treatment, besides organic pollutants, a number of hazardous inorganic pollutants may be present that affect biodegradation of other substrates or are converted into less or more harmful forms by microbial activity. Among such pollutants, metals constitute one of the most important groups.

Domestic wastewater contains heavy metals such as Cu, Zn, Hg, Pb, Cr, Ni and Cd due to leaching from several materials and food. Also stormwater runoff can be regarded as an important source of heavy metals (Margot *et al.* 2015a). Although metal-containing industrial wastewaters are strictly controlled, in many cases their discharge into domestic wastewater is still possible.

4.9.1 Metals in biological treatment systems

4.9.1.1 Speciation of metals

In a water or wastewater matrix, hardly any metal exists in the form of free ions, but is complexed in many ways:

(1) *Complexation of metals with water*: When present in water, metals are often hydrolyzed to form metal aquo complexes. These are coordination compounds containing metal ions with only water as a ligand.
(2) *Complexation of metals with organic substances (ligands)*: A large variety of organic substances found in water or wastewater, often denoted by the acronym dissolved organic matter (DOM), or expressed by the sum parameter dissolved organic carbon (DOC), can act as ligands and form complexes of variable stability.
(3) *Complexation of metals with inorganic ligands*: Metals found in water or wastewater can also be complexed with inorganic ligands such as chloride and phosphate.
(4) *Complexation of metals with EPS*: Metals also complex with the EPS surrounding the cells. The effect of metals in activated sludge and other biological systems depends also on their binding capacity to EPS.

4.9.1.2 Consequences of metal speciation for biological treatment

In most laboratory-scale studies evaluating the effects of heavy metals, speciation of metals is often disregarded and a rather simplistic approach is adopted that considers the total metal concentration only. In fact, the total metal is comprised of many fractions such as the metal found in the bulk liquid, in the EPS and inside of cells.

Speciation of metals depends on many factors such as temperature, pH and the type and amount of inorganic and organic matter in the background. Among these, the impact of background organic matter is the most difficult one to predict. On the other hand, the level and type of organic matter in a wastewater is a crucial factor affecting the behavior of metals. Organic matter was shown to affect the partitioning of heavy metals between liquid and sludge phases (Katsoyiannis & Samara, 2006). Further, the type of organic matter determines whether metals exist in the form of labile or stable complexes in the liquid phase.

Complexation of a metal with the organics in a wastewater reduces the free ion concentration in the bulk phase, thus changing the inhibitory effects of a metal on biomass. For example, compounds such as EDTA and nitrilotriacetic acid (NTA), widely used as builders in detergents, are found at relatively high concentrations in domestic wastewater (Margot *et al.* 2015b). These two compounds have the ability to form strong complexes with metals. As a result, inhibitory characteristics of most heavy metals may be greatly reduced. However, NTA is biodegradable whereas EDTA is not. Upon degradation of NTA, heavy metals may be liberated again. Once complexed with EDTA in a WWTP, however, it is likely that heavy metals will be passed on to receiving waters with the treated wastewater. For example, a former study has shown that a significant fraction of Cu and Ni discharged from WWTPs was present in a strongly complexed form (Sedlak *et al.* 1997). In such a case, these metals will also be found in an extremely stable form in seawater. Thus, complexation of metals can affect their fate in receiving waters.

Speciation of metals is highly dependent on the composition of wastewater as well as operating conditions. Often laboratory-scale experimentation is needed to study their effects (Semerci & Çeçen, 2007; Çeçen *et al.* 2010a, b). But, in the case of a high number of metals, laboratory-scale inhibition and speciation studies alone may not be sufficient to explain the behavior of metals. Therefore, complementary to them, quantitative structure-activity relationships (QSARs) are used for predicting cation toxicity, bioconcentration, biosorption, and binding strength (Sacan *et al.* 2009; Walker *et al.* 2012).

4.9.1.3 Biotransformation of metals

Although metals are inert to biodegradation, they may undergo biotransformation in the presence of microorganisms. Often, the "biotransformation" term is used interchangeably with bioreduction or biooxidation of metal or to cover four different processes such as biomineralization, biosorption, bioreduction and bioleaching. Within the scope of this chapter this term is used to indicate conversion of a substance into another form by microbial activity.

Heavy metals are discharged from many facilities. For example, metal plating facilities, tanneries, and industrial processes using catalysts discharge large amounts of Cr^{+6} and Cr^{+3}. Many species (bacteria, fungi, yeast and fungi) can remove or transform the metal ions found in water and wastewater. For example, different species of bacteria and fungi have the ability to reduce Cr^{+6} to Cr^{+3}, as reported in literature. It was shown that Cr^{+6} was effectively reduced by immobilized intact cells and the cell-free enzyme extracts of *Bacillus* sp. ES 29 in a bioreactor system (Camargo *et al.* 2004). Biotransformation of Cr^{+6} into Cr^{+3} decreases the toxicity of chromium, since the latter species settles out from solution due to low solubility. Usually, biotransformation of a metal leads to its detoxification. However, in some instances as in methylation of mercury in receiving waters, biotransformation may render metals even more toxic. It is often hard to decide whether biotransformation of a metal species occurs by direct metabolism or cometabolism.

4.9.1.4 Biosorption of metals

Once present in wastewater treatment systems, metals may sorb to primary sludge and be removed from the water phase (Wang *et al.* 2006). Then, in the biological treatment step, they may be sorbed to biological sludge whereas this type of sorption is regarded as biosorption.

A wide variety of microorganisms can sequester trace levels of heavy metal ions from water or wastewater. Metals can also reach inside of cells by internal diffusion, leading mostly to a toxic/inhibitory effect. This intracellular uptake of metals is also referred to as active uptake or bioaccumulation (Kapoor & Viraraghavan, 2003). However, in biological treatment systems, it is often observed that in addition to organic pollutants, also metals sorb to the EPS. In that way cells can be protected from adverse effects of heavy metals.

The composition of EPS is a factor affecting metal sorption. In turn, also the binding of metals to EPS can bring about a change in EPS characteristics. For example, a study has shown that binding of silver and nanosilver to EPS led to deterioration of the proteins in EPS (Geyik & Çeçen, 2016). In addition, various fractions in EPS had different uptake capacities for silver and nanosilver.

Some metals can behave in the opposite way, meaning that they leach from mineral ores or treatment sludges. This process called bioleaching takes place in the presence of specific microorganisms that produce acids. The acid production leads to solubilization and mobilization of metals that are found in the form of sulfides. In wastewater treatment, bioleaching is a method that is often used to extract metals from treatment sludges.

4.9.1.5 Inhibitory effects of metals on biological treatment

Metals inhibit organic carbon removal as well as nutrient removal processes such as nitrification. The inhibitory effect of heavy metals is usually modelled using non-competitive inhibition kinetics. Although much research has been done on simultaneous removal of organic carbon and nitrogen, few studies address inhibition of the nitrification process alone (Çeçen *et al.* 2010a, b; Semerci & Çeçen, 2007, 2009). In addition, few studies differentiate the inhibitory effects of metals on organic carbon removal from that on nitrification (Çeçen & Kılıç, 2016).

Unlike inhibitory organic compounds, metals do not biodegrade. Therefore, their total level would not change during a biological process. However, they may be biotransformed into more or less inhibitory species or precipitate out from the liquid phase. Further, the presence of background organic or inorganic matter has a great effect on their inhibitory behavior. For example, a study has shown that the inhibitory property of the silver ion (Ag^+) was drastically increased as the carbon to nitrogen (COD/TKN) ratio in a synthetic feed became lower, mainly due to enrichment of nitrifiers in sludge and increased nitrification. Also the type of organic matter in the feed had an influence on inhibition (Çeçen & Kılıç, 2016). Since peptone in the feed complexed with the silver ion, the inhibitory effect of silver was reduced substantially.

In long-term exposure of biomass to metals, the extent of inhibition may be gradually altered since the microbial ecology changes in a direction to tolerate the metal. For example, in long-term exposure of a nitrifying sludge to increased Cd loadings in a continuous-flow activated sludge reactor, the inhibitory effects of this metal became lower since a shift occurred in microbial population from the initial *Nitrosomonas* sp. to the Cd-tolerant *Nitrosospira* sp. (Semerci & Çeçen, 2009). However, most laboratory-scale experimentation is based on relatively short-term contact of sludge with metals; therefore such gradual changes taking place in microbial ecology of sludges are often not detected.

Even after aerobic treatment of wastewater, metals continue to exert their inhibitory effects, this time in anaerobic digestion of sludges. In this case too, speciation of a metal is an important factor. For example, relatively high levels of silver could be tolerated in anaerobic digestion of silver-bearing waste sludges generated from the aerobic treatment of a photoprocessing wastewater because the free silver ion concentration, the most inhibitory form of silver, was drastically reduced in the presence of sulfide (Pavlostathis & Maeng, 2000).

4.9.1.6 Nanometals in biological treatment: Speciation, fate and effects

In recent years most of the relevant studies have been devoted to the speciation and effects of nanometals such as nanosilver (AgNP) and nanozinc oxide (ZnO-NP) in biological wastewater treatment and to the disposal and utilization of treatment sludges containing them. Such metallic micropollutants usually undergo a rapid physicochemical transformation in biological treatment systems. For example, several studies report on the dissolution of AgNP in biological treatment systems to yield the silver ion. Also, other metallic nanometals such as nanozinc oxide and nanocopper may dissolve to yield ionic species. Subsequently, these species may sorb to settlable biomass, precipitate in the form of inorganic solids, or form stable aqueous complexes. In literature, more than 90% of nanomaterials was observed to attach to biomass, which is removed within the WWTP (Westerhoff *et al.* 2013).

Even after biological treatment of wastewater, speciation of a nanometal remains an important factor since it affects digestion and utilization of sludge. A model used to predict the partitioning of silver nanoparticles to biomass showed that in a treatment plant, concentrations in the effluent went down to very low values. However, at an influent silver concentration of 180 µg/L, the silver concentration in biosolids would still exceed the toxicity criteria set by the United States Environmental Protection Agency (U.S. EPA) (Benn & Westerhoff, 2008). Further, AgNPs introduced to the influent of a pilot-scale wastewater treatment system were seen to be speciated differently in aged, fresh, and incinerated biosolids. It was stated that such transformations should be accounted for in life-cycle analyses of AgNPs and in management decisions regarding the disposal of incinerated biosolids (Impellitteri *et al.* 2013).

Although many studies focus on the fate, transformation, toxicity and bioavailability of nanosilver in WWTPs, some issues are not clarified yet. Some studies have shown that nanosilver inhibits bacterial growth (Choi & Hu, 2009). However, very divergent results are also reported on AgNP inhibition. This is believed to be caused by the differences in AgNP synthesis, shape, size distribution, and coating. A study demonstrated that coating of AgNP played a very important role in determining Ag dissolution and ultimate toxicity to nitrifiers (Arnaout & Gunsch, 2012). Another study has shown that a commercial nanosilver agglomerated and had therefore no effect on activated sludge. In contrast to this, a nanoparticle synthesized according to the chemical reduction method exerted inhibitory effects on activated sludge (Geyik & Çeçen, 2016).

As in the case of heavy metal ions, the presence of inorganic or organic ligands plays a role in speciation and fate of nanometals. For example, in domestic wastewater the presence of sulfide leads to sulfidation of nanometals and reduces their inhibitory effects (Choi & Hu, 2009). In another case sulfidation of AgNP was shown to depend primarily on particle size, sulfide availability and residence times in the sewer and the WWTP. This study also confirmed that most of the AgNP was directed to the digester while only a small fraction was released to surface waters (Kaegi *et al.* 2013).

The impacts of nanometals on microbial communities are still not well understood. A study has shown that compared to Ag^+, a lower AgNP concentration resulted in a more pronounced effect on the community structure of an activated sludge, possibly reflecting a higher tendency for Ag^+ than AgNPs to be scavenged by inorganic ligands and organic matter. In addition, AgNPs decreased the abundance of nitrifying bacteria and damaged the structure of activated sludge flocs, which could affect sludge clarification and recycling (Yang *et al.* 2014). In another study, even at high doses and long-term contact of nanosilver with activated sludge, no obvious change was observed in COD removal rates whereas characteristics of protein-EPS were altered (Geyik, 2015).

In recent years, studies have also focused on the importance of nanometals in anaerobic digestion. While nanosilver is of primary interest, relatively scarce information is found on others such as nanocopper. Short-term exposure of sludge to nanocopper (1.4 mg Cu/L) was shown to cause only minor inhibition on methanogenesis. On the other hand, extended exposure caused severe toxicity and reduced the acetoclastic methanogenic activity by more than 85%. Moreover, reactor performance was completely disrupted and methane production decreased by more than 50% (Otero-González *et al.* 2014).

Nanometals can also affect the microbial processes in groundwater biotreatment. For example, a study pointed to the inhibitory effect of nanoiron (nFe0) on microbial sulfate reduction during the in situ bio-precipitation of metals (Kumar *et al.* 2014).

4.9.2 Hazardous ions in biological treatment

Besides heavy metal ions whose effects are relatively known, there are a great number of cations and anions that can bring about harm in biological treatment. For example, chloride (Cl^-), perchlorate (ClO_4^-), fluoride (F^-) and free cyanide (CN^-) are inhibitors of organic carbon and nutrient removal.

In a study, fluoride was found to be inhibitory (IC_{50}:18 to 43 mg/L) toward the species found in anaerobic digestion processes, at concentrations lower than those found in some fluoride-containing industrial effluents. In the aerobic activated sludge and in denitrification processes, very high concentrations of soluble fluoride (>500 mg/L) could be tolerated. However, such ionic species usually have a higher effect on nitrifying species. It was shown that nitrification was more sensitive to fluoride (IC_{50}:149 mg/L) than anaerobic digestion. However, nitrifiers appeared to acclimate to fluoride rapidly (Ochoa-Herrera *et al.* 2009). Therefore, it was concluded that the main problem due to the presence of fluoride would occur in anaerobic treatment.

4.10 REFERENCES

Alpaslan Kocamemi B. and Çeçen F. (2005). Cometabolic degradation of TCE in enriched nitrifying batch systems. *Journal of Hazardous Materials*, **125**(1–3), 260–265.

Alpaslan Kocamemi B. and Çeçen F. (2007a). Inhibitory effect of the xenobiotic 1,2-dichloroethane in a nitrifying biofilm reactor. *Water Science & Technology*, **55**(8–9), 67.

Alpaslan Kocamemi B. and Çeçen F. (2007b). Kinetic analysis of the inhibitory effect of trichloroethylene (TCE) on nitrification in cometabolic degradation. *Biodegradation*, **18**(1), 71–81.

Alpaslan Kocamemi B. and Çeçen F. (2009). Biodegradation of 1,2-dichloroethane (1,2-DCA) by cometabolism in a nitrifying biofilm reactor. *International Biodeterioration & Biodegradation*, **63**(6), 717–726.

Alpaslan Kocamemi B. and Çeçen F. (2010a). Biological removal of the xenobiotic trichloroethylene (TCE) through cometabolism in nitrifying systems. *Bioresource Technology*, **101**(1), 430–433.

Alpaslan Kocamemi B. and Çeçen F. (2010b). Cometabolic degradation and inhibition kinetics of 1,2-dichloroethane (1,2-DCA) in suspended-growth nitrifying systems. *Environmental Technology*, **31**(3), 295–305.

Andersen H., Siegrist H., Halling-Sorensen B. and Ternes T. A. (2003). Fate of estrogens in a municipal sewage treatment plant. *Environmental Science & Technology*, **37**(18), 4021–4026.

Arnaout C. L. and Gunsch C. K. (2012). Impacts of silver nanoparticle coating on the nitrification potential of Nitrosomonas europaea. *Environmental Science and Technology*, **46**(10), 5387–5395.

Barret M., Barcia G. C., Guillon A., Carrere H. and Patureau D. (2010). Influence of feed characteristics on the removal of micropollutants during the anaerobic digestion of contaminated sludge. *Journal of Hazardous Materials*, **181**(1–3), 241–247.

Bernhard M., Muller J. and Knepper T. P. (2006). Biodegradation of persistent polar pollutants in wastewater: comparison of an optimised lab-scale membrane bioreactor and activated sludge treatment. *Water Research*, **40**(18), 3419–3428.

Camargo F., Okeke B., Bento F. and Frankenberger W. (2004). Hexavalent chromium reduction by immobilized cells and the cell-free extract of *Bacillus* sp. ES 29. *Bioremediation Journal*, **8**(1–2), 23–30.

Çeçen F. and Aktaş Ö. (2011). Activated Carbon for Water and Wastewater Treatment: Integration of Adsorption and Biological Treatment. Wiley-VCH, Germany, 388 pages, ISBN: 978-3-527-32471-2.

Çeçen F. and Kılıç B. (2016). Inhibitory effect of silver on activated sludge: effect of organic substrate and the carbon to nitrogen ratio. *Journal of Chemical Technology & Biotechnology*, **91**(4), 1190–1198.

Çeçen F., Semerci N. and Geyik A. G. (2010a). Inhibitory effects of Cu, Zn, Ni and Co on nitrification and relevance of speciation. *Journal of Chemical Technology & Biotechnology*, **85**, 520–528.

Çeçen F., Semerci N. and Geyik A. G. (2010b). Inhibition of respiration and distribution of Cd, Pb, Hg, Ag and Cr species in a nitrifying sludge. *Journal of Hazardous Materials*, **178**(1–3), 619–627.

Choi O. K. and Hu, Z. Q. (2009). Nitrification inhibition by silver nanoparticles. *Water Science and Technology*, **59**(9), 1699–1702.

Clara M., Kreuzinger N., Strenn B., Gans O. and Kroiss H. (2005a). The solids retention time-a suitable design parameter to evaluate the capacity of wastewater treatment plants to remove micropollutants. *Water Research*, **39**(1), 97–106.

Clara M., Strenn B., Gans O., Martinez E., Kreuzinger N. and Kroiss H. (2005b). Removal of selected pharmaceuticals, fragrances and endocrine disrupting compounds in a membrane bioreactor and conventional wastewater treatment plants. *Water Research*, **39**(19), 4797–4807.

Clouzot L., Doumenq P., Roche N. and Marrot B. (2010). Kinetic parameters for 17alpha-ethinylestradiol removal by nitrifying activated sludge developed in a membrane bioreactor. *Bioresource Technology*, **101**(16), 6425–6431.

Clouzot L., Cloutier F. and Vanrolleghem P. A. (2012). Modelling 17α-ethinylestradiol in Membrane Bioreactors. In: 3rd IWA/WEF Wastewater Treatment Modelling Seminar (WWTmod2012), February 26–28, 2012, Mont-Sainte-Anne, Canada, 165–171.

Clouzot L., Choubert J. M., Cloutier F., Goel R., Love N. G., Melcer H., Ort C., Patureau D., Plosz B. G., Pomies M. and Vanrolleghem P. A. (2013). Perspectives on modelling micropollutants in wastewater treatment plants. *Water Science and Technology*, **68**(2), 448–461.

Delgadillo-Mirquez L., Lardon L., Steyer J. P. and Patureau D. (2011). A new dynamic model for bioavailability and cometabolism of micropollutants during anaerobic digestion. *Water Research*, **45**(15), 4511–4521.

Dionisi D. (2014). Potential and limits of biodegradation processes for the removal of organic xenobiotics from wastewaters. *ChemBioEng Reviews*, **1**(2), 67–82.

Dytczak M. A., Londry K. L. and Oleszkiewicz J. A. (2008). Biotransformation of estrogens in nitrifying activated sludge under aerobic and alternating anoxic/aerobic conditions. *Water Environment Research*, **80**(1), 47–52.

European Commission S.C. (1999). Opinion of the Scientific Committee on Toxicity, Ecotoxicity and the Environment (CSTEE) on a Proposed "Ready Biodegradability" Approach to Update Detergents Legislation. Adopted at the 12th CSTEE Plenary Meeting, 25 November 1999.

Falas P., Longree P., la Cour Jansen J., Siegrist H., Hollender J. and Joss A. (2013). Micropollutant removal by attached and suspended growth in a hybrid biofilm-activated sludge process. *Water Research*, **47**(13), 4498–4506.

Fernandez-Fontaina E., Omil F., Lema J. M. and Carballa M. (2012). Influence of nitrifying conditions on the biodegradation and sorption of emerging micropollutants. *Water Research*, **46**(16), 5434–5444.

Flemming H.-C., Wingender J., Griegbe C. and Mayer C. (2000). Physico-chemical properties of biofilms. In: Biofilms: Recent Advances in their Study and Control, Evans L. V. (ed.). Amsterdam: Harwood Academic Publishers, pp. 19–34.

Forrez I., Carballa M., Boon N. and Verstraete W. (2009). Biological removal of 17α-ethinylestradiol (EE2) in an aerated nitrifying fixed bed reactor during ammonium starvation. *Journal of Chemical Technology & Biotechnology*, **84**(1), 119–125.

Geyik A.G. (2015). Relationship between metal inhibition and microbial products in biological systems, Ph.D. Thesis, Boğaziçi University, Institute of Environmental Sciences, Istanbul, Turkey.

Geyik A. G. and Çeçen F. (2016). Exposure of activated sludge to nanosilver and silver ion: inhibitory effects and binding to the fractions of extracellular polymeric substances. *Bioresource Technology*, **211**, 691–697.

Grady C. P. L. J., Daigger G. T., Love N. and Filipe C. D. M. (2011). Biological Wastewater Treatment. IWA Publishing, London.

Guerra P., Kim M., Shah A., Alaee M. and Smyth S. A. (2014). Occurrence and fate of antibiotic, analgesic/anti-inflammatory, and antifungal compounds in five wastewater treatment processes. *Science of the Total Environment*, **473–474**, 235–243.

Helbling D. E., Johnson D. R., Honti M. and Fenner K. (2012). Micropollutant biotransformation kinetics associate with WWTP process parameters and microbial community characteristics. *Environmental Science and Technology*, **46**(19), 10579–10588.

Impellitteri C. A., Harmon S., Silva R. G., Miller B. W., Scheckel K. G., Luxton T. P., Schupp D. and Panguluri S. (2013). Transformation of silver nanoparticles in fresh, aged, and incinerated biosolids. *Water Research*, **47**(12), 3878–3886.

ISO (2013). ISO 14853:2005(en) Plastics – Determination of the ultimate anaerobic biodegradation of plastic materials in an aqueous system – Method by measurement of biogas production, Technical Committee ISO/TC 61, Plastics, Subcommittee SC 5, Physical-chemical properties.

Jones O. A. H., Voulvoulis N. and Lester J. N. (2005). Human pharmaceuticals in wastewater treatment processes. *Critical Reviews in Environmental Science and Technology*, **35**(4), 401–427.

Joss A., Andersen H., Ternes T., Richle P. R. and Siegrist H. (2004). Removal of estrogens in municipal wastewater treatment under aerobic and anaerobic conditions: consequences for plant optimization. *Environmental Science & Technology*, **38**, 3047–3055.

Joss A., Zabczynski S., Gobel A., Hoffmann B., Loffler D., McArdell C. S., Ternes T. A., Thomsen A. and Siegrist H. (2006). Biological degradation of pharmaceuticals in municipal wastewater treatment: proposing a classification scheme. *Water Research*, **40**(8), 1686–1696.

Kaegi R., Voegelin A., Ort C., Sinnet B., Thalmann B., Krismer J., Hagendorfer H., Elumelu M. and Mueller E. (2013). Fate and transformation of silver nanoparticles in urban wastewater systems. *Water Research*, **47**(12), 3866–3877.

Key B. D., Howell R. D. and Criddle C. S. (1998). Defluorination of organofluorine sulfur compounds by *Pseudomonas* sp. Strain D2. *Environmental Science & Technology*, **32**, 2283–2287.

Khunjar W. O. and Love N. G. (2011). Sorption of carbamazepine, 17alpha-ethinylestradiol, iopromide and trimethoprim to biomass involves interactions with exocellular polymeric substances. *Chemosphere*, **82**(6), 917–922.

Kumar N., Omoregie E. O., Rose J., Masion A., Lloyd J. R., Diels L. and Bastiaens L. (2014). Inhibition of sulfate reducing bacteria in aquifer sediment by iron nanoparticles. *Water Research*, **51**, 64–72.

Kümmerer K. (ed.) (2008). Pharmaceuticals in the Environment: Sources, Fate, Effects and Risks. Springer, Germany.

Lindblom E., Gernaey K. V., Henze M. and Mikkelsen P. S. (2006). Integrated modelling of two xenobiotic organic compounds. *Water Science & Technology*, **54**(6–7), 213.

Llorens E., Matamoros V., Domingo V., Bayona J. M. and García J. (2009). Water quality improvement in a full-scale tertiary constructed wetland: effects on conventional and specific organic contaminants. *Science of The Total Environment*, **407**(8), 2517–2524.

Margot J., Rossi L., Barry D. A. and Holliger C. (2015a). A review of the fate of micropollutants in wastewater treatment plants. *WIREs Water*, **2**, 457–487. doi: 10.1002/wat2.1090.

Margot J., Rossi L., Barry D. A. and Holliger C. (2015b). A review of the fate of micropollutants in wastewater treatment plants. *Wiley Interdisciplinary Reviews: Water*, **2**(5), 457–487.

Nam S. W., Jo B. I., Yoon Y. and Zoh K. D. (2014). Occurrence and removal of selected micropollutants in a water treatment plant. *Chemosphere*, **95**, 156–165.

Narumiya M., Nakada N., Yamashita N. and Tanaka H. (2013). Phase distribution and removal of pharmaceuticals and personal care products during anaerobic sludge digestion. *Journal of Hazardous Materials*, **260**, 305–312.

Ochoa-Herrera V., Banihani Q., Leon G., Khatri C., Field J. A. and Sierra-Alvarez R. (2009). Toxicity of fluoride to microorganisms in biological wastewater treatment systems. *Water Research*, **43**(13), 3177–3186.

Ogunlaja O. O. and Parker W. J. (2015). Impact of activated sludge process configuration on removal of micropollutants and estrogenicity. *Water Science and Technology*, **72**(2), 277–283.

Otero-González L., Field J. A. and Sierra-Alvarez R. (2014). Inhibition of anaerobic wastewater treatment after long-term exposure to low levels of CuO nanoparticles. *Water Research*, **58**, 160–168.

Pala-Ozkok I., Ubay-Cokgor E., Cakar Z. P. and Orhon D. (2014). Acute impact of erythromycin on substrate utilization by activated sludge: effect of sludge age. *Journal of Chemical Technology & Biotechnology*, **89**(7), 1091–1102.

Park H. D., Wells G. F., Bae H., Criddle C. S. and Francis C. A. (2006). Occurrence of ammonia-oxidizing archaea in wastewater treatment plant bioreactors. *Applied and Environmental Microbiology*, **72**(8), 5643–5647.

Pasquini L., Merlin C., Hassenboehler L., Munoz J. F., Pons M. N. and Gorner T. (2013). Impact of certain household micropollutants on bacterial behavior. Toxicity tests/study of extracellular polymeric substances in sludge. *Science of the Total Environment*, **463–464**, 355–365.

Pavlostathis S. G. and Maeng S. K. (2000). Fate and effect of silver on the anaerobic digestion process. *Water Research*, **34**(16), 3957–3966.

Plósz B. G., Leknes H. and Thomas K. V. (2010). Impacts of competitive inhibition, parent compound formation and partitioning behavior on the removal of antibiotics in municipal wastewater treatment. *Environmental Science & Technology*, **44**(2), 734–742.

Plosz B. G., Langford K. H. and Thomas K. V. (2012). An activated sludge modeling framework for xenobiotic trace chemicals (ASM-X): assessment of diclofenac and carbamazepine. *Biotechnology and Bioengineering*, **109**(11), 2757–2769.

Plosz B. G., Benedetti L., Daigger G. T., Langford K. H., Larsen H. F., Monteith H., Ort C., Seth R., Steyer J. P. and Vanrolleghem P. A. (2013). Modelling micro-pollutant fate in wastewater collection and treatment systems: status and challenges. *Water Science and Technology*, **67**(1), 1–15.

Pomies M., Choubert J. M., Wisniewski C. and Coquery M. (2013). Modelling of micropollutant removal in biological wastewater treatments: a review. *Science of the Total Environment*, **443**, 733–748.

Pomies M., Choubert J. M., Wisniewski C., Miege C., Budzinski H. and Coquery M. (2015a). Lab-scale experimental strategy for determining micropollutant partition coefficient and biodegradation constants in activated sludge. *Environmental Science and Pollution Research*, **22**(6), 4383–4395.

Pomies M., Choubert J. M., Wisniewski C., Miege C., Budzinski H. and Coquery M. (2015b). Lab-scale experimental strategy for determining micropollutant partition coefficient and biodegradation constants in activated sludge. *Environmental Science and Pollution Research International*, **22**(6), 4383–4395.

Rattier M., Reungoat J., Keller J. and Gernjak W. (2014). Removal of micropollutants during tertiary wastewater treatment by biofiltration: role of nitrifiers and removal mechanisms. *Water Research*, **54**, 89–99.

Ren Y. X., Nakano K., Nomura M., Chiba N. and Nishimura O. (2007). Effects of bacterial activity on estrogen removal in nitrifying activated sludge. *Water Research*, **41**(14), 3089–3096.

Rittmann B. E. (1994). In Situ Bioremediation. Noyes Publications, Park Ridge N.J., U.S.A.

Rittmann B. E. (1995). Transformation of organic micropollutants by biological processes. In: Water Pollution: Drinking Water and Drinking Water Treatment. Hrubec J. (ed.), pp. 31–60, doi:10.1007/978-3-540-48468-4. Springer, Berlin Heidelberg.

Rittmann B. E. and McCarty P. L. (2001). Environmental Biotechnology:Principles and Applications. McGraw-Hill Book Co, Singapore.

Roh H., Subramanya N., Zhao F., Yu, C. P., Sandt J. and Chu K. H. (2009). Biodegradation potential of wastewater micropollutants by ammonia-oxidizing bacteria. *Chemosphere*, **77**(8), 1084–1089.

Sacan M. T., Çeçen F., Erturk M. D. and Semerci N. (2009). Modelling the relative toxicity of metals on respiration of nitrifiers using ion characteristics. *SAR and QSAR in Environmental Research*, **20**(7–8), 727–740.

Sathyamoorthy S., Chandran K. and Ramsburg C. A. (2013). Biodegradation and cometabolic modeling of selected beta blockers during ammonia oxidation. *Environmental Science and Technology*, **47**(22), 12835–12843.

Sedlak D. L., Phinney J. T. and Bedsworth W. W. (1997). Strongly Complexed Cu and Ni in Wastewater Effluents and Surface Runoff. *Environmental Science & Technology*, **31**(10), 3010–3016.

Semerci N. and Çeçen F. (2007). Importance of cadmium speciation in nitrification inhibition. *Journal of Hazardous Materials*, **147**(1–2), 503–512.

Semerci N. and Çeçen F. (2009). Effect of continuous Cd feeding on the performance of a nitrification reactor. *Biodegradation*, **20**(2), 155–164.

Shi J., Fujisawa S., Nakai S. and Hosomi M. (2004). Biodegradation of natural and synthetic estrogens by nitrifying activated sludge and ammonia-oxidizing bacterium Nitrosomonas europaea. *Water Research*, **38**(9), 2322–2329.

Stasinakis A. S. (2012). Review on the fate of emerging contaminants during sludge anaerobic digestion. *Bioresource Technology*, **121**, 432–440.

Suarez S., Lema J. M. and Omil F. (2010). Removal of pharmaceutical and personal care products (PPCPs) under nitrifying and denitrifying conditions. *Water Research*, **44**(10), 3214–3224.

Tadkaew N., Hai F. I., McDonald J. A., Khan S. J. and Nghiem L. D. (2011). Removal of trace organics by MBR treatment: the role of molecular properties. *Water Research*, **45**(8), 2439–2451.

Tan D. T., Arnold W. A. and Novak P. J. (2015). Effects of estrone and organic carbon exposure on the transformation of estrone. *Environmental Science: Water Research & Technology*, **1**(4), 457–464.

Ternes T. A., Herrmann N., Bonerz M., Knacker T., Siegrist H. and Joss A. (2004). A rapid method to measure the solid–water distribution coefficient (Kd) for pharmaceuticals and musk fragrances in sewage sludge. *Water Research*, **38**(19), 4075–4084.

Tunkel J., Howard P., Boethling R. S., Stiteler W. and Loonen H. (2000). Predicting ready biodegradability in the Japanese Ministry of International Trade and Industry Test. *Environmental Toxicology and Chemistry*, **19**(10), 2478–2485.

Urase T. and Kikuta T. (2005). Separate estimation of adsorption and degradation of pharmaceutical substances and estrogens in the activated sludge process. *Water Research*, **39**(7), 1289–1300.

Vader J. S., van Ginkel C. G., Sperling F. M. G. M., de Jong J., de Boer W., de Graaf J. S., van der Most M. and Stokman P. G. W. (2000). Degradation of ethinyl estradiol by nitrifying activated sludge. *Chemosphere*, **41**, 1239–1243.

van Beelen E. S. (2007). Municipal Waste Water Treatment Plant (WWTP) Effluents: a Concise Overview of the Occurrence of Organic Substances. Association of River Waterworks – RIWA, The Netherlands.

van Leeuwen J. C. and Vermeire T. G. (eds) (2007). Risk Assessment of Chemicals: An Introduction. Springer, Dordrecht, The Netherlands.

Verlicchi P., Al Aukidy M. and Zambello E. (2012). Occurrence of pharmaceutical compounds in urban wastewater: removal, mass load and environmental risk after a secondary treatment – a review. *Science of the Total Environment*, **429**, 123–155.

Walker J. D., Newman M. C. and Enache M. (eds) (2012). Fundamental QSARs for Metal Ions. CRC Press, Boca Raton, FL.

Wang J., Huang C. P. and Allen H. E. (2006). Predicting metals partitioning in wastewater treatment plant influents. *Water Research*, **40**(7), 1333–1340.

Westerhoff P. K., Kiser A. and Hristovski K. (2013). Nanomaterial removal and transformation during biological wastewater treatment. *Environmental Engineering Science*, **30**(3), 109–117.

Yang Y., Quensen J., Mathieu J., Wang Q., Wang J., Li, M., Tiedje J. M. and Alvarez P. J. J. (2014). Pyrosequencing reveals higher impact of silver nanoparticles than Ag+ on the microbial community structure of activated sludge. *Water Research*, **48**, 317–325.

Yi T. and Harper W. F. (2007). The link between nitrification and biotransformation of 17α-ethinylestradiol. *Environmental Science & Technology*, **41**(12), 4311–4316.

Zhang D., Gersberg R. M., Ng, W. J. and Tan S. K. (2014). Removal of pharmaceuticals and personal care products in aquatic plant-based systems: a review. *Environmental Pollution*, **184**, 620–639.

Chapter 5

Experimental assessment of the inhibitory effect and biodegradation of hazardous pollutants

Ferhan Çeçen[1]*, Ulaş Tezel*[1] *and Bilge Alpaslan Kocamemi*[2]

[1]*Boğaziçi University, Institute of Environmental Sciences, 34342 Bebek, Istanbul, Turkey; E-Mail: cecenf@boun.edu.tr; ulas.tezel@boun.edu.tr.*

[2]*Marmara University, Department of Environmental Engineering, Istanbul, Turkey; E-Mail: bilge.alpaslan@marmara.edu.tr.*

5.1 INTRODUCTION

Hazardous pollutants pose a risk to the water environment. In biological treatment systems, mainly biotransformation/biodegradation characteristics and inhibitory effects of hazardous pollutants make them a focus of attention. Biotransformation is a structural modification of a substance by the enzymes of microorganisms. Complete oxidation or reduction of a substance into end products such as CO_2, NH_4^+, N_2, CH_4 and H_2O is called biodegradation. Inhibition is defined as impairment of the enzymatic system of the living cell or direct damage to the cell structure in the presence of a substance, resulting in the slowing down of cell activity (Henze *et al.* 2008).

In industrial wastewaters, hazardous pollutants are often present at high concentrations. Yet, most of them are found at trace concentrations in domestic wastewater. In comparison to bulk organic matter, most hazardous organic pollutants are characterized by their slow or poor biodegradability. This is often attributed to the physicochemical characteristics of pollutants, inability of microorganisms to assimilate them or to a lack of specific microorganisms in

Ferhan Çeçen, Ulaş Tezel
doi: 10.2166/9781780407708_183

biological sludge. Since conventional biological treatment units are not specifically designed to remove hazardous organic pollutants, hazardous organic pollutants are often found in the effluent stream or in waste sludge. Compared to them, hazardous inorganic pollutants are often known for their inhibitory effects on biological processes. Although they cannot undergo biodegradation, microorganisms can convert them into ionic or non-ionic species.

Complementary to the knowledge provided in other chapters, this chapter provides information on the experimental procedures used to determine the inhibitory effects and biodegradation potential of hazardous pollutants under different Terminal Electron Accepting Processes (TEAPs) of wastewater treatment. Some of those processes shown in Table 5.1 also apply to biological drinking water treatment.

Table 5.1 General classification of Terminal Electron Accepting Processes (TEAPs) in wastewater treatment.

	Aerobic	Anoxic	Anaerobic
e^- acceptor	O_2	NO_2^-, NO_3^-	Organic compounds
TEAPs	Oxygenic respiration Nitrification	Denitrification Anammox	Methanogenesis Fermentation
End products	CO_2, NO_2^-, NO_3^-	N_2	CH_4, CO_2, H_2

Anammox: Anaerobic ammonium oxidation.

In biological removal of hazardous pollutants, multiple criteria such as biomass source, substance characteristics, and environmental and operational conditions play a role. Therefore, it is often impossible to predict the removal and behavior of hazardous pollutants without experimenting. As stated in literature, wastewater engineering itself is in fact an extremely experiment-based area and there is always a need to develop and standardize the methods used in experimenting (Loosdrecht *et al.* 2016). Similarly, there is also a need for experimenting in drinking water treatment.

Determining the inhibitory effects of hazardous pollutants plays a crucial role in understanding their fate in biological treatment systems. Also in biodegradation experiments, it is important to know the inhibitory behavior of a pollutant. It is of great importance whether a hazardous pollutant inhibits its own removal (self- or substrate inhibition) or the removal of other substances. Thus, the present chapter first discusses the inhibition experiments and the standard tests used for this purpose. In inhibition as well as biodegradation experiments, without an initial model in mind, it often becomes impossible to interpret the results of experiments. Therefore, along with experimental procedures the chapter also tackles inhibition models, inhibition kinetics and data evaluation. Then shortcomings of existing inhibition tests are discussed and suggestions are made for their improvement.

Following inhibition experiments, standard biodegradation tests are discussed in detail. In biological systems it is important to know whether a hazardous pollutant is biodegraded under aerobic, anoxic or anaerobic conditions. In biodegradation of hazardous pollutants, the rates and pathways are also important. Yet, experimenting alone is often unsatisfactory in determination of biodegradation. Therefore, the chapter first provides detailed information on the databases, assessment methods and predictive models that can be used prior to conduction of biodegradation tests. In that regard, the chapter connects experimental and non-experimental methods and highlights that they should complement each other.

The chapter further distinguishes between two types of tests: (1) Standard tests, and (2) Freely-designed tests. Standard tests are those that have been set by the agencies such as the Organisation for Economic Co-operation and Development (OECD). Since the presence of hazardous pollutants is of great concern in the aerobic stages of domestic wastewater treatment plants (WWTPs), most of those tests deal with aerobic conditions. Freely-designed tests are determined by a researcher depending on the objectives of a study. An overview of numerous papers reveals that the latter procedure is more common in examination of biodegradability and inhibition.

The chapter also addresses specific issues which have not yet received enough recognition in standard in inhibition and biodegradation tests. In that regard, attention is paid to growth-linked and cometabolic biodegradation, microbial ecology of sludge, inoculum source, acclimation, and feeding regimes for hazardous pollutants.

5.2 DECIDING ON EXPERIMENTAL SYSTEMS

5.2.1 Closed versus open bioreactors

The choice between a closed or open biological reactor depends on the physicochemical properties of target substances, the TEAP and the substances to be monitored during the course of the experiment. Aerobic systems are often open systems whereas closed bioreactors are chosen for anoxic or anaerobic systems to avoid the intrusion of O_2. The closure of the bioreactor may also be necessary for monitoring of the reaction products such as CO_2, CH_4, N_2 and volatile organics.

5.2.2 Batch versus continuous-flow bioreactors

Although most laboratory-scale inhibition and biodegradation experiments are performed in batch reactors, the use of continuous-flow bioreactors with long-term operation is also possible.

In batch operation, there is a concentration gradient of the pollutant with respect to time. In the initial periods of operation, the biomass is exposed to a rather high concentration which then decreases towards the end. This concentration gradient may also change the kinetic order of a reaction such as from zero- to first-order. In addition, in the case of highly inhibitory pollutants, the biomass will be first exposed to a high level of pollutant which may decrease its activity.

The physicochemical characteristics of the tested substances determine also the selection of reactor material. For example, an important factor is the sorbability of the pollutant. The bioreactor may be constructed from glass or plexiglass while the former is usually preferred to avoid an abiotic loss due to sorption to reactor walls. Another important factor may be the penetration of light into a reactor, which in some cases may induce photochemical degradation or unwanted growth of organisms such as photosynthetic bacteria or algae. Thus, amber material may be chosen in such cases. If pure cultures are used in bioreactors, the maintenance of sterile conditions is necessary to avoid contamination. This can more easily be accomplished in closed reactors than in open ones.

In continuous-flow operation, there is a steady flux of pollutants into the bioreactor. In a continuous-flow stirred tank reactor (CSTR), at steady-state operation, the pollutant concentration is lowered to a specific value which is equal to the effluent. Thus, the removal rate of the pollutant inside the reactor takes a single value. Since the biomass is not exposed to concentration gradients, it will not be exposed to highly inhibitory concentrations. Under non-steady (i.e. dynamic) conditions, however, which actually represent reality, both the flow rate or concentration of a pollutant may vary; therefore fluctuations are expected in effluent concentration.

CSTRs are normally operated for a much longer period of time than batch reactors. Therefore, in contrast to batch reactors, in these reactors microorganisms may acclimate or evolve upon exposure to the tested pollutant. For example, when the inhibitory effect of Cd on nitrification was investigated in a continuous-flow system with a sludge in which nitrifying bacteria were enriched, the inhibitory effect of Cd on nitrification became lower since metal-tolerant ammonia oxidizing species became dominant in long-term operation (Semerci & Çeçen, 2009).

A plug-flow reactor (PFR) is similar to a batch reactor; this time the distance from the inlet to the outlet of the PFR is comparable to the timescale in a batch reactor. However, PFRs are hardly used in standard or freely-designed tests. This can mainly be attributed to the difficulty of maintaining non-dispersed flow under real conditions.

Another type of bioreactor used in experiments is the fed-batch reactor (FBR) reactor which is rather applied in testing the removal and inhibitory kinetics of specific pollutants in activated sludge. In this type of operation, a concentrated wastewater is dosed into the reactor with a small volumetric flow rate so that the maximum removal rate of the pollutant is exceeded while the volume of the reactor remains the same, leading to a build-up of the pollutant. Comparison of the pollutant feed rate and pollutant response lines allows the calculation of pollutant utilization rate. If inhibition occurs, utilization rates will decrease. But, in standard tests this type of operation is not included whereas there are a lot of data in freely-designed experimental systems.

5.2.3 Suspended- versus attached-growth (biofilm) reactors

Inhibition and biodegradation studies can be conducted in both suspended- and attached-growth reactors. It is a known fact that most pollutants exert a lesser effect on biomass due to the reasons discussed in **Chapter 4** if the biomass is retained in the form of a biofilm. However, due to the ease of working with suspended-growth cultures, most biodegradability and inhibition experiments reported in Sections 5.3 and 5.4 are conducted in suspended-growth bioreactors.

5.2.4 Evaluation of abiotic removal

When required, in biodegradation and inhibition tests, abiotic removal of the test substance should also be checked. The main abiotic mechanisms leading to the removal of a pollutant in a bioreactor are sorption, abiotic degradation and volatilization.

As discussed in **Chapter 4**, biosorption or sorption is the second most important process after biodegradation. However, it is surprising to find out that in biological treatment studies, sorption experiments delineating the distribution of a hazardous pollutant between the solid and liquid phases of a reactor have not been standardized yet. If the sorption of a pollutant onto biomass is studied, usually a sample is taken from the bioreactor which is then contacted with the test substance in the absence of a feed. However, different results might be achieved in the presence of a feed since it can change the sorption behavior. Another option is to study the removal of the hazardous pollutant inside the bioreactor. In that case, to distinguish between biodegradation and sorption, control bioreactors may be used in which the activity of biomass is inhibited with some chemicals. However, external addition of an inhibitor may also change the conditions in the sludge such as pH, ionic strength etc., finally causing a change in the sorption/biosorption ability of tested pollutant.

Abiotic loss of hazardous substances can also occur by volatilization which is much more difficult to control than sorption. Hydrolysis of the tested substance is also possible. Photolysis is unlikely to occur in most biological wastewater treatment systems since the penetration of light is hindered, as discussed in **Chapter 4**.

5.3 INHIBITION EXPERIMENTS

Organic hazardous substances in a wastewater may be generally categorized as inhibitory and noninhibitory as shown in Figure 5.1. Inhibitory and noninhibitory organic hazardous substances can be further sub-categorized as biodegradable and nonbiodegradable. Inorganic hazardous substances can also be inhibitory and noninhibitory, but they are nonbiodegradable.

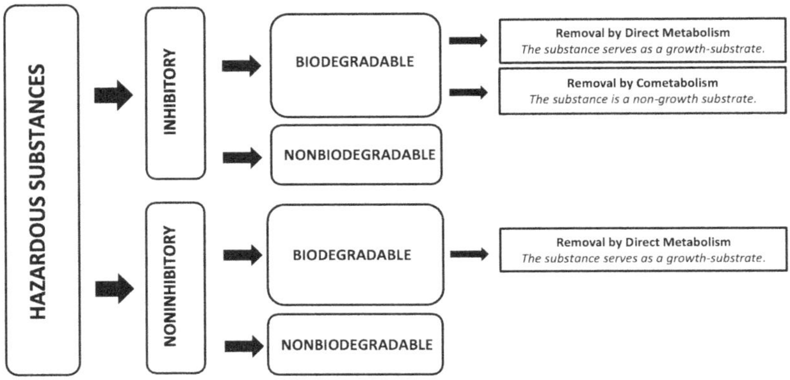

Figure 5.1 Classification of hazardous substances for inhibition assays.

In wastewater treatment biodegradable and inhibitory pollutants are the most common. Among the inhibitory nonbiodegradable pollutants, metals and metal halides constitute the major group of inorganics inhibiting biological reactions. Most of the polyhalogenated xenobiotic organic compounds (XOCs) and intermediate products generated from cometabolic degradation of some xenobiotics also belong to the inhibitory and nonbiodegradable class.

Biodegradable hazardous compounds may serve as a growth-substrate in one of the TEAPs and be degraded by direct metabolism. In addition, there are hazardous compounds which do not support growth but are degraded cometabolically as a non-growth substrate in the presence of a growth-substrate, as discussed in detail in **Chapter 4**. Cometabolically degradable compounds may cause inhibition on TEAPs depending on concentration and exposure time. Competitive inhibition is the most commonly observed inhibition type. In addition, biodegradable hazardous pollutants may inhibit their own degradation via self-inhibition if found at very high concentrations. An example of this is the inhibitory effect of phenol on its oxidation under aerobic conditions. Nonbiodegradable and inhibitory pollutants inhibit several TEAPs without being transformed; thus they appear in effluent streams unless they are sorbed to biological sludge.

5.3.1 Why do we need inhibition tests?

Hazardous pollutants may enter WWTPs either continuously or as a result of shock loads. The potential inhibitory effect(s) of these pollutants on different TEAPs can be assessed in the design as well as operation stages of WWTPs by performing short-term (batch) or long-term (bench-scale continuous-flow) experiments. Inhibition tests evaluating the synergistic effect of more than one hazardous substance on a microbial population need to be conducted to draw conclusions regarding their behavior in a treatment plant. For biodegradable and inhibitory pollutants, inhibition tests have to be carried out prior to biodegradability tests (Section 5.4) in order to identify the noninhibitory level of the pollutant.

Inhibition tests are designed according to the specific aim. Mainly, they serve to clarify the following points:

(a) Evaluation of the extent of inhibition.
(b) Identification of inhibition type.
(c) Estimation of kinetic parameters in the presence of hazardous substances.
(d) Monitoring the change in microbial population in the long term.

Inhibition tests are mostly performed over a short term to estimate 50% reduction in microbial activity (IC_{50}) upon exposure to an individual hazardous substance. However, IC_{50} values give a rough idea about the extent of acute inhibition. IC_{50} values can also be utilized to predict the sudden response of a treatment process against shock loads of a hazardous substance. These values may also be used to compare the hazardous effects of different hazardous substances on the same microbial population. In any case, kinetic parameters have to be determined in the presence of an inhibitor and the type of inhibition has to be assessed to predict the effect on any TEAP. In this way measures can be taken to minimize the negative effects. IC_{50} values cannot be directly utilized to determine the change in kinetic parameters. Further, IC_{50} values cannot provide information on the long-term effect of a hazardous substance. Each species in a microbial community may be affected in a different way if exposed to a hazardous substance in the long term. Some microbial species may adapt themselves to a hazardous substance while others will disappear from the community. Such information about the dynamics of a microbial population is quite helpful in operation of treatment plants.

5.3.2 Strategies for the design of inhibition experiments

The main aim in the design of an inhibition experiment would be to obtain reliable data in a short period of time by conduction of the least number of experiments. To do this, the researcher may first collect relevant background information about the test substance. The fact sheets prepared by governmental agencies are quite helpful, not only to search about the physicochemical properties of a test substance but also in learning about its environmental fate.

In inhibition studies, often preliminary tests are conducted in order to identify (i) the possible abiotic losses, (ii) acute inhibitory range, and (iii) biodegradability potential (by direct or cometabolism) of the tested substance. They are usually performed in a short term in batch reactors.

The checklist in Table 5.2 summarizes the major variables in inhibition studies. Inhibition studies can follow standard test protocols (e.g. OECD, ISO) or they can be freely designed by the researcher. In both cases, the variables shown in Figure 5.2 may directly affect results. In standard tests there is almost no chance to change the variables. This is one of the major challenges limiting the use of these tests in inhibition studies. Most of the factors shown in Figure 5.2 may also apply to biodegradation tests.

Table 5.2 Relevant background information in the design of inhibition experiments.

Is the pollutant biodegradable?

If yes, the following questions should be asked:

- Is it a growth- or non-growth (cometabolic) substrate?
- For which TEAPs is it known to be a growth-substrate?
- For which TEAPs is it known to be a non-growth substrate?
- What are the TEAPs that possibly lead to cometabolic removal of the pollutant?

Which abiotic mechanism(s) leads to the removal of this pollutant?

e.g. sorption, abiotic degradation and volatilization

In which cases is the pollutant found at macro- or microlevels?

Is it a macro- or micropollutant in your designed system?

Which TEAP might be threatened in the presence of this pollutant?

e.g. aerobic carbon degradation, fermentation, nitrification, denitrification, Anammox

Does the pollutant exist alone (not in a matrix with other hazardous pollutants)?

If yes, the following questions should be asked:

- Is there any reported antagonistic/synergistic inhibition effect of pollutants?
- Are other pollutants present in the matrix biodegradable?

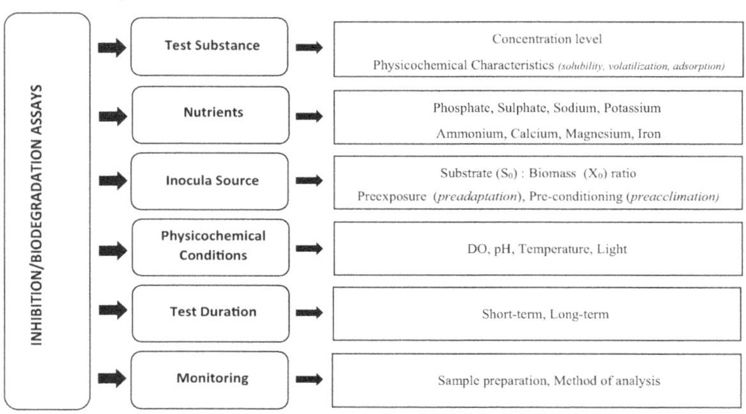

Figure 5.2 Major variables affecting inhibition and biodegradation assays. (DO: dissolved oxygen).

Testing the effect of each variable alone requires much time, since many experimental runs are needed. The response surface methodology (RSM), a well-known combined mathematical and statistical approach used in experimental design, is the optimum way to evaluate the influence of a single or several independent variables (factors) on the response variable. Hence, use of RSM allows the prediction of the degradation rate (response) in a TEAP in the presence of an inhibitory substance, by accounting for one or more of the independent variables shown in Figure 5.2. RSM can particularly simplify experimental design when an experiment has to be conducted to evaluate the synergistic/antagonistic effect of several hazardous pollutants. The study of Zhang *et al.* (2016) is a good example demonstrating the use of RSM in the design of an experiment which aims to examine the interactive effects of the heavy metals Cu and Zn on Anammox activity.

5.3.3 Monitoring methods in inhibition experiments

In inhibition tests the following items, shown in Sections 5.3.3.1–5.3.3.7, can be monitored. Such tests may also give an idea about the noninhibitory level of a test substance in biodegradation studies (Section 5.4).

5.3.3.1 Utilization of terminal electron acceptor

Tests based on the measurement of terminal electron acceptor are usually referred to as respirometric. Respirometry is a particularly attractive tool to determine the effect of inhibitory but nonbiodegradable hazardous substances that do not degrade themselves but inhibit the removal of the growth-substrate in a TEAP (Figure 5.1). Respirometric methods are widely applied in aerobic and anoxic systems. Since in aerobic systems the terminal electron accepter is oxygen, these tests continuously monitor the change in the oxygen uptake rate (OUR) and/or cumulative oxygen consumption in a biological reactor. Such aerobic respirometric tests are mainly classified based on (i) location of oxygen measurement (liquid/gas), and (ii) state of gas and liquid phases (flowing/static) (Loosdrecht *et al.* 2016).

Since the late 1980s, several short-term aerobic respiration tests have been developed that measure inhibition while the tests OECD 209 (OECD, 2010), EPA 712-C-014 (Environmental Protection Agency [EPA], 2012) and ISO 8192 (International Standard Organisation [ISO], 2007) can be listed among the most common ones. The OECD 209 test (OECD, 2010) measures the respiration rate of activated sludge OUR to give an idea about the extent of organic carbon removal and/or ammonium oxidation under defined conditions in the presence of an inhibitor. Inhibition of both heterotrophic and autotrophic bacteria, as well as inhibition of heterotrophs and autotrophs alone can be determined by this procedure. Thus, this allows the researcher to differentiate between inhibition of organic carbon removal and nitrification. For this purpose, a specific nitrification inhibitor is used in these tests. The study of Volskay and Grady (1988) is a good example demonstrating the use of the modified OECD 209 (OECD, 2010) test in studying the inhibitory effect of 33 organic substances that are found in domestic wastewaters and appear in the list of the Resource and Recovery Act (RCRA) (e.g. chlorinated aliphatic organic compounds, xylene, phenolic compounds) (Rücker & Kümmerer, 2012). Anoxic respirometric tests are performed by measuring the nitrate or nitrite uptake rate (NUR) of facultative heterotrophs. But, unlike OUR tests, standardized anoxic tests are not available yet.

Both aerobic or anoxic respirometric methods usually report the IC_{50} value which is generally defined as the concentration of an inhibitory pollutant causing 50% reduction in the specific oxygen/nitrite/nitrate uptake rate of activated sludge. As discussed in Section 5.3.1, IC_{50} values can be utilized for quick assessment of the inhibitory effect of a pollutant. But, these results cannot be used to predict the effect of an inhibitor on the performance of a bioreactor since they cannot be directly translated into a quantitative effect on Monod kinetic parameters (Grady *et al.* 2011).

Moreover, these tests can be used in order to evaluate the response of biomass to occasional inhibitor pulses from point sources (Pala-Ozkok & Orhon, 2013). However, such tests do not give an idea about the response of a microbial community to the tested substance in the long term. They can also be used to assess the combined inhibitory effect (synergism or antagonism) of multiple substances on activated sludge. Usually, inhibitory effects of structurally-related substances can be quickly compared via respirometric experiments.

5.3.3.2 Product formation

Product formation rates are usually measured in anaerobic systems. Such tests mainly consider the products of anaerobic respiration, namely methane (CH_4), carbon dioxide (CO_2) or hydrogen (H_2). The gas production can be quantified with manometric methods by the measurement of pressure increase in headspace or with volumetric methods, also known as the serum bottle or Marriotte's bottle test, based on the displacement of water with gas. The International Standard Organisation (ISO) proposes two methods to assess the inhibition of gas production by anaerobic microorganisms. In one of them concentrated anaerobic digester sludge is employed (ISO, 2003), while in the other, diluted digester sludge at low biomass concentrations is used (ISO, 2003). The study of Gartiser *et al.* (2007) demonstrates the application of ISO 13641-1 (ISO, 2003) to find the inhibitory impact of 16 different antibiotics on anaerobic systems. The test OECD 224 (OECD, 2007) determines the inhibition of anaerobic microorganisms by the manometric measurement of gas production.

Another test, namely OECD 311 (OECD, 2006b) monitors the gas production in anaerobic systems. However, this test offers limited opportunity in determination of inhibition. This is because the sludge is diluted to one tenth of its normal concentration and a relatively high concentration of a test substance is used. Therefore, a longer test period is needed than the retention time in an anaerobic digester. The test is applicable to substances which are soluble or insoluble in water, including volatile substances. But special care is necessary in the case of substances with low water-solubility and high volatility. Additionally, erroneous results may be obtained in the case of inhibitory but biodegradable test substances.

For the anoxic TEAPs, there is yet no standard inhibition test protocol that considers the rate of product formation. However, there are some freely-designed experimental methodologies. A manometric test methodology developed by Dapena-Mora *et al.* (2007) for Anammox systems, called the Specific Anammox Activity (SAA) test, is a good example of an anoxic respirometric test. The methodology is based on the measurement of pressure in the headspace. As the Anammox reaction proceeds, different forms of nitrogen (ammonia and nitrite) are converted into N_2 gas. Hence, the gas pressure is directly proportional to the nitrogen gas produced. The study of Lotti *et al.* (2012) demonstrates the use of SAA in the assessment of the inhibitory effect of antibiotics (oxytetracycline and sulfathiazole) and heavy metals (copper and zinc) on Anammox.

For nitrification too, there are standard test protocols that are used to monitor product formation. The test ISO 9509 (ISO, 2006) and the modified test ISO 9509 quantify the inhibitory effect of a test substance on nitrification by monitoring the nitrate production in the short- (6–9 hours) and long term (about 72 hours), respectively.

5.3.3.3 Degradation of growth-substrate

In a real operation, the main interest lies in how the main substrate (or the growth-substrate) is removed in the presence of an inhibitor. Therefore, a methodology should be selected that mimics the conditions in a real plant. For this purpose, long-term experiments can be conducted with continuous dosing of inhibitory substance and the changes taking place in the removal of the main growth-substrate can be monitored. Usually, in wastewater the main substrate to be measured is bulk organic matter. But, in a nitrifying sludge, changes taking place in ammonia (ammonium) utilization can also be followed. Not only the inhibitory effect of a single hazardous pollutant, but also the cumulative impact of multiple low-concentration pollutants, can be determined by this procedure.

There are cases in which the growth-substrate definitely needs to be measured. In a mixed microbial consortium, a hazardous organic compound inhibiting one group of bacteria might be biodegradable by another group, whereas these different groups may use the same terminal electron acceptor. Therefore, it is impossible to determine the extent of inhibition by monitoring the decrease in the terminal electron acceptor alone. In such cases inhibition can be successfully determined only if the decrease in substrate degradation rate is measured. A good example of this is phenol, which is known to be biodegradable under aerobic conditions by phenol degraders. However, at the same time, this compound is inhibitory to nitrifying microorganisms. A researcher aiming to evaluate the inhibitory effect of phenol on nitrification in a mixed consortium has to follow the degradation rate of ammonium, namely the growth-substrate in nitrification, and not the oxygen uptake rate. An example of this is the study conducted by Amor *et al.* (2005) in which the inhibitory effect of phenol on nitrification was determined by measuring ammonium utilization rates.

It is also very common to monitor the inhibitory effect of a cometabolically degradable hazardous substance on a growth-substrate. However, for this purpose no standard test protocols have been developed yet. But, such inhibitory effects have been examined in many freely-designed studies in literature. The example in Box 5.1 shows such a study that evaluates the inhibitory effect of trichloroethylene (TCE) on nitrification.

BOX 5.1 INHIBITORY EFFECT OF TCE ON A NITRIFYING ACTIVATED SLUDGE

TCE is a man-made chemical that is widely used in industrial applications. In the present example the removal of TCE was examined in nitrifying batch activated sludge reactors. Ammonia (ammonium) served as a growth-substrate whereas TCE was cometabolized in the presence of ammonia. At the same time, TCE had an inhibitory effect on nitrification. As shown below, specific ammonium degradation rates (qNH_4-N) decreased with increasing TCE concentration. Analysis of data indicated that TCE was a competitive inhibitor of nitrification (Alpaslan Kocamemi, 2005; Alpaslan Kocamemi & Çeçen, 2005).

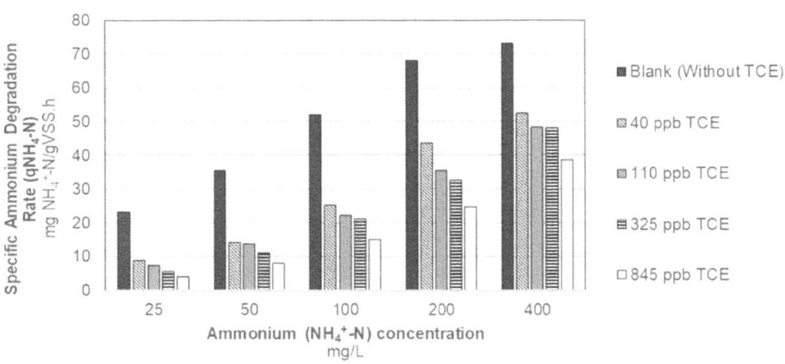

Concentration of the Cometabolite TCE (ppb)	% Decrease in Ammonium Degradation Rate (qNH_4^+-N)				
	25 mg/l NH_4^+-N	50 mg/l NH_4^+-N	100 mg/l NH_4^+-N	200 mg/l NH_4^+-N	400 mg/l NH_4^+-N
0	n/a	n/a	n/a	n/a	n/a
40	62	60	52	36	28
110	68	61	57	48	34
325	75	70	60	52	34
845	82	78	71	63	46

5.3.3.4 Degradation of inhibitory and biodegradable pollutants

This type of monitoring is especially important for hazardous substances that can also be removed by cometabolism. Hazardous organic substances, especially the chlorinated ones, are usually degraded as cometabolites in the different TEAPs of wastewater/water treatment. In other words, they cannot serve as an energy or carbon source, but are removed as a non-growth substrate (cometabolic substrate) in the presence of a growth-substrate (primary substrate).

However, in such systems competition is often observed between growth- and cometabolic substrates. The presence of the cometabolic substrate, which is the hazardous pollutant of interest, may decrease the removal rate of the growth-substrate and vice versa. Additionally, cometabolic degradation of a hazardous substance may lead to generation of by-products or products which can even be more inhibitory than the original substance. Therefore, for a reliable assessment of inhibition, the concentrations of cometabolic substrate, its by- or end-products as well as the concentration of growth-substrate need to be monitored (Section 5.3.3.3). A decrease in cometabolic degradation rate and/or an increase in by-product formation rate would indicate that the hazardous pollutant is not mineralized yet, indicating an inhibition in the presence of growth-substrate. As explained in **Chapter 4**, this happens usually in the treatment of domestic wastewater when micropollutants are degraded as cometabolites in the presence of bulk organic matter or ammonia that act as growth-substrates. The latter two are found at much higher concentrations than micropollutants. If a competitive inhibition exists between them and micropollutants, generally micropollutant removal rates are reduced.

5.3.3.5 Enzyme activity

Enzyme activity tests can be used to evaluate the inhibitory effects of hazardous substances on substrate degradation. Dehydrogenases are enzymes catalyzing the dehydration of the substrate (oxidation of the substrate). They transfer protons and electrons to further stages of the respiratory chain (Makuch *et al.* 2010). They are the most widely used enzymes in toxicity testing (Bitton & Koopman, 1992). Other enzyme activity tests utilize ATPases, esterases, phosphatases, urease, luciferase, beta-galactosidase, protease, amylase, or beta-glucosidase (Bitton & Koopman, 1992). Yet, no standard OECD or other test protocols are available that assess the inhibitory effect of hazardous substances using enzymes. However, in numerous freely-designed studies, dehydrogenase activity tests were frequently employed. The paper of Makuch *et al.* (2010) summarizes some of those tests and demonstrates a study evaluating the inhibitory effect of pharmaceuticals (sulphonamides) on activated sludge via the measurement of dehydrogenases activity.

5.3.3.6 Bacterial luminescence

To measure the inhibitory effects of a test substance, bioluminescent bacteria can be used. Bacterial luminescence is linked to respiration and at increased concentrations hazardous substances reduce the light emitted by such cultures (Henze *et al.* 2008). The change in luminescence rate can be measured via a luminometer.

Vibrio fischeri (Photobacterium phosphoreum), a marine bioluminescent bacterium, is the most commonly used microorganism in bioluminescence tests. These bacteria are provided by manufacturers in a lyophilized form and can be stored for several months to be used "on demand" (Rizzo, 2011). However, *V. fischeri* is not representative of microorganisms found in activated sludge since it is not likely to be found in a WWTP. Despite this fact, it is a widely used bacterium in testing the effects of various inhibitors due to some advantages like sensitivity, reproducibility and ease of application (Tobajas *et al.* 2016).

The bioluminescence inhibition test conducted with the *V. fischeri* species has become a standard test (ISO, 1998) which has many commercially available modifications (Rizzo, 2011). In this category the most widely used one is the Microtox Acute Toxicity test developed according to ISO 11348-3 (ISO, 1998). Hoffmann and Christofi (2001) modified the test by incorporating *V. fischeri* into a sludge taken from a real activated sludge plant. Lappalainen *et al.* (2001) developed a methodology to eliminate the interference caused by colored substances in the *V. fischeri* assay.

Since 2000, luminescent genetically engineered microorganisms (GEMs) have also been used for wastewater toxicity testing (Ren, 2004). An in vitro bioluminescence assay aims to quantify acute inhibition. It assumes that bioluminescence depends on the amount of adenosine triphosphate (ATP) extracted from activated sludge (Dalzell & Christofi, 2002).

5.3.3.7 Response of microbial population to an inhibitor

ISO 15522 (ISO, 1999) is a well-known long-term inhibition test that monitors the increase in the number of microbial cells upon exposure to an inhibitory test substance. Recently, in freely-designed long-term studies also adaptation and/or resistance of a microbial community to a selected inhibitor as well as the shifts occurring in microbial composition were examined (Cetecioglu *et al.* 2012; Pala-Ozkok & Orhon, 2013). The newly developed molecular biology tools, metagenomics, proteomics, and metabolomics provide understanding of complex changes in microbial community functions, and detection and identification of existing and novel proteins (enzymes) and metabolites involved in the biodegradation of the hazardous substance of interest (Kowalczyk *et al.* 2015). A thorough discussion of such molecular biology tools can be found in **Chapter 7**.

5.3.4 Inhibition kinetics and data analysis

The effect of an inhibitor on any TEAP requires identification of inhibition type and quantification of kinetic parameters. As discussed in **Chapter 4**, the type of inhibition can be determined by determining the changes in the two kinetic parameters, namely the maximum specific substrate degradation rate (q_m) and the half-velocity constant

(K_m) in the Michaelis-Menten equation shown in Table 5.3. The apparent kinetic constants found in the presence of an inhibitor (q_m^{app} and K_m^{app}) are then compared with those under uninhibited conditions (q_m and K_m) to determine the type of inhibition.

Table 5.3 Change in kinetic constants in different types of inhibition.

Michaelis-Menten Expression $q = q_m \frac{S}{K_m + S}$

Inhibition Type	New Expression	Change in q_m and K_m
Competitive	$q = \frac{q_m^{app} S}{K_m^{app} + S} = \frac{q_m S}{K_m(1 + (I/K_{ic})) + S}$	$q_m^{app} = q_m$ $K_m^{app} > K_m$
Uncompetitive	$q = \frac{q_m^{app} S}{K_m^{app} + S} = \frac{q_m\left[1/(1 + I/K_{iu})\right] \cdot S}{K_m\left[1/(1 + I/K_{iu})\right] + S}$	$q_m^{app} < q_m$ $K_m^{app} < K_m$
Mixed	$q = \frac{q_m^{app} S}{K_m^{app} + S} = \frac{q_m\left[1/(1 + I/K_{iu})\right] \cdot S}{K_m\left[(1 + I/K_{ic})/(1 + I/K_{iu})\right] + S}$	$q_m^{app} < q_m$ $K_m^{app} < K_m$ or $K_m^{app} > K_m$
Non-competitive	$q = \frac{q_m^{app} S}{K_m^{app} + S} = \frac{q_m\left[1/(1 + I/K_{iu})\right] \cdot S}{K_m + S}$	$q_m^{app} < q_m$ $K_m^{app} = K_m$

q: specific substrate degradation rate.
q_m: maximum specific substrate degradation rate.
q_m^{app}: apparent maximum specific substrate degradation rate.
S: substrate concentration.
I: concentration of inhibitor.
K_m: half-velocity constant (affinity of substrate and enzyme).
K_m^{app}: apparent half-velocity constant (affinity of substrate and enzyme in the presence of inhibitor).
K_{ic}: competitive inhibition constant demonstrating the affinity of the free enzyme for the inhibitor.
K_{iu}: uncompetitive inhibition constant demonstrating the affinity of the enzyme substrate complex (ES) for the inhibitor.

5.3.4.1 Estimation of the kinetic parameters in degradation of growth-substrate

Before an inhibition study is started, preliminary experiments are necessary to find the kinetic parameters pertaining to degradation of the growth-substrate under uninhibited conditions. This is done in a substrate concentration range extending from 0.2 to 10 K_m (Cornish-Bowden, 2012). In aerobic systems measurements such as oxygen uptake rate may help decide quickly about this concentration range.

Kinetic experiments yield data that show the specific degradation rate of a substrate with respect to its concentration. In analysis of such data, mainly two approaches are followed:

(a) *Linearization of data*: The nonlinear Michaelis-Menten equation shown in Table 5.3 is brought into a linear form, followed by estimation of the kinetic parameters, the maximum specific substrate degradation rate (q_m) and the half-velocity constant (K_m). The linear regression techniques, namely the Lineweaver-Burk, Hanes-Woolf, and Eadie Hoftsee bring data to the form of $1/q_m$ versus $1/S$; S/q_m versus S; and q_m versus q_m/S, respectively. Among these techniques, the Lineweaver-Burk is by far the most widely used one. However, at small values of maximum specific substrate degradation rates (q_m) small errors in q_m lead to enormous errors in $1/q_m$; but for large values of q_m the same small errors in q_m lead to barely noticeable errors in $1/q_m$. In the Hanes-Woolf technique, over a fair range of S, the errors in S/q_m provide a faithful reflection of those in q_m. For this reason, the Hanes-Woolf technique is preferred over others for most purposes. The Eadie-Hofstee technique gives fairly good results in practice, despite the fact that q_m appears in both coordinates. This means that errors in q_m affect both of them and cause deviations towards or away from the origin rather than parallel with the ordinate (Cornish-Bowden, 2012).

(b) *Analysis of data by nonlinear techniques*: Also, the nonlinear least square regression (NLSR) can be used for estimation of kinetic parameters. Nowadays, several user-friendly computer programs are available for nonlinear fitting. Therefore, the difficulties in application of NLSR that existed in the past have been overcome (Smith *et al.* 1997). NLSR has been shown to be a better method for estimating the kinetic parameters than linearization (Kim *et al.* 2002) because data points are more likely to be normally distributed about a curve representing the model (Smith *et al.* 1997).

5.3.4.2 Determination of inhibition kinetics and inhibition type

For a rapid estimation of the inhibitory range of a hazardous substance, either the utilization rate of the terminal electron acceptor such as oxygen (Section 5.3.3.1) can be measured or bacterial luminescence (Section 5.3.3.6) can be determined. Later, experiments can be performed in a wide concentration range to find the impact of the inhibitor on the degradation of the growth-substrate. In this way, at various inhibitor levels, curves can be generated that show the degradation rate of the growth-substrate with respect to its concentration.

Another approach for identifying the type of inhibition is visual comparison of either linearized or nonlinear plots with sample plots given in literature for each type of inhibition (Bailey & Ollis, 1986; Cornish-Bowden, 2012).

The inhibition constants K_{ic} and K_{iu} shown in Table 5.3 can be determined with three different approaches (Kakkar *et al.* 2000): (i) Dixon plot, (ii) direct method (simultaneous nonlinear regression, SNLR), and (iii) Replot method (K_m^{app} Method). In the Dixon method, the reciprocal of substrate degradation rate is plotted against the substrate concentration at each concentration of the inhibitory substance. This method results in unreliable estimates of K_i (K_{ic} or K_{iu}), especially at high substrate concentrations (Kakkar *et al.* 2000). The SNLR and K_m methods provide more accurate and reliable estimates of K_i compared to the Dixon method (Kakkar *et al.* 2000). The inhibition constants K_{ic} and K_{iu} can also give an idea about the behavior of an inhibitor having mixed inhibitor characteristics. If K_{ic} is greater than K_{iu}, the mixed inhibition is predominantly of competitive type. If K_{ic} is less than K_{iu}, the mixed inhibition is predominantly of uncompetitive type. If K_{ic} is equal to K_{iu}, this indicates a special type of mixed inhibition, called non-competitive inhibition.

An example of the inhibitory effects of two organic hazardous compounds is shown in Box 5.2. The study was performed to estimate the apparent nitrification constants (q_m^{app} and K_m^{app}) in the presence of trichloroethylene (TCE) and 1,2-dichloroethane (1,2-DCA), respectively. Comparing the results of the study with the information presented in Table 5.3 shows that TCE and 1,2-DCA acted as competitive and mixed inhibitors, respectively.

5.3.5 Critical appraisal of standard inhibition tests and suggestions

The following limitations exist in standard inhibition tests:

(1) *Limited number of tests*: Standard inhibition tests issued by OECD or other international organizations were initially developed to account for inhibition of substrate degradation in sewage treatment plants (STPs). These tests provide a rapid screening of the effect of a hazardous substance. The synergistic or antagonistic behavior of several hazardous substances can also be quickly assessed with these tests. However, compared to the diversity of standard biodegradation tests (Section 5.4), standard inhibition tests are still very limited in number.

(2) *Absence of anoxic tests*: Standard inhibition tests have been mainly developed for aerobic and anaerobic systems. Currently no tests are available for anoxic systems (e.g. denitrification, Anammox systems).

(3) *Lack of tests for biofilm systems*: Current inhibition tests are developed for suspended-growth systems like activated sludge. Hence, they are not applicable to biofilm (attached-growth) systems. On the other hand, biofilm systems receive great attention, in particular for nutrient removal processes such as nitrification and Anammox that include slowly-growing microorganisms.

BOX 5.2 DETERMINATION OF INHIBITION TYPE IN A NITRIFYING ACTIVATED SLUDGE EXPOSED TO TCE AND 1,2-DCA

TCE and 1,2-DCA are hazardous compounds generated from industrial applications. The effect of each compound on an enriched nitrifying activated sludge was investigated. Initially, the baseline specific ammonium degradation rates were determined at various NH_4^+-N concentrations. Data were evaluated by NLSR and the kinetic constants, the maximum specific ammonium degradation rate q_m and the half-velocity constant K_m, belonging to uninhibited nitrification were determined. Then, in the presence of either TCE or 1,2-DCA, specific ammonium degradation rates were determined. To do this, first the inhibitory ranges of these compounds were determined using short-term oxygen uptake measurements.

The figures below show the specific ammonium degradation rates with respect to NH_4^+-N/K_m for various concentrations of TCE or 1,2-DCA, respectively. As seen in the tables below, with an increase in TCE, q_m^{app} remained the same as the original value (q_m), but K_m shifted to a higher one (K_m^{app}). Thus, TCE was identified as a competitive inhibitor of nitrification with an inhibition constant (K_{ic}) of 666 ppb. However, with an increase in 1,2-DCA, the apparent q_m (q_m^{app}) became lower compared to the uninhibited case whereas K_m shifted to a higher value. Thus, 1,2-DCA was identified as a mixed inhibitor of nitrification. For this pollutant, the inhibition constants K_{ic} and K_{iu} were determined as 8045 ppb and 187500 ppb, respectively (Alpaslan Kocamemi, 2005; Alpaslan Kocamemi & Çeçen, 2005; Alpaslan Kocamemi & Çeçen, 2010).

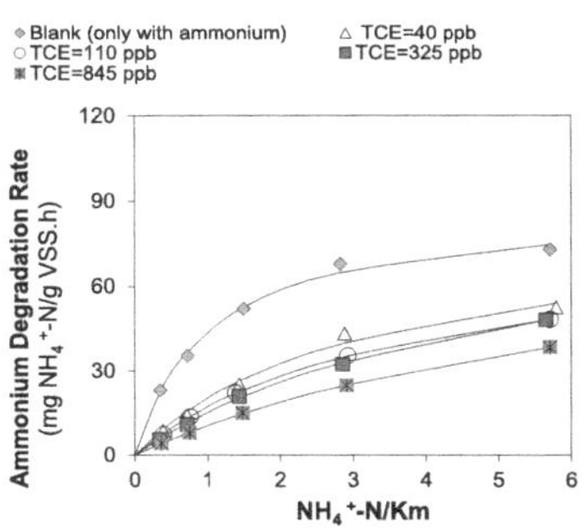

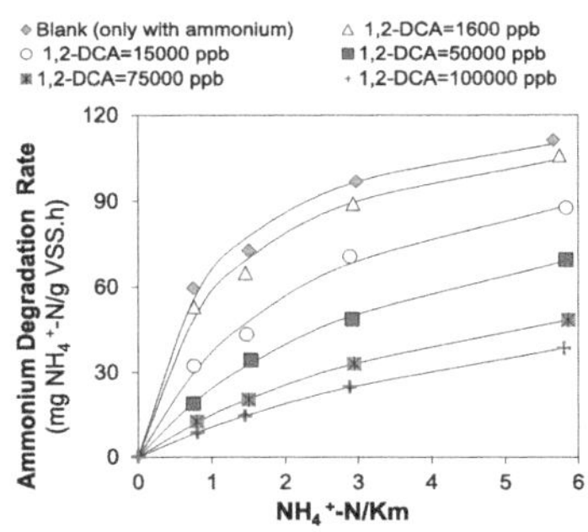

TCE	q_m^{app}	K_m^{app}
ppb	mgNH$_4$-N / gVSS.h	mg/L
0 (Blank)	87.80 ± 3.4	70.09 ± 8.1
40	82.73 ± 8.8	215.59 ± 47.3
110	78.87 ± 1.1	251.85 ± 7.1
325	90.45 ± 3.4	348.55 ± 23.3
845	88.97 ± 2.1	523.06 ± 19.8

1,2-DCA	q_m^{app}	K_m^{app}
ppb	mgNH$_4$-N / gVSS.h	mg/L
0 (Blank)	129.26 ± 7.7	68.54 ± 13.5
1600	126.41 ± 7.9	82.61 ± 15.6
15000	124.5 ± 11.5	163.99 ± 34.8
50000	112 ± 5.5	249.16 ± 24
75000	90.50 ± 2.1	349.66 ± 14.8
100000	85.68 ± 6	486.52 ± 53.8

(4) *Pollutant concentration and type*: Standard inhibition tests are performed with a synthetic sewage to which hazardous substances are added in the order of mg/L. While these concentration levels may often be encountered in industrial wastewaters, in domestic wastewaters most hazardous pollutants of interest today, called micropollutants, exist at ng/L or μg/L levels. If tests are conducted at high concentrations and these data are extrapolated to low concentrations, this may not reflect the reality. Thus, the "high concentration" is one of the shortcomings of inhibition tests.

Further, two different cases need to be considered here, namely the case of biodegradable and nonbiodegradable hazardous pollutants. The conduction of a test is more straightforward in the case of nonbiodegradable pollutants such as metals or nonbiodegradable organic pollutants. In addition, interpretation of results is easier since the concentrations of such substances usually do not change during a test. On the other hand, tests conducted with biodegradable hazardous pollutants are more difficult to interpret. Yet, such inhibition tests may provide information for further biodegradation tests since they give an idea about the noninhibitory concentration. If present at high concentrations, an essentially biodegradable pollutant might exert an inhibitory effect on its own removal and other substrates. Thus, it would not biodegrade, but would probably inhibit TEAPs. However, nowadays most pollutants encountered in domestic wastewater treatment exist at extremely low concentrations. At these concentrations, inhibition is not very likely. At very low concentrations, the degradation rate of biodegradable hazardous organics usually follows first-order kinetics as discussed in **Chapter 4**.

(5) *Disregard of cometabolism in existing tests*: A biodegradable hazardous organic test substance may serve as a growth-substrate itself. Another possibility is that the substance is removed as a cometabolite (non-growth substrate) in the presence of a growth-substrate. In treatment works, a significant number of hazardous substances are degraded by cometabolism nowadays. However, standard inhibition tests do not consider the removal and effects of such substances yet. A hazardous substance that is removed by cometabolism, may also inhibit the degradation of growth-substrate. It may also be inhibitory to its own biodegradation at very high concentrations, a phenomenon known as substrate inhibition.

For any hazardous substance that is a candidate for cometabolic degradation in any aerobic TEAP, simple respirometric experiments may provide evidence of cometabolism (Horvath, 1972). If only the hazardous substance is present in the medium and a higher oxygen consumption is recorded than endogenous respiration, it is a sign that the substance degrades by cometabolism under resting cell conditions. However, the exact evidence will be provided only if the test substance disappears with production of by- or end-products.

(6) *Fixed test conditions*: The standard inhibition tests were developed for fixed test conditions (i.e. constant pH, DO, temperature). Therefore, the response of TEAPs to a hazardous substance under variable environmental conditions cannot be assessed with the use of these tests.

(7) *Inoculum source*: Standard tests consider the use of an inoculum taken from the activated sludge unit of a domestic WWTP. Since in each plant operating conditions differ, the diversity and number of microorganisms in the inoculum may vary greatly. These variations may result in spurious results.

(8) *Inoculum concentration*: In standard tests, the inoculum concentration varies in a wide range. As shown in Figure 5.2 and explained in detail in Section 5.4.4.4, in a batch test, for example, the initial amount of substrate per biomass, namely the S_0/X_0 ratio, may have a great impact on biodegradation. The same may apply to inhibition.

(9) *Pretreatment and/or preadaptation of inoculum*: Standard inhibition tests do not permit pretreatment or preadaptation of inocula to the test substance. However, these two factors directly affect inhibition and biodegradation kinetics. For example, preconditioning of inoculum by means of filtration, centrifugation etc. may affect test results since it changes the diversity in an inoculum.

(10) *Lack of detailed information about inhibition*: Standard tests do not provide information about the pathway and rate of inhibition, and the inhibition caused by accumulation of by- and/or end-products.

(11) *Abiotic losses*: In standard inhibition tests, there is no clear experimental methodology to determine the loss of a hazardous substance by abiotic mechanisms.

(12) *Lack of simulation of long-term operation*: The duration in standard inhibition tests may not be sufficient to predict the real behavior of hazardous test substances. Experiments usually have to be conducted in the long term, ideally in continuous-flow bioreactors if the idea is to simulate the real operating conditions in a treatment plant. Ideally a real wastewater should be used in such cases. During the entire test period, continuous monitoring of the following parameters is strongly recommended: (i) major environmental factors (i.e. pH, DO, temperature), (ii) the concentration of growth-substrate, (iii) the concentration of hazardous substance, (iv) the concentration of by-products, and (v) the concentrations of major end products.

(13) *Disregard of changes in microbial ecology*: Due to their short duration, standard inhibition tests do not allow adaptation of inoculum to the test substance. Adaptation may, however, possibly lower the inhibitory effects. Changes in microbial diversity and species quantity can be followed by regular sampling of sludge and performance of molecular analyses (e.g. Denaturing Gradient Gel Electrophoresis (DGGE), quantitative Polymerase Chain Reaction (qPCR), cloning, sequencing and phylogenetic

analysis). The microscopic analysis of microbial consortium (e.g. Scanning Electron Microscopy (SEM), Transmission Electron Microscopy (TEM) is also recommended for observing the changes taking place in the taxonomy upon exposure to a hazardous substance.

(14) *Disregard of recovery from inhibition*: Existing standard tests do not allow the examination of recovery from inhibition. The example shown in Box 5.3 demonstrates some part of a long-term study that examined the inhibitory effect of 1,2-dichloroethane (1,2-DCA) in a submerged nitrifying biofilm reactor and the recovery from inhibition (Alpaslan Kocamemi, 2005; Alpaslan Kocamemi & Çeçen, 2009).

BOX 5.3 INHIBITORY EFFECT OF 1,2-DCA ON A NITRIFYING BIOFILM REACTOR AND RECOVERY FROM INHIBITION

1,2-DCA is a hazardous substance that is commonly used in industrial applications. The inhibitory effect of 1,2-DCA as well as recovery from inhibition were investigated in a continuous-flow submerged nitrifying biofilm reactor (Alpaslan Kocamemi, 2005; Alpaslan Kocamemi & Çeçen, 2009). Initially, an ammonium solution alone was continuously fed to the reactor. The loading and degradation rates were expressed per surface area of packing medium. Then, also 1,2-DCA was dosed into the reactor to observe its inhibitory effect on nitrification. Each time steady-state conditions were reached. Following the inhibition period, 1,2-DCA dosing was terminated and the reactor was fed again with ammonium alone to monitor the recovery of nitrification activity. As seen from the figure below, at influent 1,2-DCA concentrations of 4728 ppb and 8087 ppb, ammonium degradation decreased to 84% and 60%, respectively. However, when 1,2-DCA feeding was stopped, nitrifying bacteria regained their activity and ammonium degradation increased back to 90%. Repeatedly, higher concentrations of 1,2-DCA were dosed into the reactor. It was seen that the recovery period lasted longer in such cases.

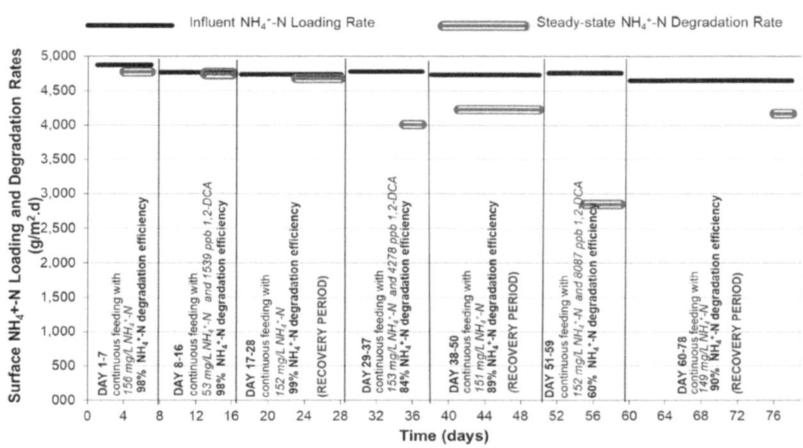

5.4 BIODEGRADATION EXPERIMENTS

5.4.1 Why do we need biodegradation experiments?

Biodegradation experiments are performed for three reasons:

(a) If no *a priori* biodegradability information exists about a pollutant.
(b) To determine the potential of a biological system to remove a biodegradable pollutant.
(c) To determine the time needed for a pollutant to be degraded in a biological treatment system.

BOX 5.4 BIODEGRADATION OF BISPHENOL A (BPA)

BPA is a chemical, extensively used for the production of polymers, such as epoxy resins and polycarbonate plastics. These plastics are used in many food and drink packaging applications. As a result, BPA is one of the most detected chemicals in wastewater. Guerra *et al.* (2015) measured BPA in 499 liquid and 347 solid samples collected from 25 WWTPs (Guerra *et al.* 2015). Median BPA concentration in the sewage was 400 ng/L. Given the fact that BPA is a probable endocrine disrupting compound (EDC), it should be removed from wastewater. However, median removal efficiencies in treatment plants ranged from 1 to 77% (Guerra *et al.* 2015). Moreover, median BPA levels in primary sludge, secondary sludge and digester sludge were 230, 260, and 460 ng/g, respectively, suggesting that BPA has an affinity to sorb on the organic/inorganic particles in wastewater. So, based on these occurrence data, what can be concluded? Is BPA biodegradable? Obviously, there is no clear answer to this, since the findings above do not differentiate between biological or other types of removal.

A closer look at the microbial degradation of BPA yields the following results: A limited number of bacteria that can utilize BPA has been identified to date.

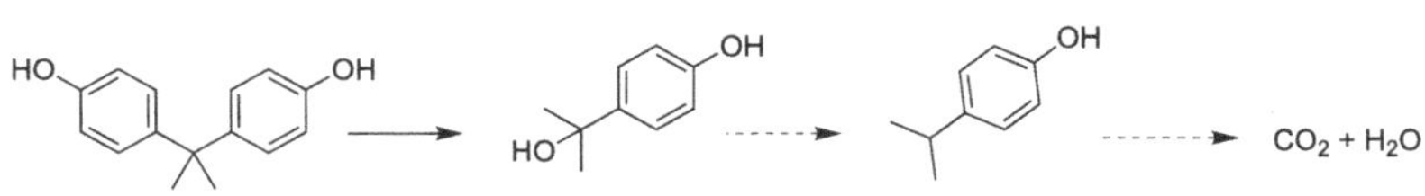

bisphenol A — 4-(2-Hydroxypropan-2-yl)phenol — 4-iso-Propylphenol

***Sphingomonas* sp. TTNP3** — ***Sphingobium fuliginis* TIK-1**

A strain called *Sphingomonas* sp. TTNP3 can convert BPA up to 4-iso-propylphenol (Kolvenbach *et al.* 2007). However, 4-iso-propylphenol cannot be utilized by the strain TTNP3 (Kolvenbach *et al.* 2007). On the other hand, 4-iso-propylphenol can be mineralized to CO_2 by *Sphingobium fuliginis* TIK-1 (Kolvenbach *et al.* 2007; Toyama *et al.* 2010). As a result, biodegradation of BPA needs at least two different species to be present. Moreover, sorption on biomass does not account for biodegradation even though it is often called "preliminary degradation". Therefore, simple disappearance from the wastewater does not truly mean that BPA is degraded. In addition, biodegradation rate of BPA is important.

It is very difficult to determine if a pollutant is "biodegradable" or not using the concentrations of the pollutant in the influent and effluent of a biological treatment system. To clarify this, the example regarding Bisphenol A (BPA) is shown in Box 5.4. As seen in this example, occurrence data alone have no explanatory power. To correctly judge the biodegradation potential of a hazardous substance in any treatment system, we need a systematic testing of biodegradation that accounts for biotransformation, biodegradation and abiotic removal rates.

5.4.2 Types of biodegradation experiments

In wastewater treatment studies, there are standard methods to test biodegradation of substances by microbial communities in the different TEAPs shown in Table 5.1. Among those, the tests developed by OECD receive the most attention and application. These tests are divided into three as:

(1) Screening (ready biodegradability) tests
(2) Inherent biodegradability tests
(3) Simulation tests

Briefly, screening tests (ready biodegradability tests) assume that the test substance is rapidly mineralized. Therefore, most of them target inorganic end-products such as CO_2, CH_4, N_2 as the endpoint of biodegradation. These tests primarily aim to determine whether the substance is potentially readily biodegradable, rather than to predict its actual rate of biodegradation (Ahtiainen *et al.* 2003). Particularly, for hazard classification in the water environment, data obtained from these screening tests are taken into consideration. Inherent biodegradability tests aim at assessing whether the test substance has any potential for primary or ultimate biodegradation. Simulation tests follow either the actual test substance or radiolabeled end-products associated with the test substance. In the new OECD guidelines, simulation tests have largely replaced inherent biodegradability tests (Boethling & Lynch, 2007). In simulation tests, the concentration of the test substance is kept low, generally around the expected concentrations in a wastewater, whereas biodegradation at high concentrations is tested in screening tests. Such differences in concentration levels may also result in differences in the probability of biodegradability and biodegradation rates under the indicated redox condition.

The design of biodegradation tests is complicated. Prior information on the biodegradability of the test substance is always helpful. Further, the design of biodegradability tests depends also on the availability of analytical methods to measure the test substance, duration of the tests and extent of the information needed as an output. In Figure 5.3, we offer a pipeline for designing of biodegradation tests and describe how each step is performed.

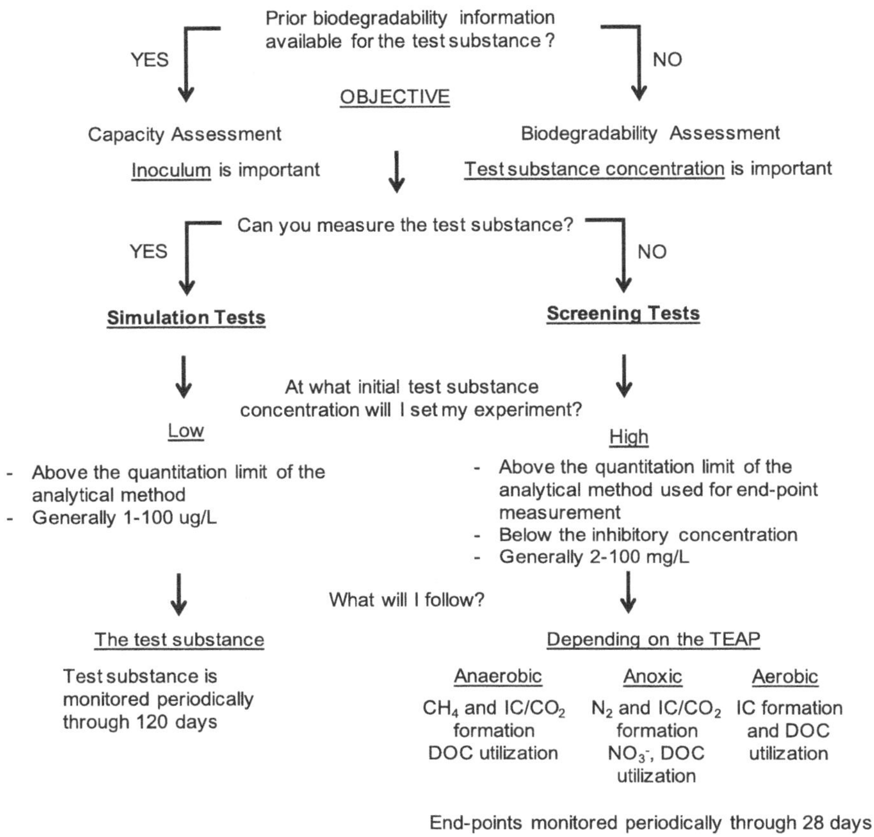

Figure 5.3 Pipeline of designing biodegradation tests.

5.4.3 Prior information on biodegradability of a test substance

Prior information on the biodegradability of a test substance is necessary to set the objectives of biodegradation tests. The databases and tools helping in conduction of biodegradation experiments are shown in Figure 5.4.

Currently, there is a large database called METACyc to identify if the test substance can be biotransformed by a microorganism, its biotransformation pathway, and genes/enzymes involved. Presence of a test substance in the database confirms that the substance can be biotransformed by a microorganism present in the environment, otherwise it suggests that either the substance is not degradable or has not been studied yet. In the latter case, biodegradation potential of a test substance can be estimated by calculating if the transformation reaction is energetically favorable or using the EPI Suite BIOWIN module which predicts biodegradation probability using group contribution method. EPI Suite™ is the EPI (Estimation Programs Interface) Suite™, a Windows®-based set of physical/chemical properties and environmental fate estimation program developed by OPPT to screen new chemicals lacking experimental data.

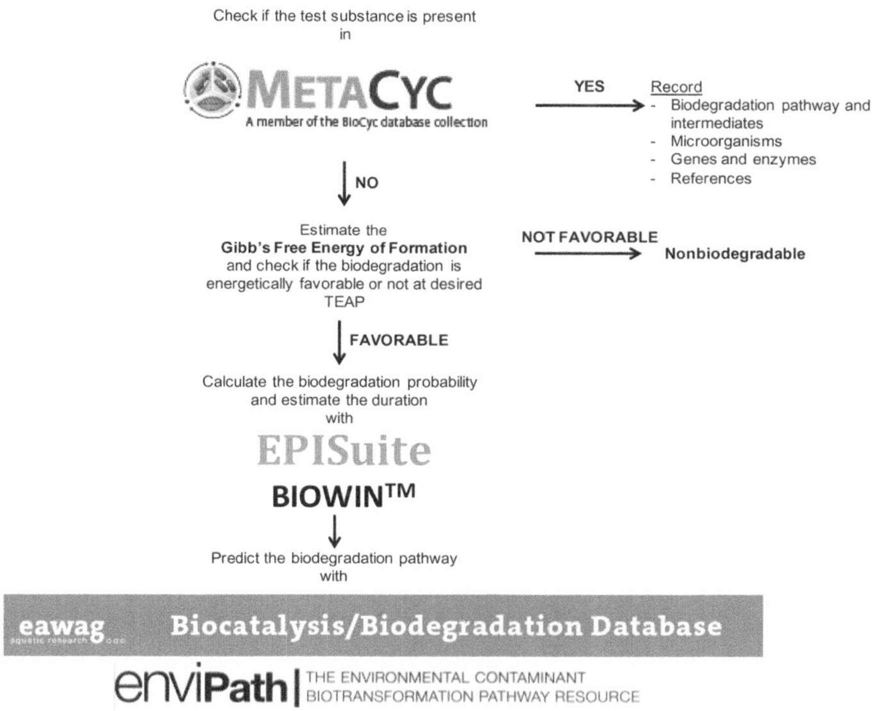

Figure 5.4 Pipeline of collecting prior information on biodegradation of a test substance.

The biodegradation pathway of a test substance which has not been studied to date can be predicted by the database of the Swiss Federal Institute of Aquatic Science and Technology, namely the EAWAG BBD (EAWAG Biocatalysis/Biodegradation Database), and enviPath, using biotransformation reactions of similar compounds having similar functional groups (Figure 5.4). The tool enviPath has been designed as a more modern and flexible successor system of EAWAG BBD-Pathway Prediction System (EAWAG-BBD/PPS), to extend its functionality (Wickert *et al.* 2016).

METACyc is an extensive curated database of experimentally elucidated metabolic pathways from all domains of life. METACyc contains 2491 pathways from 2816 different organisms. Pathways include both primary and secondary metabolism, as well as associated metabolites, reactions, enzymes, and genes. One can search the database simply by entering the name of the test substance. The output includes the complete recorded pathway by an organism, intermediates observed, enzymes involved, the test conditions and the references related to the pathway (Caspi *et al.* 2016). As an example of this, Figure 5.5 illustrates the biodegradation of linuron, a phenylurea herbicide that is widely used in agriculture.

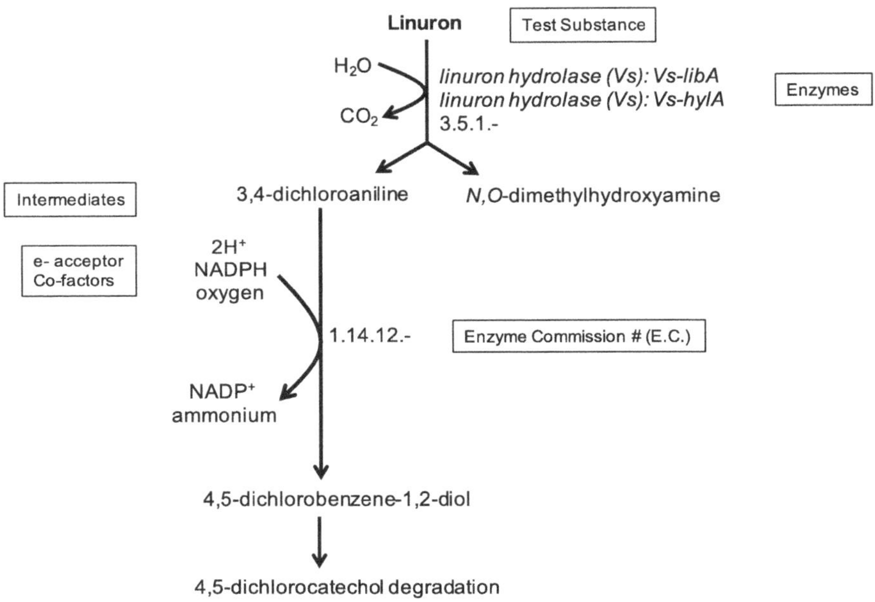

Figure 5.5 The output of METACyc for biodegradation of linuron.

If a biodegradation pathway is present in the METACyc database this suggests that the test substance can be utilized by a microorganism in the environment. However, it should be noted that the pathway and the organisms may not be limited to what is listed in the output of the database. If the potential of a certain treatment system in biodegradation of a test substance is determined, the METACyc output would offer if a screening or simulation test is appropriate and suggest endpoints to monitor.

If there is no record on the biodegradation of a test substance, first we need to check if biodegradation of the substance is energetically favorable under a certain TEAP. The standard Gibb's free energies of formation ($\Delta G_f^{0\prime}$) of many hazardous pollutants have not been published. A group contribution method developed by Mavrovouniotis (1990) was used to calculate $\Delta G_f^{0\prime}$ values of Quaternary Ammonium Compounds (QACs) and potential QAC transformation products using Equation 5.1 (Mavrovouniotis, 1990):

$$\Delta G_f^{0\prime} = \sum_{1}^{i} a_i \Delta G_{f,i}^{0\prime} \tag{5.1}$$

where, a is the number of occurrence of group i and $\Delta G_{f,i}^{0\prime}$ is the standard Gibb's free energy of formation of group i.

As an example, as shown in Table 5.4, $\Delta G_f^{0\prime}$ of benzyldimethyldodecyl ammonium (C_{12}BDMA), a quaternary ammonium disinfectant and an emerging pollutant, is calculated by adding the contribution of each group present in the molecule as described in Equation 5.1.

Table 5.4 Molecular structure, groups, their occurrences and contribution to Gibb's Free Formation Energy of C_{12}BDMA.

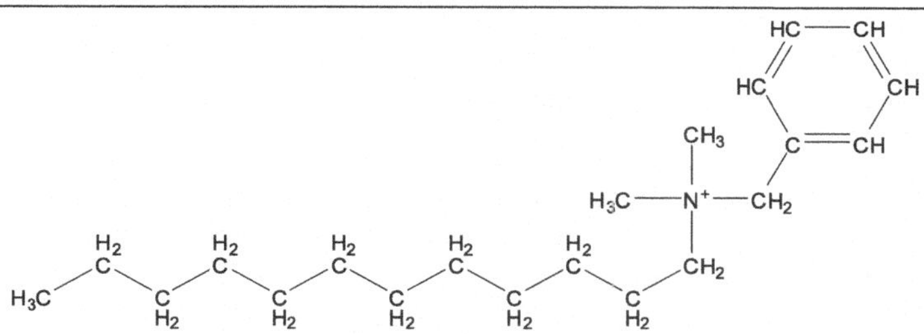

Benzyldimethyldodecyl ammonium (C_{12}BDMA)

Group/Correction	**# of occurrence**	**Contribution (kcal/mol)**	**Total Contribution (kcal/mol)**
Origin	1	–24.7	–24.7
Aromatic ring	1	–6.0	–6.0
$-CH_3$	3	8.5	25.5
$-CH_2-$	12	1.7	20.4
$>N^+<$	1	8.9	8.9
>CH–benzene	5	8.6	43.0
>C–benzene	1	1.1	1.1
$\Delta G_f^{0\prime}$			**68.2 kcal/mol (285.4 kJ/mol)**

$\Delta G_f^{0\prime}$ values of other compounds involved in reactions of interest can be obtained from the literature (Thauer *et al.* 1977; Rittmann & McCarty, 2001). The standard

Gibb's free energies of half-reactions ($\Delta G^{0\prime}$) and oxidation-reduction reactions ($\Delta G_{reac}{}^{0\prime}$) without cell synthesis can be calculated by using Equation 5.2 according to the method described by Rittmann and McCarty (2001).

$$\Delta G^{0\prime}(\text{or } \Delta G^{0\prime}_{reac}) = \sum_{1}^{j} p_j \Delta G^{0\prime}_{f,p_j} - \sum_{1}^{k} r_k \Delta G^{0\prime}_{f,r_k} \tag{5.2}$$

where, p is the molar fraction of product j, and r is the molar fraction of reactant k, whereas $\Delta G_{f,p}{}^{0\prime}$ and $\Delta G_{f,r}{}^{0\prime}$ are the standard Gibb's free energies of formation of products and reactants, respectively.

The standard Gibb's free energies of half-reactions of C_{12}BDMA and various TEAPs as well as free energy of C_{12}BDMA biodegradation are given in Table 5.5. As in the example, energetics calculations suggest that biodegradation of C_{12}BDMA-Cl is energetically favorable under aerobic and denitrifying conditions (Tezel & Pavlostathis, 2009; Tezel *et al.* 2012), but limited under methanogenic conditions which is also observed in reality (Tezel *et al.* 2006). In addition, biodegradation of C_{12}BDMA under aerobic and denitrifying conditions may require co-enzymes such as flavin adenine dinucleotide (FADH), cytochromes or rubredoxin, but not nicotinamide adenine dinucleotide (NADH) (Figure 5.6).

Table 5.5 Half-reactions of C_{12}BDMA and various TEAPs and their Gibb's free energies as well as free energies of C_{12}BDMA biodegradation under those TEAPs at pH = 7 and 298°K.

Half-reaction	**$\Delta G^{0\prime}$ (kJ/e⁻eq.)**
e- acceptor	
$\frac{1}{4}O_2 + H^+ + e^- \leftrightarrow \frac{42}{118}H_2O$	–78.72
$\frac{1}{5}NO_3 + \frac{6}{5}H^+ + e^- \leftrightarrow \frac{1}{10}N_2 + \frac{3}{5}H_2O$	–72.20
$\frac{1}{8}CO_2 + H^+ + e^- \leftrightarrow \frac{1}{8}CH_4 + \frac{1}{4}H_2O$	23.53
e- donor	
$\frac{1}{118}C_{21}H_{38}N^+ + \frac{42}{118}H_2O \leftrightarrow \frac{21}{118}CO_2 + \frac{1}{118}NH_4^+ + H^+ + e^-$	–28.70
Energetics of C_{12}BDMA Biodegradation	**$\Delta G_{reaction}{}^{0}$**
$\Delta G_{O2\ respiration}{}^{0\prime}$	–107.42
$\Delta G_{Denitrification}{}^{0\prime}$	–100.90
$\Delta G_{Methanogenesis}{}^{0\prime}$	–5.17

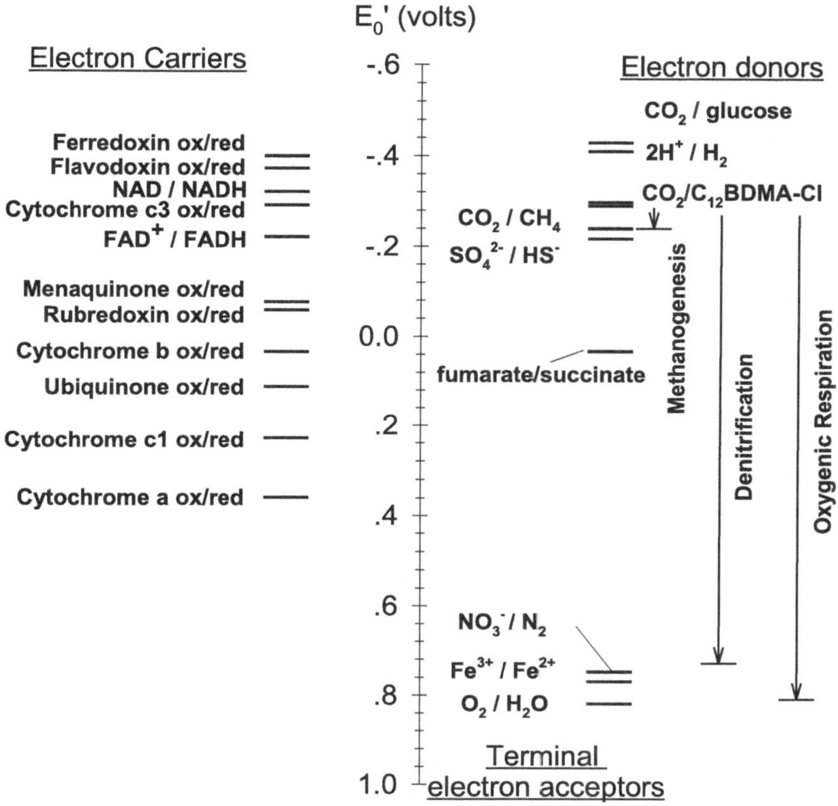

Figure 5.6 Schematic showing the positions of $CO_2/C_{12}BDMA$, TEAP and co-enzyme redox couples on the electron tower.

If biodegradation of a test substance is energetically favorable, its biodegradation probability and duration under aerobic and anaerobic conditions can be predicted using BIOWIN models. These models are implemented in EPI Suite and OECD QSAR Toolbox Software which are freely available.

BIOWIN contains seven individual group contribution models to predict the aerobic or anaerobic biodegradation potential and extent of a test substance at screening level. Out of six aerobic biodegradation models, four are qualitative whereas two (BIOWIN 3 and 4) are semi-quantitative (Boethling *et al.* 2007).

BIOWIN 1 and 2 were developed using a BIODEG database which includes qualitative biodegradation information on 264 chemicals. Similarly, BIOWIN 5 and 6 which are analogous to BIOWIN 1 and 2, estimate the likelihood of rapid biodegradation based on a set of Japanese Ministry of International Trade and

Industry (MITI) ready biodegradation data for 884 compounds (Boethling *et al.* 2007). On the other hand, semi-quantitative estimates of rates of primary and ultimate biodegradation, gathered from a survey of experts who analyzed 200 carefully selected substances, were used to develop BIOWIN 3 and BIOWIN 4. In addition, BIOWIN 7 is a screening-level model for the prediction of anaerobic ultimate degradation in sludge digesters (Boethling *et al.* 2007). BIOWIN runs these seven models simultaneously using the user input of test substance in SMILES format. The output contains seven tables. Each table includes the contribution of the functional groups in the test substance to predict the overall probability of degradation.

For example, as shown in Figure 5.5, when the biodegradation probability of linuron, a phenylurea herbicide that can be degraded by *Lysinibacillus sphaericus*, *Variovorax* sp. PBS-H4, *Variovorax* sp. SRS1 and *Variovorax* sp. WDL1, is estimated by BIOWIN models, the predictions show that its degradation is not fast, but may take weeks or months under aerobic conditions. Moreover, it is not biodegradable under anaerobic conditions (Table 5.6).

Table 5.6 Biodegradation probability of linuron under aerobic and anaerobic conditions estimated by seven BIOWIN models.

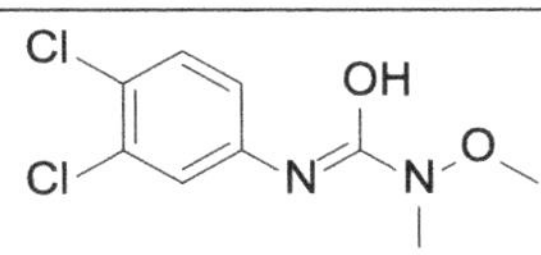

Linuron

SMILES: CN(/C(=N\c1ccc(c(c1)Cl)Cl)/O)OC

Model	Value	Criteria
BIOWIN 1	0.2641	≥0.5: Degrades FAST
BIOWIN 2	0.0104	<0.5: Does NOT degrade FAST
BIOWIN 3	2.2355	5→hours, 4→days, 3→weeks, 2→months,
BIOWIN 4	3.1577	1→longer (the values indicate the relative time span of biodegradation)
BIOWIN 5	0.0089	≥0.5: Readily degradable
BIOWIN 6	0.0090	<0.5: NOT readily degradable
BIOWIN 7	−0.4139	≥0.5: Degrades FAST <0.5: Does NOT degrade FAST

Thus, the results provided by BIOWIN models can be used to decide on the type (ready biodegradability vs. inherent biodegradability) and the duration of the biodegradation experiment that will be performed for the test substance.

The major shortcoming of the BIOWIN models is the limited number of substances that the models use as the training set. Therefore, not all fragments of significance for biodegradation are included or adequately represented in the training sets. This is the reason why BIOWIN lacks coefficients for phosphonate (C-P bond), imidazole ring, pyrimidine ring, ethoxylate ether, alicyclic rings, and quaternary nitrogen (Boethling *et al.* 2007). Therefore, the biodegradation probability of some emerging contaminants having these groups, such as QACs and organophosphate pesticides including parathion and dichlorvos, cannot be predicted accurately.

When BIOWIN models are limited, other prediction tools can supply information on the probability of biodegradation as well as its pathway. One of these tools is EAWAG Biodegradation/Biocatalysis Database (BBD).

EAWAG BBD contains information on microbial biocatalytic reactions and biodegradation pathways, primarily for xenobiotic chemical compounds (Gao *et al.* 2010). Currently, EAWAG BBD provides information on 1503 microbial enzyme-catalyzed reactions involved in the 219 pathways for biotransformation of 1396 compounds by 993 enzymes and 543 microorganisms that are important for biotechnology. One can search the database by simply entering the name, molecular formula or the Chemical Abstracts Service (CAS) number of the test substance. Biotransformation pathway including individual reactions, intermediate chemical compounds, organisms transforming the compound, enzymes, and genes are then displayed.

Main disadvantages of the EAWAG BBD system are as follows: Available data remain still limited although the structural diversity of xenobiotic chemicals is high. Further, most of the biotransformation data in EAWAG BBD is obtained from experiments conducted mainly with pure or enrichment cultures, which are of limited relevance for actual environmental situations. Moreover, conditions of the experiments are generally not described. Therefore, METACyc described above is a more comprehensive database than EAWAG BBD.

However, EAWAG BBD has a pathway prediction system (PPS). EAWAG-PPS predicts microbial catabolic reactions using substructure searching, a rule-base, and atom-to-atom mapping. The system is able to recognize the organic functional groups in a compound and predict transformations based on biotransformation rules. The biotransformation rules are based on the reactions found in the EAWAG BBD database or in scientific literature. EAWAG-PPS pathway predictions are mostly accurate. For instance, linuron whose biodegradation pathway is present in METACyc is not present in the EAWAG database. However, EAWAG-PPS is able to predict the major intermediates of linuron biodegradation, which have been identified experimentally as correct (Bers *et al.* 2011; Bers *et al.* 2013). Predictions indicate that formation of 3,4-dichloroaniline is "very likely" whereas the probability of formation of 4,5-dichlorobenzene-1,2-diol from 3,4-dichloroaniline is "neutral" (Figure 5.7). It has been shown that conversion of 3,4-dichloroaniline to 4,5-dichlorobenzene-1,2-diol is a slow reaction; thus the accumulation of

3,4-dichloroaniline was observed in the experiments performed with *Variovorax* sp. WDL1 (Breugelmans *et al.* 2010). EAWAG-PPS predicts that the formation of N,O dimethylhydroxyl amine is "unlikely" since reduction of methoxy(methyl) carbamic acid requires reductive conditions (Figure 5.7). Probably, the reduction reaction yielding N,O dimethylhydroxyl amine is abiotic.

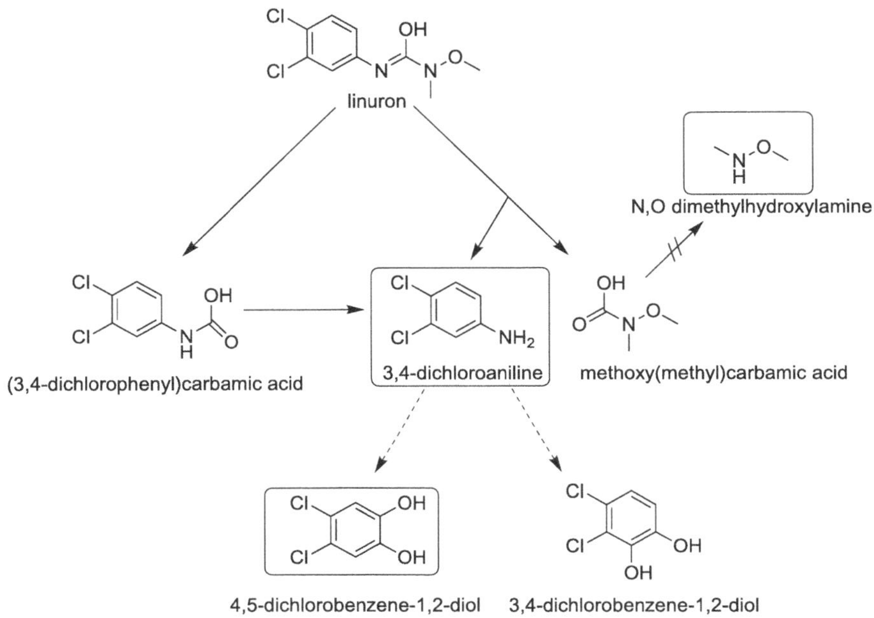

Figure 5.7 Pathway of linuron biodegradation by EAWAG-PPS. Compounds in the boxes are intermediates observed in actual experiments. Solid, dashed and blocked arrows indicate whether the probability of the reaction is "Very Likely", "Neutral" or "Unlikely", respectively.

In summary, prior information on the biodegradability of a test substance can be acquired using software and online tools freely available. The information is important to decide on the type of biodegradation experiment as well as the endpoints to monitor and duration of experiments.

5.4.4 Screening (ready biodegradability) tests

Screening tests which are also referred to as "ready biodegradability tests" are performed at high concentrations of the test substance ranging from 2 to 100 mg/L. The endpoints for these tests are non-specific parameters like Dissolved Organic Carbon (DOC), Biochemical Oxygen Demand (BOD), CO_2, CH_4, and N_2 production.

Activated sludge, anoxic sludge from a Biological Nutrient Removal (BNR) system or an anaerobic sludge taken from a digester are used as the source of microorganisms (inoculum) in tests conducted under aerobic, anoxic and anaerobic conditions, respectively.

The inoculum should not have been preadapted to degradation of the test substance by previous exposure to the substance itself or structurally related chemicals. The duration of the test is generally about 30 days. Indication of 60% degradation within 10 days of incubation is the pass criteria for the test and suggests that the test substance undergoes rapid and ultimate degradation in treatment plants. Before biodegradation tests are performed for a test substance, inhibition tests should be conducted and the minimum inhibitory concentration of the test substance on microbial inoculum should be found. A reference substance which is known to degrade should be used to verify the activity of the inoculum.

OECD offers a number of standard screening tests for aerobic and anaerobic biodegradability. However, a protocol for testing biodegradability under anoxic conditions is currently missing. In this section, we are going to discuss standard tests and describe new protocols for testing biodegradability.

5.4.4.1 Aerobic biodegradability of a test substance

The following standard OECD tests are available to determine ready biodegradability of organic chemicals (OECD, 1992, 2014):

- DOC Die-Away Test (301A)
- CO_2 Evolution Test (301B)
- Modified MITI Test I (301C)
- Closed Bottle Test (301D)
- Modified OECD Screening Test (301E)
- Manometric Respirometry Test (301F)
- CO_2 in sealed vessels (Headspace Test) (310)

In all of these tests, the test substance serves as the sole carbon and energy source, thus it is a growth- or primary-substrate. Relatively low concentration of biomass is used to avoid excessive adsorption of the test substance as well as to limit release of carbon due to endogenous decay. The test substance is not measured during the course of the experiment, but non-specific analytical methods are used to indirectly verify biodegradation. The advantage of using non-specific methods is that they are applicable to all organic substances. Thus, there is no need to develop specific analytical procedures. In addition, these methods also measure any biodegradation residues or transformation products. Complete removal of DOC is indicative of ultimate biodegradation. However, concentration of the test substance should be high enough to be measured by the mentioned non-specific methods. As a result, the initial concentration of a test substance in the screening test is generally kept at a very high level and does not represent the actual concentrations in a wastewater.

An activated sludge sample diluted with a mineral medium to maintain a biomass concentration between 4–30 mg/L as suspended solids (SS) is spiked with a solution or suspension of the test substance in a bottle or a flask. The reactors are incubated under aerobic conditions in the dark or in diffuse light at 22–25°C (Table 5.7). The pH of the medium is kept around neutral. The amount of DOC stemming from the inoculum should be as low as possible compared to the organic carbon of the test substance which ranges between 2–40 mg DOC/L (Table 5.7). An extra reactor amended with a reference compound, i.e. aniline, sodium acetate or sodium benzoate, which meet the criteria for ready biodegradability, is set as a control for inoculum activity and analytical measurements. In addition, a reactor containing inoculum alone is prepared to correct for the carbon released during endogenous decay of biomass.

Table 5.7 Composition and environmental conditions of OECD biodegradation screening tests.

OECD Test	301A	301B	301C	301D	301E	301F	310
Test Substance mg/L			100	2–10		100	
mg DOC/L	10–40	10–20			10–40		2–40
mg ThOD/L				5–10		50–100	
Inoculum							
mg SS/L	≤30	≤30	30			≤30	4
cells/L	10^7–10^8	10^7–10^8	10^7–10^8	10^4–10^6	10^5	10^7–10^8	10^5–10^8
pH	7.4 ± 0.2	7.4 ± 0.2	7.0	7.4 ± 0.2	7.4 ± 0.2	7.4 ± 0.2	7.4 ± 0.2
Temperature	22 ± 2	22 ± 2	25 ± 1	22 ± 2	22 ± 2	22 ± 2	22 ± 2

Biodegradation of the test substance is followed by measuring DOC, CO_2 production and oxygen uptake of the inoculum. Other than 301D and 310, every test is performed in loosely covered open bottles or flasks for protection from evaporation and cross contamination. Reactors are monitored frequently to identify the beginning and end of biodegradation within an incubation time of 28 days.

In order to describe a test substance as readily biodegradable, the following pass levels should be satisfied at the end of 10 days after the start of biodegradation:

- 70% DOC removal in tests 301A and 301E.
- 60% theoretical carbon dioxide ($ThCO_2$) production in test 301B.
- 60% theoretical oxygen demand (ThOD) in tests 301C, 301D and 301F.
- 60% of theoretical maximum inorganic carbon (IC) or CO_2 production (ThIC or $ThCO_2$) in test 310.

The ready biodegradability test 310 (CO_2 in sealed vessels-headspace test) represents an alternative to the CO_2 Evolution Test (301B). In this test, the CO_2 evolution resulting from the ultimate aerobic biodegradation of the test substance is determined by measuring the IC or CO_2 produced in sealed test bottles, and the pass level has been defined as 60% of theoretical values (ThIC or $ThCO_2$). The pass levels of either 60% ThOD (or $ThCO_2$) or 70% DOC removal practically represent complete ultimate degradation of the test substance as the remaining 30–40% of the test substance is assumed to be used in biosynthesis.

In order to select the most appropriate biodegradability test, information is necessary on the molecular structure, formula, solubility, vapor pressure and the octanol-water partitioning coefficient (K_{ow}). K_{ow}, also denoted as the P value, is a physicochemical parameter that defines the hydrophobicity of a test substance, and is used not only to calculate theoretical values and/or check measured values of parameters, e.g. ThOD, $ThCO_2$, DOC and TOC, but also to decide on the reactor type, i.e. closed or open, as mentioned in Section 5.2. All of these chemical properties can be acquired using an online database called ChemSpider (www.chemspider.com). In addition, information on the purity of the tested material is required in order to interpret the results obtained, especially when the result lies close to the pass level. Noninhibitory concentration of the test substance to the inoculum should be identified using OECD 209 "Activated Sludge Respiration Inhibition Test" for selecting appropriate test concentrations, as described in Section 5.3.

All the methods listed above can be applied for test substances whose solubility is higher than 100 mg/L, given that they are non-volatile and non-adsorbing. However, 301A and 301E tests are preferential due to their simplicity compared to other tests. For those chemicals which are poorly soluble in water, volatile or adsorbing, suitable methods are indicated in Table 5.8. Overall, test methods based on DOC measurements such as 301A and 301E are not preferred for the substances which are poorly soluble or have a high affinity for sorption on biomass since lower aqueous concentrations are measured in such cases. This results then in incorrect interpretation of biodegradation (Table 5.8). The tests performed in open or loosely sealed bottles such as 301A, B, C, E and F are not preferred for assessing biodegradability of volatile substances due to unattained loss by volatilization (Table 5.8).

Tests such as 301B and 310 basing on the evolution of CO_2 or IC give the most direct evidence of oxidation of organic carbon during biodegradation. The removal of DOC or the uptake of oxygen are indirect measures of biodegradability and may give misleading results. Therefore, these methods need strict control reactors.

The DOC die-away and modified OECD screening methods are limited to soluble chemicals with low K_{ow} and low vapor pressure since loss of test substance due to adsorption and vaporization would be misleadingly accounted as biodegradation.

The respirometric methods can be used for soluble, insoluble and volatile chemicals while the CO_2 evolution method can deal with soluble and insoluble but not volatile chemicals. On the contrary, the OECD test 310 can eliminate the volatilization of organics, since tightly sealed bottles are used. The only limitation of the closed vessel test is the limited agitation which elongates the period of ultimate degradation of compounds that adsorb on biomass. In 301D and 310 tests performed in closed bottles, the concentration of the test substance is limited to 2 to 5 mg/L because of the low solubility of oxygen in water, i.e. 8.9 mg/L at 22°C. These tests rely on the consumption of oxygen from the dissolved phase.

Table 5.8 Applicability of OECD ready biodegradability tests based on substance characteristics.

Test	Parameter to be Monitored	Test Substance Property		
		Poorly Soluble	Volatile	Adsorbing
301A	Dissolved organic carbon	–	–	+/–
301B	Respirometry: CO_2 evolution	+	–	+
301C	Respirometry: Oxygen consumption	+	+/–	+
301D	Respirometry: Dissolved oxygen	+/–	+	+
301E	Dissolved organic carbon	–	–	+/–
301F	Oxygen consumption	+	+/–	+
310	Dissolved inorganic carbon or CO_2	+	+	+

The accuracy and precision of the methods used in measuring the performance parameters are also important. The determination of DOC can be made with a precision of 0.25 mg/L with a lower detection limit of about 0.5 mg/L. The detection limit of O_2 using automatic respirometers and probes is about 1 mg DO/L. In the closed bottle test, the concentration of DO can be determined with a precision of 0.05 mg/L. However, measurements of CO_2 in the CO_2 evolution test (301B) are considered to be less precise than DOC and oxygen in other tests because the relatively small amounts of CO_2 evolved in the period between measurements is determined from the difference between two relatively high titration values. On the other hand, this limitation is overcome using sealed vessels and measuring headspace CO_2 after acidification of the test medium by gas chromatography or IC after the medium is treated with sodium hydroxide.

Other than measuring the consumption of dissolved oxygen in water, other configurations are also possible. There are automated respirometers that constantly supply air into a closed bottle. They measure the O_2 consumed in the gas phase and the release of CO_2 into the gas phase. Using such apparatus, it is possible to monitor the oxygen uptake/CO_2 release at any time and also the cumulative O_2 consumption/CO_2 evolution.

In general, every method can be used to test biodegradability if the test substance is soluble. Preferentially 301A and 301E should be selected because of their simplicity and high precision of DOC measurements (Figure 5.8). For poorly soluble test substances, 301C and 301F are preferable. Volatile test substances should be tested in closed and sealed bottles such as 310 and 301D (Figure 5.8). The test 310 is preferred for all types of substances, especially at very low concentrations since it employs sealed bottles and precise measurement techniques such as infrared measurement of IC and gas chromatographic measurement of CO_2 (Figure 5.8).

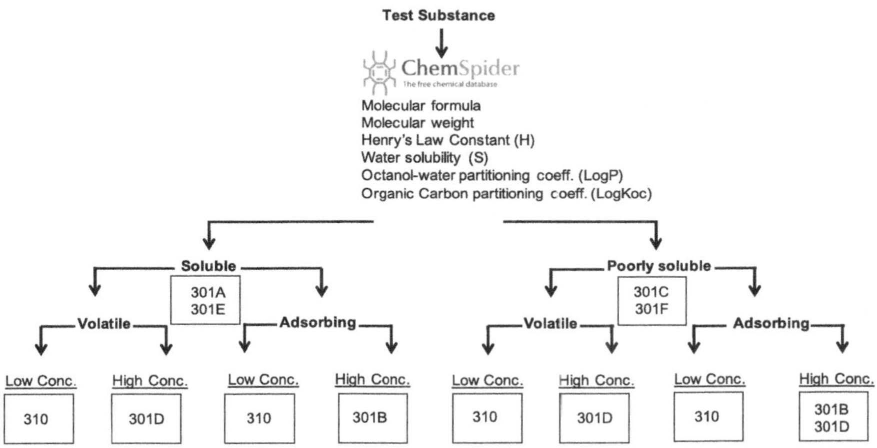

Figure 5.8 Selection of OECD ready biodegradability tests based on characteristics of the test substance.

In Box 5.5, a hypothetical biodegradability test on linuron is shown. Linuron is a pesticide which is soluble in water up to 75 mg/L. Its low Henry's law constant and moderately high log P (log K_{ow}) value indicate that this substance is a non-volatile but an adsorbing chemical. The inhibitory concentration of linuron on the microbial community in an activated sludge is not known. The best way to determine the biodegradability of linuron is to use either test 301B in the case where it is not toxic around 75 mg/L or test 310 if it is toxic above at least 2 mg/L (Box 5.5).

In Box 5.5 we illustrate the test results that a researcher would obtain in the biodegradability tests 301A, 301D and 310 which use different endpoints such as DOC utilization, oxygen uptake and CO_2 evolution, respectively. Initial concentration of the linuron was set at 10 mg/L, lower than its solubility. This is equivalent to 4.34 mg/L DOC which is higher than the DOC detection limit. Within the course of 28 days, at least 10 samples have to be taken until DOC drops below the detection limit.

BOX 5.5 LINURON BIODEGRADATION UNDER AEROBIC CONDITIONS

Properties

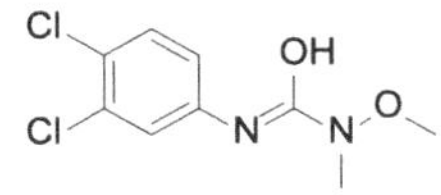

Molecular Formula: $C_9H_{10}Cl_2N_2O_2$; Molecular Weight: 249.1 g/mole; Solubility: 75 mg/L; Henry's Law Constant: 6.25 10^{-9} atm-m^3/mole; Log P = 3.20

Oxidation Reaction

Donor: $C_9H_{10}Cl_2N_2O_2 + 16H_2O \leftrightarrow 9CO_2 + 2NH_3 + 2HCl + 34H^+ + 34e^-$

Acceptor: $O_2 + 4H^+ + 4e^- \leftrightarrow 2H_2O$

Overall: $C_9H_{10}Cl_2N_2O_2 + 8.5O_2 \leftrightarrow 9CO_2 + 2NH_3 + 2HCl + H_2O$

Theoretical Values

$$DOC = \frac{9 \times 12}{249.1} = 0.434 \text{ g C/g Linuron}$$

$$ThCO_2 = \frac{9 \times 44}{249.1} = 1.590 \text{ g } CO_2\text{/g Linuron or } 9 \text{ mol} \times 82.06 \frac{\text{mL} \cdot \text{atm}}{\text{mol} \cdot \text{K}} \times \frac{295\text{ K}}{1\text{ atm}} \div 249.1\text{ g Linuron} = 875 \text{ mL/g Linuron}$$

$$ThOD \text{ or } COD = \frac{8.5 \times 32}{249.1} = 1.09 \text{ g } O_2\text{/g Linuron}$$

Three ready biodegradability tests were set up for an initial linuron concentration of 10 mg/L and 100 mL liquid volume was used. The outputs of the tests 301A, 301D and 310, assuming first-order biodegradation of linuron with a rate constant of 0.4/day, show that linuron passes test 301A within 3 days. DOC measurements correlate directly with the test substance concentration. Linuron is also biodegraded completely in 10 days in tests 301D and 310. However, a seemingly lower removal is recorded in these tests than in 301A. This arises because in these two tests, DO decrease and CO_2 release reflect the substrate consumed in the energy reaction only.

This example clearly shows the importance of method selection. To avoid this bias within test methods, the pass criteria might be lowered down to 50% (OECD, 2006a). Increasingly, direct quantification of the test substance by advanced instrumental methods is preferred in decision making about important hazardous compounds.

301A: DOC Die-away Test **DOC_0 = 4.34 mg/L** — **301D: Closed Bottle Test** **ThOD = 10.9 mg-O_2/L** — **310: Headspace Test** **$ThCO_2$ = 15.9 mg/L or 0.875 mL**

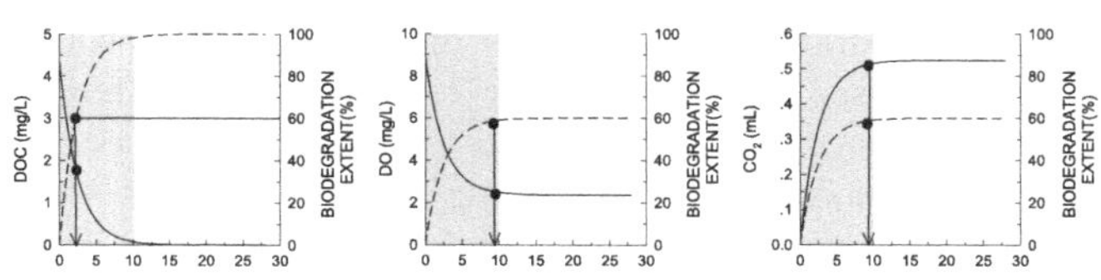

While designing biodegradation tests in closed bottles/sealed vessels, the concentration of the test substance should be chosen so that its theoretical oxygen demand (ThOD) should not exceed the DO present in the medium. ThOD of 10 mg/L linuron is 10.9 mg O_2/L. Given the fact that biodegradability assessment is performed in closed bottles in 301D and the solubility of oxygen in water at 22°C is 8.9 mg O_2/L, the oxygen demand of linuron is higher than the available DO. Theoretically, the test would be performed under an oxygen deficit environment. However, the actual oxygen demand is 6.54 mg/L since 30–40% of the organic test substance is used for biomass synthesis. Therefore, O_2 would be sufficient during the entire course of biodegradation (Box 5.5).

Like 301D, O_2 is limited in 310 since it is performed in a sealed vessel. However, vessels have headspaces which can supply more oxygen than in 301D. For instance, a 125-mL sealed bottle containing 100 mL liquid medium and 25 mL headspace has 7.82 mg O_2 whereas the same bottle containing 100 mL medium without a headspace has only 0.89 mg O_2 that is you can test biodegradability of test substance at 10 times higher concentration in 310 than 301D. However, if headspace CO_2 is measured during test 310, having a large volume of headspace would decrease the sensitivity of the test since the partial pressure of the CO_2 produced from the same amount of test substance is lower in a large headspace than in a small one.

5.4.4.2 Anoxic biodegradability of a test substance

Nitrate reduction occurs in two dissimilatory pathways. One is reduction of nitrate and nitrite to ammonia which is called dissimilatory nitrate reduction to ammonia (DNRA). DNRA preferentially occurs at very low redox potential values (Knowles, 1982), in the presence of sulfide (Brunet & Garcia-Gil, 1996) or in the presence of organic matter provided that the COD/N ratio is high (Akunna *et al.* 1992), depending on the nature of the carbon source (Akunna *et al.* 1993). In wastewater treatment systems designed for BNR, denitrification is the dominant nitrate reduction process which is shown as the sequential reduction of nitrate (NO_3^-) to nitrite (NO_2^-), nitric oxide (NO), nitrous oxide (N_2O) and dinitrogen (N_2). During denitrification, in total 5 electrons are transferred to the electron donor in a stepwise manner:

$$NO_3^- + 2e^- \rightarrow NO_2^- + e^- \rightarrow NO + e^- \rightarrow N_2O \rightarrow N_2 \quad (5.3)$$

There is no standard protocol for testing the biodegradability of hazardous substances under nitrate-reducing conditions. Since in denitrification tests with domestic or other types of wastewater the main aim is the removal of nitrate, usually a more or less readily biodegradable carbon source is supplied along with nitrate (Loosdrecht *et al.* 2016). However, in the case of hazardous test substances, one is usually faced with a slowly degradable carbon source. The biodegradation rate of the hazardous substance would correlate with the reduction rate of nitrate. Therefore, in assessment of its biodegradability, the nitrate used as an electron acceptor as well as the N_2 gas produced can be measured.

A procedure suggested for ready anoxic biodegradability may be as follows. It would be simply a modification of aerobic ready biodegradability tests. Ideally, the inoculum is taken from the anoxic tank of a BNR plant. Experiments are performed in sealed bottles which are preferentially degassed with He which is free of N_2 and CO_2. The reaction volume will be not less than 100 mL. At least half the reaction volume is filled with the diluted inoculum whereas the rest consists of the nutrient medium, nitrate solution and the test substance. The inoculum concentration may be about 30 mg SS/L in bottles as in aerobic tests. In accordance with Table 5.7, the concentration of the test substance may be in the range of 20–100 mg C/L. The initial nitrate concentration is set accordingly.

When determining the concentration of NO_3^- used in tests, it should be kept in mind not to overdose nitrate since only a fraction of the added organic test substance will be used in denitrification of nitrate. Literature values show that heterotrophic anoxic growth yield in denitrification varies between 0.18–0.66 mg cell COD/substrate COD (Loosdrecht *et al.* 2016). The value used may be assumed as 0.4 mg cell COD/substrate COD, suggesting that 40% of the organic substance is used for biomass synthesis whereas the rest is used in reduction of nitrate.

The biodegradation of the test substance can be measured indirectly by monitoring the utilization of NO_3^- and production of N_2. NO_3^- and N_2 are measured using ion chromatography and gas chromatography which are sensitive enough to monitor those N-species at very low concentrations. The duration of the test should be 30 days which is comparable to tests performed under aerobic conditions. Complete utilization of nitrate or reaching a plateau of NO_3^- at less than 10% of the initial NO_3^- is required. As opposed to a 10-day criterion for reaching 60% biodegradation under aerobic conditions, ultimate utilization of NO_3^- within about 15 days might be a reasonable pass criterion for ready biodegradability under anoxic conditions.

Box 5.6 presents the outputs of a hypothetical ready biodegradability test under nitrate-reducing conditions.

5.4.4.3 Anaerobic biodegradability of a test substance

OECD Test No. 311 "Anaerobic Biodegradability of Organic Compounds in Digested Sludge: by Measurement of Gas Production" is offered as a standard method to test the biodegradability of organic compounds under methanogenic conditions. Briefly, digested sludge obtained from an anaerobic digester of a WWTP is diluted to a total solids (TS) concentration of 1–3 g/L TS and incubated at 35 ± 2°C in sealed vessels with the test substance at 20 to 100 mg C/L for up to 60 days. The increase in headspace pressure resulting from the production of

BOX 5.6 LINURON BIODEGRADATION UNDER NITRATE-REDUCING CONDITIONS

Oxidation Reaction

Donor: $C_9H_{10}Cl_2N_2O_2 + 16H_2O \leftrightarrow 9CO_2 + 2NH_3 + 2HCl + 34H^+ + 34e^-$

Acceptor: $NO_3^- + 6H^+ + 5e^- \leftrightarrow 0.5N_2 + 3H_2O$

Overall: $C_9H_{10}Cl_2N_2O_2 + 6.8NO_3^- + 6.8H^+ \leftrightarrow 3.4N_2 + 9CO_2 + 2NH_3 + 2HCl + 4.4H_2O$

Theoretical Values

$$ThNO_3^- \text{ needed} = \frac{6.8 \times 62}{249.1} = 1.69 \text{ g } NO_3\text{/g Linuron or } 0.38 \text{ g } NO_3\text{-N/g Linuron}$$

$$ThN_2 = 3.4 \text{ mol} \times 82.06 \frac{\text{mL} \cdot \text{atm}}{\text{mol} \cdot \text{K}} \times \frac{295 \text{ K}}{1 \text{atm}} \div 249.1 \text{g Linuron} = 330.4 \text{ mL/g Linuron}$$

A ready biodegradability test was set up for an initial linuron concentration of 50 mg/L and 100 mL liquid volume and 25 mL headspace were used in the test.

It is assumed that biodegradation of linuron follows first-order degradation kinetics with a rate constant of 0.4/day. While the aerobic tests 301D and 310 determine the substrate consumed in the energy reaction by measuring the DO consumption and CO_2 evolution, respectively; the proposed anoxic test measures the decrease in the electron acceptor NO_3^- and evolution of N_2 gas. Here, the assumption is made that nitrate is not used in cell synthesis. Therefore, it should be assured that sufficient ammonium is present in the medium to be used as a source of nitrogen in cell synthesis.

At an initial linuron concentration of 50 mg/L, theoretically 84.5 mg/L NO_3^- (19 mg/L NO_3^--N) would be needed if all of the linuron were directed to the energy reaction only. In fact some fraction of this organic test substance will be used in cell synthesis. If heterotrophic anoxic growth yield in denitrification is assumed as 0.4 mg cell COD/ substrate COD, only 60% of the linuron, namely 30 mg/L linuron will be used in the energy reaction to denitrify 11.4 mg/L NO_3^--N. Per 100 mL this would result in about 1 mL N_2 as seen in the lower figure. Thus, in an anoxic test, by considering the formation of N_2, one can determine the extent of linuron removal as well as it biodegradation rate.

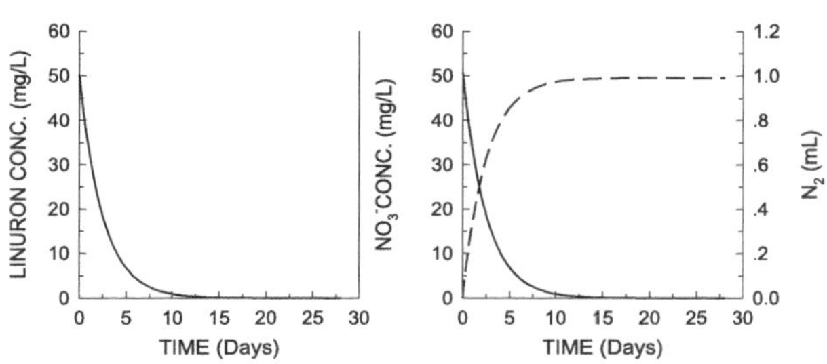

CO_2 and CH_4 is measured to judge the biodegradation. The amount of carbon (inorganic plus methane) resulting from the biodegradation of the test substance, is calculated from the net gas production and net inorganic carbon (IC) formation in the liquid phase, in excess of blank control. The extent of biodegradation is calculated as the total IC and methane-C produced per the measured or calculated amount of carbon present in the test substance. The disadvantages of this method are: (1) it does not simulate the actual conditions in a digester where complex organics are also present, and (2) it assumes that the test substance is mineralized and does not account for biotransformation since it does not measure the test substance itself.

Before the experiment starts, an inoculum put into a closed bottle is degassed with He and "starved" till no gas is produced, to eliminate any residual organics present in the sludge. The experiment is set up in either 125 or 160-mL air-tight serum bottles. Bottles are prepared to contain 100 mL total liquid volume, at least 50% of which should be occupied with the inoculum at 1–3 g/L as TS. The rest contains a nutrient medium composed of K, N, Mg, Fe, Ca, P salts, and trace metals, especially Ni which is essential for methanogens, and resazurine as a redox indicator. Air tight bottles are first degassed with He. Then inoculum, nutrient medium and the hazardous substance are added to the bottle in that order. Finally, the total volume is adjusted to 100 mL. The headspace is degassed with He and the bottles are incubated at 35°C. Agitation is necessary, but generally satisfied with manual tumbling every day. Another bottle containing sludge alone serves as a reference.

Total gas, methane and carbon dioxide are measured to determine the extent of biodegradation. Under methanogenic conditions, the heterotrophic yield is usually lower than 0.2; thus it may be assumed that all of the substance is used in the energy reaction. Based on this assumption, one can calculate the theoretical amount of methane produced if all of the initial substance were completely converted into end products. The total gas produced during the incubation is measured using either the gas/water displacement method or a pressure transducer. Methane and carbon dioxide are generally measured as a fraction of produced total gas using gas chromatography (GC) with a thermal conductivity detector. Also, cumulative methane and carbon dioxide production curves are generated. Gas measurements have to be done more frequently, i.e. once in two days during first 10 days of incubation, to capture the acute impact of the test substance on anaerobic metabolism. The frequency of gas sampling can be once in five days after 10 days till the end of the experiment which generally lasts 60 days.

The pass criterion in OECD test 311 is that 60% of the carbon present in the test substance is converted into methane and carbon dioxide. Box 5.7 shows a hypothetical ready biodegradability test with linuron under methanogenic conditions.

BOX 5.7 LINURON BIODEGRADATION UNDER METHANOGENIC CONDITIONS

Oxidation Reaction

Donor: $C_9H_{10}Cl_2N_2O_2 + 16H_2O \leftrightarrow 9CO_2 + 2NH_3 + 2HCl + 34H^+ + 34e^-$
Acceptor: $CO_2 + 8H^+ + 8e^- \leftrightarrow CH_4 + 2H_2O$
Overall: $C_9H_{10}Cl_2N_2O_2 + 7.5H_2O \leftrightarrow 4.25CH_4 + 4.75CO_2 + 2NH_3 + 2HCl$

Theoretical Values

$ThCO_2 =$

$$4.75\,\text{mol} \times 82.06\frac{\text{mL}\cdot\text{atm}}{\text{mol}\cdot\text{K}} \times \frac{308\,\text{K}}{1\,\text{atm}} \div 249.1\,\text{g Linuron} = 482\,\text{mL/g Linuron}$$

$ThCH_4 =$

$$4.25\,\text{mol} \times 82.06\frac{\text{mL}\cdot\text{atm}}{\text{mol}\cdot\text{K}} \times \frac{308\,\text{K}}{1\,\text{atm}} \div 249.1 = \text{g Linuron} = 431\,\text{mL/g Linuron}$$

A ready biodegradability test was set up for linuron under methanogenic conditions. The initial concentration of linuron was 50 mg/L. 100 mL liquid volume and 25 mL headspace were used in the test. Since initially 5 mg linuron was present this would theoretically yield 2.41 mL $ThCO_2$ and 2.155 mL $ThCH_4$ as shown in the figure below. The curves were generated assuming first-order biodegradation kinetics for linuron with a rate constant of 0.4/day. The results show that linuron passes the test within 3 days, as in 301A, since almost all of the initial substance (more than 60% as required in the test) is converted into end products. This happens because biomass synthesis is negligible under methanogenic conditions.

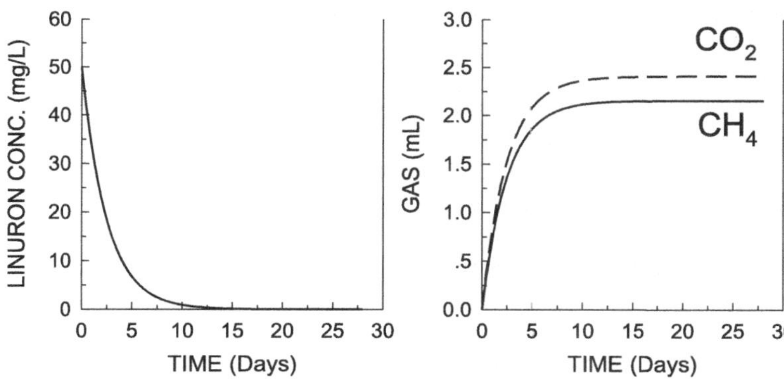

5.4.4.4 Shortcomings of screening (ready biodegradability) tests

Factors related with inoculum properties

Inoculum source: Inocula used in biodegradation tests may stem from different sources such as activated sludge, sewage effluent, surface waters, soils etc. (Kowalczyk *et al.* 2015). Although biodegradation strongly depends on the types of microorganisms, the inoculum used in tests is often unidentified, more so in the tests freely designed by the researcher. Yet, the source of inoculum appears to be an even more important factor affecting biodegradation rates than the concentration level of the pollutant (Ahtiainen *et al.* 2003).

The inoculum used in ready biodegradability tests is usually obtained from biological units of WWTPs. Given the fact that every treatment plant operates at different conditions and receives wastewater with different characteristics, community structure and composition of inocula may vary (Forney *et al.* 2001). The variations in inoculum may include presence of a specific microorganism that can degrade the test substance, pre-exposure of the inoculum to the test substance and the cell density. The presence and abundance of a microorganism that can degrade a certain hazardous chemical in an inoculum is related to the origin and history of that sample. This is especially evident for samples taken from treatment plants that are treating wastewater containing the test substance. A test prepared with an inoculum taken from such a source would have a better chance to pass biodegradability than an inoculum that was never exposed to it before. Thus, variations in the inoculum are the main challenge for the reproducibility of OECD tests. Such variations finally result in misinterpretation of the "ready biodegradability" criterion.

Concentration of inoculum: Another factor that affects the reproducibility of test results is abundance of the microorganism(s) that can degrade the test substance in the inoculum. The cell density in aerobic ready biodegradability tests is less than 30 mg SS/L (10^4–10^8 CFU/L). Given the fact that the actual biomass concentration of activated sludge is above 3000 mg SS/L, in such a case only 1% of the inoculum is used in the test. In a case where the abundance of the microorganisms that can degrade the test substance is less than 1%, the biodegradation test would fail even if a degrader is present in the original inoculum before dilution. In short, using low density inoculum would decrease the probability of encountering rare organisms that may degrade the test substance (Thouand *et al.* 2011).

Overall, when biodegradability of a hazardous substance is assessed, the test should be designed so that the probability of relatively rare specific degraders is high in the inoculum and its fraction should be large enough to enable the transformation of the test substance at the applied concentration during the test duration.

Microbial ecology of inoculum: Determination of microbial community structure and composition is not an easy procedure. However, recent advancements in next generation sequencing techniques enable rapid characterization of microbial communities based on the 16S rDNA gene. In addition, as described in **Chapter 7**, PhyloChips are used to rapidly identify the presence of more than 8000 strains of bacteria in complex samples based on microarray technology (DeSantis *et al.* 2007). However, functional characterization of the communities is still a challenge.

Given the fact that inoculum composition is the main reason for irreproducibility in OECD ready biodegradability tests, standardization of the inoculum is suggested to overcome this problem (Paixao *et al.* 2006). Paxiao and co-workers compared the performance of two commercial inocula and an inoculum prepared as a mixture of *Acinetobacter* sp., *Enterobacter* sp., *Staphylococcus capitis*, *Pseudomonas putida*, *Bacillus subtilis*, and *Bacillus cereus* with an activated sludge community, on biodegradation of reference chemicals used in OECD ready biodegradability tests (Paixao *et al.* 2006). The designed inoculum was as efficient as the activated sludge community suggesting that a reference inoculum can be designed and used for ready biodegradability tests.

On the other hand, a reference inoculum should be designed so that it contains species that can degrade a wide range of xenobiotic organic compounds (XOCs) and are abundant in the environment. Such an inoculum should contain *Pseudomonas* spp., *Arthrobacter* spp. followed by *Rhodococcus* spp. and *Burkholderia* spp. since as of November 2016, 20% of the microorganisms in the EAWAG-BBD are affiliated to the genus *Pseudomonas* which was involved in 90 pathways. The compounds metabolized represent aliphatic and aromatic hydrocarbons; aliphatic and aromatic acids; aliphatic and aromatic compounds with chloro, cyano, nitro, and phosphoryl groups substituted; and cycloaliphatic compounds. The genus with the second largest set of substrates metabolized is *Arthrobacter* followed by *Rhodococcus* and *Burkholderia* (Wackett & Hershberger, 2001).

Enrichment of nitrifiers: As shown in **Chapter 4**, in activated sludge or biofilm systems most of the pharmaceuticals or hormones are not removed directly by heterotrophs, but by the cometabolism provided by nitrifiers. This is because the AMO enzyme induced by nitrifiers leads to indirect oxidation of some micropollutants. However, in sludges taken from the biological units of treatment plants or existing laboratory-scale bioreactors, the fraction of nitrifiers is highly variable due to the differences in feeding and operating conditions. As known, the fraction of nitrifiers in a biological sludge depends significantly on parameters such as sludge age and temperature. In standard or freely-designed tests, the raw sludge is initially taken from a domestic/industrial WWTP and the fraction and activity of nitrifiers is often unknown. Therefore, the ability of sludge to degrade hazardous organics may vary greatly from one study to another as well as during the course of the same study, since experimental conditions are decisive for enrichment of such bacteria. Thus, if the removal of specific pollutants such as pharmaceuticals and hormones are of interest under aerobic conditions, it should be checked whether experimental conditions favor the growth of nitrifiers or not. The same idea may apply to (micro)pollutants that are directly or cometabolically removed under anoxic or anaerobic conditions.

Consideration of cometabolism: Although recent models dealing with biological removal of micropollutants account also for mechanisms other than primary utilization (**Chapter 4**), surprisingly, no standard methods have been established yet that take into account the removal as a secondary substrate or cometabolite.

On the other hand, most of the micropollutants studied today, particularly those found in domestic wastewater, may be removed by these mechanisms.

The key point here is how standard biodegradation experiments should be designed in future to distinguish between different mechanisms. For example, one of the issues to be studied is how to distinguish cometabolism or secondary substrate utilization from direct metabolism. An important factor here is the presence of bulk organic matter since the strength of the organic matter strongly affects biodegradation of hazardous pollutants. For example, researchers studied, in a Membrane Bioreactor (MBR), the exposure of biomass to estrone (E1) and alternate organic substrates where the aim was to determine whether cometabolism or multiple substrate utilization is an operating mechanism for the transformation of E1 and if feeding intervals affect the selection of E1-degrading bacteria. The results indicated that the presence of other organic substrates sped up the transformation of E1. The research demonstrated that E1 was removed by multiple substrate utilization (Tan *et al.* 2015).

Factors related with substrate properties

Concentration of test substance: One of the most important shortcomings of screening tests is that the concentration of the test substance is extremely high compared to the actual concentration of many hazardous pollutants in wastewater which is frequently below 10 μg/L (Gavrilescu *et al.* 2015; Kowalczyk *et al.* 2015; Petrie *et al.* 2015). Therefore, results obtained from screening tests performed at high concentrations may not be representative for systems receiving pollutants at very low concentrations. First of all, concentration of the test substance affects its biodegradation kinetics. The biodegradation rate is higher at high concentrations compared to low concentrations. In addition, induction of genes that synthesize the enzymes responsible for biodegradation of a chemical is achieved at a concentration above a certain level which is specific for the enzyme and its substrate. For instance, a certain test substance can be degraded in screening tests, but that substance may not be degraded in biological treatment units due to its low concentration (Kowalczyk *et al.* 2015).

Co-presence of hazardous organic pollutants and bulk organics: When biodegradation and/or inhibitory properties of hazardous substances are studied, mainly two different types of approaches are adopted:

(a) Screening tests are generally performed using a hazardous organic substance at high concentrations as the sole source of carbon. In most cases, however, hazardous substances are often used as cometabolites in the presence of primary substrates. Feeding a hazardous pollutant alone disregards this probability and may result in misinterpretation of biodegradability. Therefore, the fate of a hazardous substance in a wastewater treatment system cannot be adequately predicted with this approach. Also, in freely-designed experiments, hazardous organic substances are often fed to the bioreactor alone, excluding any other organic carbon source. This results usually in a low COD loading in a bioreactor, leading to a very high endogenous respiration due to the lack of sufficient carbon, making a correct evaluation difficult.

(b) Hazardous organic substances are fed to the bioreactor along with biogenic organic substrates. This ensures that a high organic loading rate is kept in the bioreactor. The presence of biogenic material may lead to transformation of hazardous compounds either as a secondary substrate or a cometabolite. Considering a real water and wastewater treatment system, this experimental approach is more realistic since in any case hazardous substances coexist with biogenic organic material. However, coexistence of hazardous and biogenic substances brings about a disadvantage too. The hazardous organic substance is less biodegradable than bulk organic matter and is often found at much lower concentrations. It may not be preferred as a primary substrate when other readily biodegradable organics are present at high concentrations. Therefore, its biodegradation may experience a lag phase or may take place at a reduced rate. As explained in **Chapter 4**, this is then observed as a kind of competition between hazardous organics and bulk organic matter.

Ratio of hazardous pollutants to inoculum: Standard or freely-designed tests carried out in activated sludge units also enable the calculation of kinetic constants belonging to biodegradation of the test substance. But, often variable kinetic parameters are obtained for individual substances. This variability is often attributed to the changes taking place in the physiological state of inoculum (biomass), as discussed by Grady and co-workers (McAvoy *et al.* 1998).

Depending on the type of biodegradation assay employed and on the physiological state of bacteria, two types of kinetic parameter estimates, namely intrinsic and extant, may be obtained in the case of XOCs (Grady *et al.* 2011). As shown in Table 5.9 intrinsic parameters are obtained when the initial substrate to biomass (S_0/X_0) ratio is high and bacteria grow under unrestricted conditions using an XOC as the sole carbon and energy source. In such cases the estimated kinetic parameters are unaffected by the history of biomass. This might, however, not represent the real conditions, since the microorganism's physiological conditions will differ from the initial state. On the other hand, extant conditions are present when the S_0/X_0 is kept rather low, more closely representing the conditions in a real biological treatment system. For example, in an activated sludge tank, the mixed liquor would have a concentration of 3000 mg SS/L whereas the hazardous pollutant might exist at 100 μg/L. Thus, this would result in an extremely low S_0/X_0 ratio.

Table 5.9 Characteristics of intrinsic and extant biodegradation tests.

Initial Substrate to Biomass Ratio (S_0/X_0) (COD Basis)	Definition	Characteristics
High: 20	Intrinsic (McAvoy *et al.* 1998)	• High substrate concentration. • Unrestricted biomass growth, physiological change in biomass. • Reflecting the inherent (potential) capacity of microorganisms to degrade the substrate. • Calculated kinetic parameters are independent of culture history.
Low: 0.025	Extant	• Low substrate concentration. • Biomass does not synthetize additional enzymes or other proteins, reflection of the original physiological status. • Substrate removal depends on growth history and culture conditions. • Calculated kinetic parameters reflect the culture history.

Thus, not only the concentration range of an organic pollutant is important, but also its ratio to biomass. However, in most standard biodegradability tests there is no indication under which S_0/X_0 conditions the biodegradation parameters have to be found. As discussed in Section 5.4.4. the initially developed ready biodegradation tests suggest that the initial S_0/X_0 is kept at a rather high level, more in the direction to favor intrinsic conditions. By analyzing the ATP content, in a study it was shown that high concentration regimes (20 mg C/L) in standard test conditions with activated sludge appeared to yield exponential growth of degraders (Ahtiainen *et al.* 2003). However, in newer simulation tests, this shortcoming seems to be overcome. In simulation tests the S_0/X_0 is likely to be low, favoring more extant conditions. Yet, in most freely-designed experiments conducted in batch or continuous-flow reactors, not enough attention is paid to the S_0/X_0 ratio when kinetic parameters are found. This is often the reason why great discrepancies are observed from one study to another.

Respirometry is often used as a tool in determination of kinetic parameters. In that regard, Ellis and co-workers developed a respirometry method which measures the kinetic parameters under the conditions close to a real activated sludge system, namely under extant conditions (Ellis *et al.* 1996). This is accomplished by allowing the respirometer to operate at conventional food to microorganism (F/M) ratios (Ellis *et al.* 1996; Magbanua *et al.* 2003; Magbanua *et al.* 2004). In recent years extant kinetic tests have also been applied to the micropollutants such as EE2 found in domestic wastewater (Khunjar *et al.* 2011).

5.4.5 Inherent biodegradability tests

Inherent tests are designed for testing the primary or ultimate biodegradability of test chemicals (Kowalczyk *et al.* 2015). In these tests, specific and non-specific analyses are employed to assess primary and ultimate biodegradation, respectively. The test procedures allow prolonged exposure of microorganisms to a test substance and a low ratio of the test substance to biomass, which offer a better chance to obtain a positive result compared to tests for ready biodegradability. Some of these tests may be conducted using microorganisms that have previously been exposed to the test substance, which frequently results in adaptation, leading to a significant increase of degradation rate.

5.4.6 Simulation tests

Simulation tests aim to assess the rate and extent of biodegradation in a laboratory system that represents either the aerobic treatment stage of a WWTP or environmental compartments, such as fresh or marine surface water (Kowalczyk *et al.* 2015).

5.4.6.1 Simulation tests in suspended-growth systems

An example of suspended-growth tests is the simulation test OECD 303A which determines the biodegradation of water-soluble organic compounds in a continuously operated test system simulating the activated sludge process. In this test the aim is to determine the elimination and the primary and/or ultimate biodegradation of water-soluble organic compounds by aerobic microorganisms. The test substance is added at a dissolved organic carbon concentration of 10–20 mg DOC/L in addition to the easily biodegradable organic matter in synthetic sewage which has a strength of 100 mg DOC/L (OECD Simulation Test - Aerobic Sewage Treatment: 303 A-Activated Sludge Units). This test was initially designed to examine the biodegradability of surfactants which were found in the indicated range. However, as stated in OECD guidelines, actually in sewage few substances are present at these concentrations. In the 303A test, synthetic sewage, domestic sewage or a mixture of both is permissible as the organic medium. However, the use of domestic sewage is preferred over synthetic medium since this was shown to lead to a better biodegradability. The test also takes into consideration the adsorption of the test substance onto activated sludge. In the case of poorly water-soluble and volatile test substances, the application of those simulations is quite problematic and might yield variable results.

OECD offers also other simulation tests for aerobic (314B) and anaerobic (314C) biodegradability of chemicals. Some of the shortcomings mentioned for ready biodegradability tests, in particular those related with substrate, are overcome in these tests. These tests simulate the biodegradation of a test substance in activated sludge and anaerobic digesters using native biomass at the concentrations present in the reactors, and media at a typical temperature which represents the particular environment. A low concentration of test substance is used in tests designed to determine its biodegradation. Due to the low concentration, the degradation of most hazardous organic pollutants follows first-order kinetics. On the other hand, higher concentrations are normally used for identification and quantification of major transformation products for analytical reasons.

In these tests, the initial concentration of the test substance ranges between 1 μg/L and 100 μg/L which is low enough to ensure that the biodegradation kinetics obtained in the test reflect those expected in the environment being simulated as well as being higher than the limit of detection of the instrument used for the measurement of the chemical or the by-products. Biodegradation is measured either by radiolabeling techniques or using specific instrumental methods. Shortcomings of screening tests (ready biodegradability), like low biomass density, high initial test substance concentration and indirect measurement of biodegradation by non-specific parameters like DOC (Section 5.4.4), are eliminated in simulation tests.

5.4.6.2 Simulation tests in attached-growth (biofilm) systems

There are few tests evaluating the biodegradability of test substances in biofilms like the OECD simulation test 303B. These tests are applied when information is required about the behavior of a test substance in wastewater systems that involve biofilm reactors such as trickling filters, rotating biological contactors, or fluidized beds.

Compared to suspended-growth systems, the start-up of biofilm systems is more tedious since a longer time is needed for the initial development of a biofilm layer and a steady-operation. The main difficulty in these tests seems to be that conditions leading to effective biofilm formation are not well defined. Different feed conditions would lead to the development of biofilm layers that are occupied by different microorganisms. In suspended-growth systems, like activated sludge, there is no stratification of microorganisms whereas in biofilms microorganisms occupy different layers. For example, in aerobic biofilms, nitrifiers are usually found next to heterotrophs that occupy the uppermost layer. Therefore, suspended- and attached-growth systems may also respond differently to inhibitory hazardous pollutants. Variations in biofilm thickness may be another important factor. In addition, analysis of experimental results is more difficult in biofilm systems due to the added complexity of pollutant diffusion and reaction kinetics inside the biofilm.

5.5 REFERENCES

Ahtiainen J., Aalto M. and Pessala P. (2003). Biodegradation of chemicals in a standardized test and in environmental conditions. *Chemosphere*, **51**(6), 529–537.

Akunna J. C., Bizeau C. and Moletta R. (1992). Denitrification in anaerobic digesters: Possibilities and influence of wastewater COD/N-NOX ratio. *Environmental Technology*, **13**(9), 825–836.

Akunna J. C., Bizeau C. and Moletta R. (1993). Nitrate and nitrite reductions with anaerobic sludge using various carbon sources: glucose, glycerol, acetic acid, lactic acid and methanol. *Water Research*, **27**(8), 1303–1312.

Alpaslan Kocamemi B. (2005). Cometabolic Degradation of Trichloroethylene (TCE) and 1, 2-Dichloroethane (1, 2-DCA) in Nitrification Systems. Ph.D.Thesis, Boğaziçi University, Istanbul.

Alpaslan Kocamemi B. and Çeçen F. (2005). Cometabolic degradation of TCE in enriched nitrifying batch systems. *Journal of Hazardous Materials*, **125**(1–3), 260–265.

Alpaslan Kocamemi B. and Çeçen F. (2009). Biodegradation of 1,2-dichloroethane (1,2-DCA) by cometabolism in a nitrifying biofilm reactor. *International Biodeterioration & Biodegradation*, **63**(6), 717–726.

Alpaslan Kocamemi B. and Çeçen F. (2010). Cometabolic degradation and inhibition kinetics of 1, 2-dichloroethane (1, 2-DCA) in suspended-growth nitrifying systems. *Environmental Technology*, **31**(3), 295–305.

Amor L., Eiroa M., Kennes C. and Veiga M. C. (2005). Phenol biodegradation and its effect on the nitrification process. *Water Research*, **39**(13), 2915–2920.

Bailey J. E. and Ollis D. F. (1986). Biochemical Engineering Fundamentals. McGraw-Hill, New York, USA.

Bers K., Leroy B., Breugelmans P., Albers P., Lavigne R., Sorensen S. R., Aamand J., De Mot R., Wattiez R. and Springael D. (2011). A novel hydrolase identified by genomic-proteomic analysis of phenylurea herbicide mineralization by Variovorax sp. strain SRS16. *Applied and Environmental Microbiology*, **77**(24), 8754–8764.

Bers K., Batisson I., Proost P., Wattiez R., De Mot R. and Springael D. (2013). Hyla, an alternative hydrolase for initiation of catabolism of the phenylurea herbicide linuron in Variovorax sp. strains. *Applied and Environmental Microbiology*, **79**(17), 5258–5263.

Bitton G. and Koopman B. (1992). Bacterial and Enzymatic Bioassays for Toxicity Testing in the Environment. In: Reviews of Environmental Contamination and Toxicology, Ware G. W. (ed.), Vol. 125, Springer Verlag, New York Inc, pp. 1–22.

Boethling R. S. and Lynch D. G. (2007). Biodegradation of US premanufacture notice chemicals in OECD tests. *Chemosphere*, **66**(4), 715–722.

Boethling R. S., Sommer E. and DiFiore D. (2007). Designing small molecules for biodegradability. *Chemical Reviews*, **107**(6), 2207–2227.

Breugelmans P., Leroy B., Bers K., Dejonghe W., Wattiez R., De Mot R. and Springael D. (2010). Proteomic study of linuron and 3,4-dichloroaniline degradation by Variovorax sp. WDL1: evidence for the involvement of an aniline dioxygenase-related multicomponent protein. *Research in Microbiology*, **161**(3), 208–218.

Brunet R. C. and Garcia-Gil L. J. (1996). Sulfide-induced dissimilatory nitrate reduction to ammonia in anaerobic freshwater sediments. *FEMS Microbiology Ecology*, **21**(2), 131–138.

Caspi R., Billington R., Ferrer L., Foerster H., Fulcher C. A., Keseler I. M., Kothari A., Krummenacker M., Latendresse M., Mueller L. A., Ong Q., Paley S., Subhraveti P., Weaver D. S. and Karp P. D. (2016). The MetaCyc database of metabolic pathways and enzymes and the BioCyc collection of pathway/genome databases. *Nucleic Acids Research*, **44**(D1), D471–D480.

Cetecioglu Z., Ince B., Orhon D. and Ince O. (2012). Acute inhibitory impact of antimicrobials on acetoclastic methanogenic activity. *Bioresource Technology*, **114**, 109–116.

Cornish-Bowden A. (2012). Fundamentals of Enzyme Kinetics, 4th edn, Wiley-VCH. Weinheim, Germany.

Dalzell D. J. and Christofi N. (2002). An ATP luminescence method for direct toxicity assessment of pollutants impacting on the activated sewage sludge process. *Water Research*, **36**(6), 1493–1502.

Dapena-Mora A., Fernandez I., Campos J. L., Mosquera-Corral A., Mendez R. and Jetten M. (2007). Evaluation of activity and inhibition effects on Anammox process by batch tests based on the nitrogen gas production. *Enzyme and Microbial Technology*, **40**(4), 859–865.

DeSantis T. Z., Brodie E. L., Moberg J. P., Zubieta I. X., Piceno Y. M. and Andersen G. L. (2007). High-density universal 16S rRNA microarray analysis reveals broader diversity than typical clone library when sampling the environment. *Microbial Ecology*, **53**(3), 371–383.

Ellis T. G., Barbeau D. S., Smets B. F. and Grady Jr. C. P. L. (1996). Respirometric technique for determination of extant kinetic parameters describing biodegradation. *Water Environment Research*, **68**(5), 917–926.

EPA (2012). Ecological effects test guidelines. OCSPP 850.3300: modified activated sludge, respiration. Inhibition test.

Forney L. J., Liu W. T., Guckert J. B., Kumagai Y., Namkung E., Nishihara T. and Larson R. J. (2001). Structure of microbial communities in activated sludge: potential implications for assessing the biodegradability of chemicals. *Ecotoxicology and Environmental Safety*, **49**(1), 40–53.

Gao J., Ellis L. B. and Wackett L. P. (2010). The University of Minnesota Biocatalysis/Biodegradation Database: improving public access. *Nucleic Acids Research*, **38**(Database issue), D488–D491.

Gartiser S., Urich E., Alexy R. and Kümmerer K. (2007). Anaerobic inhibition and biodegradation of antibiotics in ISO test schemes. *Chemosphere*, **66**(10), 1839–1848.

Gavrilescu M., Demnerova K., Aamand J., Agathos S. and Fava F. (2015). Emerging pollutants in the environment: present and future challenges in biomonitoring, ecological risks and bioremediation. *N Biotechnology*, **32**(1), 147–156.

Grady Jr. C. L., Daigger G. T., Love N. G. and Filipe C. D. (2011). Biological Wastewater Treatment. CRC press, Boca Raton, FL.

Guerra P., Kim M., Teslic S., Alaee M. and Smyth S. A. (2015). Bisphenol-A removal in various wastewater treatment processes: operational conditions, mass balance, and optimization. *Journal of Environmental Management*, **152**, 192–200.

Henze M., van Loosdrecht M. C., Ekama G. A. and Brdjanovic D. (2008). Biological Wastewater Treatment. IWA Publishing, London.

Hoffmann C. and Christofi N. (2001). Testing the toxicity of influents to activated sludge plants with the Vibrio fischeri bioassay utilising a sludge matrix. *Environmental Toxicology*, **16**(5), 422–427.

Horvath R. S. (1972). Microbial co-metabolism and the degradation of organic compounds in nature. *Bacteriological Reviews*, **36**(2), 146.

ISO (1998). 11348–1-Water quality – Determination of the inhibitory effect of water samples on the light emission of Vibrio fischeri (Luminescent bacteria test) – Part 1: Method using freshly prepared bacteria.

ISO (1999). 15522-Water quality-Determination of the inhibitory effect of water constituents on the growth of activated sludge microorganisms.

ISO (2003). 13641–1 Water quality – Determination of inhibition of gas production of anaerobic bacteria.

ISO (2006). 9509-Water quality – Toxicity test for assessing the inhibition of nitrification of activated sludge microorganisms.

ISO (2007). 8192-Water quality – Test for inhibition of oxygen consumption by activated sludge for carbonaceous and ammonium oxidation.

Kakkar T., Pak Y. and Mayersohn M. (2000). Evaluation of a minimal experimental design for determination of enzyme kinetic parameters and inhibition mechanism. *Journal of Pharmacology and Experimental Therapeutics*, **293**(3), 861–869.

Khunjar W. O., Mackintosh S. A., Skotnicka-Pitak J., Baik S., Aga D. S. and Love N. G. (2011). Elucidating the relative roles of ammonia oxidizing and heterotrophic bacteria during the biotransformation of 17alpha-Ethinylestradiol and Trimethoprim. *Advances in Environmental Science and Technology*, **45**(8), 3605–3612.

Kim Y., Arp D. J. and Semprini L. (2002). A combined method for determining inhibition type, kinetic parameters, and inhibition coefficients for aerobic cometabolism of 1, 1, 1-trichloroethane by a butane-grown mixed culture. *Biotechnology and Bioengineering*, **77**(5), 564–576.

Knowles R. (1982). Denitrification. *Microbiology Reviews*, **46**(1), 43–70.

Kolvenbach B., Schlaich N., Raoui Z., Prell J., Zuhlke S., Schaffer A., Guengerich F. P. and Corvini P. F. (2007). Degradation pathway of bisphenol A: does ipso substitution apply to phenols containing a quaternary alpha-carbon structure in the para position? *Applied and Environmental Microbiology*, **73**(15), 4776–4784.

Kowalczyk A., Martin T. J., Price O. R., Snape J. R., van Egmond R. A., Finnegan C. J., Schafer H., Davenport R. J. and Bending G. D. (2015). Refinement of biodegradation tests methodologies and the proposed utility of new microbial ecology techniques. *Ecotoxicology and Environmental Safety*, **111**, 9–22.

Lappalainen J., Juvonen R., Nurmi J. and Karp M. (2001). Automated color correction method for Vibrio fischeri toxicity test. Comparison of standard and kinetic assays. *Chemosphere*, **45**(4), 635–641.

Loosdrecht M. C. M. V., Nielsen P. H., Lopez-Vazquez C. M. and Brdjanovic D. (eds). (2016). Experimental Methods in Wastewater Treatment. IWA Publishing, London, UK.

Lotti T., Cordola M., Kleerebezem R., Caffaz S., Lubello C. and van Loosdrecht M. (2012). Inhibition effect of swine wastewater heavy metals and antibiotics on Anammox activity. *Water Science and Technology*, **66**(7), 1519–1526.

Magbanua B. S. J., Smets B. F., Bowyer R. L., Rodieck A. G., Sanders R. W., Sowers W. W., Stolze S. B. and Grady C. P. L. (2003). Relative efficacy of intrinsic and extant parameters for modeling biodegradation of synthetic organic compounds in activated sludge: steady-state systems. *Water Environment Research*, **75**(2), 126–137.

Magbanua J. B. S., Stanfill J. C., Fehniger S. M., Smets B. F., Farkas F. and Grady Jr. C. P. L. (2004). Relative efficacy of intrinsic and extant parameters for modeling biodegradation of synthetic organic compounds in activated sludge: dynamic systems. *Water Environment Research*, **76**(3), 256–267.

Makuch A., Medrzycka K. and Plaza E. (2010). The effect of sulphonamides on activated sludge dehydrogenase activity. In: Research and Application of New Technologies in Wastewater Treatment and Municipal Solid Waste Disposal in Ukraine, Sweden and Poland: Proceedings of a Polish-Swedish-Ukrainian Seminar, September 23–25, 2010. Plaza E. and Levlin E. (eds), pp. 113–122.

Mavrovouniotis M. L. (1990). Group contributions for estimating standard gibbs energies of formation of biochemical-compounds in aqueous-solution. *Biotechnology and Bioengineering*, **36**(10), 1070–1082.

McAvoy D. C., Grady Jr. C. P. L., Blok J., Feijtel T. C. J., Federle T. W. and Larson R. J. (1998). A simplified modeling approach using microbial growth kinetics for predicting exposure concentrations of organic chemicals in treated wastewater effluents. *Chemosphere*, **36**(10), 2291–2304.

OECD (1992). Test No. 301: Ready Biodegradability. OECD Publishing, Paris.

OECD (2006a). Revised Introduction to the OECD Guidelines for Testing of Chemicals, Section 3. OECD Publishing, Paris.

OECD (2006b). Test No. 311: Anaerobic Biodegradability of Organic Compounds in Digested Sludge: By Measurement of Gas Production. OECD Publishing, Paris.

OECD (2007). Test No. 224: Determination of the Inhibition of the Activity of Anaerobic Bacteria. OECD Publishing, Paris.

OECD (2010). Test No. 209: Activated Sludge, Respiration Inhibition Test (Carbon and Ammonium Oxidation). OECD Publishing, Paris.

OECD (2014). Test No. 310: Ready Biodegradability – CO_2 in Sealed Vessels (Headspace Test). OECD Publishing, Paris.

Paixao S. M., Saagua M. C., Tenreiro R. and Anselmo A. M. (2006). Biodegradability testing using standardized microbial communities as inoculum. *Environmental Toxicology*, **21**(2), 131–140.

Pala-Ozkok I. and Orhon D. (2013). Chronic effect of erythromycin on substrate biodegradation kinetics of activated sludge. *Biochemical Engineering Journal*, **81**, 29–39.

Petrie B., Barden R. and Kasprzyk-Hordern B. (2015). A review on emerging contaminants in wastewaters and the environment: current knowledge, understudied areas and recommendations for future monitoring. *Water Research*, **72**, 3–27.

Ren S. (2004). Assessing wastewater toxicity to activated sludge: recent research and developments. *Environment International*, **30**(8), 1151–1164.

Rittmann B. E. and McCarty P. L. (2001). Environmental Biotechnology: Principles and Applications. ISBN-13: 978–0072345537. McGraw Hill, New York.

Rizzo L. (2011). Bioassays as a tool for evaluating advanced oxidation processes in water and wastewater treatment. *Water Research*, **45**(15), 4311–4340.

Rücker C. and Kümmerer K. (2012). Modeling and predicting aquatic aerobic biodegradation – a review from a user's perspective. *Green Chemistry*, **14**(4), 875.

Semerci N. and Çeçen F. (2009). Effect of continuous Cd feeding on the performance of a nitrification reactor. *Biodegradation*, **20**(2), 155–164.

Smith L. H., Kitanidis P. K. and McCarty P. L. (1997). Numerical modeling and uncertainties in rate coefficients for methane utilization and TCE cometabolism by a methane-oxidizing mixed culture. *Biotechnology and Bioengineering*, **53**(3), 320–331.

Tan D. T., Arnold W. A. and Novak P. J. (2015). Effects of estrone and organic carbon exposure on the transformation of estrone. *Environmental Science: Water Research & Technology*, **1**(4), 457–464.

Tezel U. and Pavlostathis S. G. (2009). Transformation of benzalkonium chloride under nitrate reducing conditions. *Environmental Science and Technology*, **43**(5), 1342–1348.

Tezel U., Pierson J. A. and Pavlostathis S. G. (2006). Fate and effect of quaternary ammonium compounds on a mixed methanogenic culture. *Water Research*, **40**(19), 3660–3668.

Tezel U., Tandukar M., Martinez R. J., Sobecky P. A. and Pavlostathis S. G. (2012). Aerobic biotransformation of n-tetradecylbenzyldimethylammonium chloride by an enriched *Pseudomonas* spp. community. *Environmental Science and Technology*, **46**(16), 8714–8722.

Thauer R. K., Jungermann K. and Decker K. (1977). Energy conservation in chemotrophic anaerobic bacteria. *Bacteriology Reviews*, **41**(1), 100–180.

Thouand G., Durand M., Maul A., Gancet C. and Blok H. (2011). New concepts in the evaluation of biodegradation/persistence of chemical substances using a microbial inoculum. *Frontiers in Microbiology*, **2**(164).

Tobajas M., Verdugo V., Polo A. M., Rodriguez J. J. and Mohedano A. F. (2016). Assessment of toxicity and biodegradability on activated sludge of priority and emerging pollutants. *Environmental Technology*, **37**(6), 713–721.

Toyama T., Momotani N., Ogata Y., Miyamori Y., Inoue D., Sei K., Mori K., Kikuchi S. and Ike M. (2010). Isolation and characterization of 4-tert-butylphenol-utilizing Sphingobium fuliginis strains from Phragmites australis rhizosphere sediment. *Applied and Environmental Microbiology*, **76**(20), 6733–6740.

Volskay Jr. V. and Grady Jr. C. L. (1988). Toxicity of selected RCRA compounds to activated sludge microorganisms. *Journal (Water Pollution Control Federation)*, **60**(10), 1850–1856.

Wackett L. P. and Hershberger C. D. (2001). Biodegradation and Biocatalysis. ASM Press, 1752 N St., N.W., Washington, DC 20036–2904, U.S.A.

Wickert J., Lorsbach T., Gutlein M., Schmid E., Latino D., Kramer S. and Fenner K. (2016). enviPath – The environmental contaminant biotransformation pathway resource. *Nucleic Acids Research*, **44**(D1), D502–D508.

Zhang Z. Z., Zhang Q. Q., Xu J. J., Deng R., Ji Z. Q., Wu Y. H. and Jin R. C. (2016). Evaluation of the inhibitory effects of heavy metals on anammox activity: a batch test study. *Bioresource Technology*, **200**, 208–216.

Chapter 6

Removal of hazardous pollutants in full-scale wastewater treatment plants

Wenqing Xu and Metin Duran

Villanova University, Civil and Environmental Engineering Department, 800 Lancaster Ave., Villanova, 19085, Pennsylvania, USA;
E-mail: wenqing.xu@villanova.edu; metin.duran@villanova.edu

6.1 INTRODUCTION

The advancements in analytical chemistry including extraction, chromatography, and detection technologies have enabled environmental engineers to detect and quantify a wide range of hazardous pollutants at concentration levels as low as nanograms per litre (ng/L). Although municipal wastewater treatment plants (WWTPs) are not necessarily designed to remove these pollutants, some of them can be removed through biological and physicochemical processes that include transformation and phase change. In this chapter, the fate of major classes of hazardous pollutants through full-scale WWTPs are reviewed. The classes of the hazardous pollutants reviewed include pharmaceuticals and personal care products, volatile organic compounds, agricultural pollutants, and surfactants. Other hazardous pollutants, such as nanomaterials and X-ray contrast medium, are also reviewed under the "others" category, mainly due to the limited literature on these pollutants in full-scale WWTPs.

The emphasis of this chapter was primarily on the degree of pollutant removal during the treatment processes in full-scale WWTPs. Both influent and effluent concentrations of pollutants were summarized when data were available. Only limited literature is available where the removal mechanisms are investigated and such information is included in this chapter. Moreover, reported investigations

 Hazardous Pollutants in Biological Treatment Systems: Fundamentals and a Guide to Experimental Research
Ferhan Çeçen, Ulaş Tezel
doi: 10.2166/9781780407708_239

focused on the activated sludge (AS) process in most cases. A limited body of literature has reported data on waste stabilization ponds and constructed wetlands and those are reviewed in the corresponding subcategory of hazardous pollutants. The reported results that are reviewed in this chapter are summarized in Table 6.1.

A critical analysis of the reported data was intentionally avoided due to significant discrepancies among the data reported by different researchers. Similarly, a quantitative assessment of the process types and operating conditions on the fate of these pollutants was not possible due to lack of detailed information on the processes and operating conditions under which the data were reported.

6.2 PHARMACEUTICALS AND PERSONAL CARE PRODUCTS

Pharmaceuticals and personal care products (PPCPs) are a group of hazardous organic compounds that are introduced into the environment via anthropogenic activities. High concentrations of PPCPs are often observed in wastewater influent, including estrogens and endocrine disrupting compounds, antibiotics, and antibacterial sanitizers. Herein, we review the occurrence and removal efficiencies of pharmaceuticals and hormones in full-scale WWTPs using conventional treatment technologies. Alternative treatment technologies are also discussed, including AS with nitrogen removal and membrane bioreactors.

6.2.1 Estrogens and other endocrine disrupting compounds

Endocrine disrupting compounds (EDCs) are a group of hazardous organic compounds with estrogenic or androgenic activity. They are believed to elicit negative effects on the endocrine systems and can cause adverse reproductive and developmental effects in wildlife and humans (Crisp *et al.* 1997; Knobil *et al.* 1999). EDCs include both natural products (e.g. estrone (E1), 17 β-estradiol (E2), and estriol (E3) and synthetic chemicals such as 17 α-ethinylestradiol (EE2) (Blair *et al.* 2000). During the past decade or so, significant research endeavors have been made to better understand the occurrence and behavior of EDCs in the environment. It has been reported that EDCs are often present in the environment at very low concentrations (μg/L and ng/L), including in wastewaters, rivers, streams, and groundwater (Ternes *et al.* 1999; Snyder *et al.* 2003; Fernandez *et al.* 2007). Numerous evidences have suggested that WWTPs are one of the main sources for EDCs (Kolpin *et al.* 2002). Therefore, it is important to understand the relationship between WWTPs treatment processes and removal efficiencies of EDCs in order to reduce and eliminate the occurrence of EDCs in the environment. Herein, we reviewed the data from full-scale WWTPs to identify the most persistent and/or potent EDCs, the current available best practice, and the optimal treatment conditions for the removal of EDCs.

Table 6.1 Summary of hazardous pollutants through full-scale municipal wastewater treatment plants.

Pollutant	Category	Process	Concentration (ng/L)		Removal (%)	Reference
			Influent	Effluent		
estrone	PPCP	AS*			83	Ternes *et al.* (1999)
		AS*	48		75–98	Servos *et al.* (2005)
		AS*			61	Baronti *et al.* (2000)
		AS			98	Anderson *et al.* (2003)
		AS*			90	Joss *et al.* (2004)
17 β-estradiol	PPCP	AS*			99.9	Ternes *et al.* (1999)
		AS*	15.6		0–98	Servos *et al.* (2005)
		AS*			87	Baronti *et al.* (2000)
		AS			98	Anderson *et al.* (2003)
		AS*			90	Joss *et al.* (2004)
estriol	PPCP	AS*			95	Baronti *et al.* (2000)
ethinyl estradiol	PPCP	AS*			78	Ternes *et al.* (1999)
		AS*			90	Joss *et al.* (2004)
ciprofloxacin	PPCP	AS	3,800	600		Watkinson *et al.* (2007)
		AS*			18–55	Li and Zhang (2011)
lincomycin	PPCP	AS	60	50		Watkinson *et al.* (2007)
sulfamethoxazole	PPCP	AS	360	270		Watkinson *et al.* (2007)
		AS*			62–92	Li and Zhang (2011)
trimethoprim	PPCP	AS	340	50	13–42	Watkinson *et al.* (2007)
roxithromycin	PPCP	AS*			40–46	Li and Zhang (2011)
erythromycin-H_2O	PPCP	AS*			15–26	Li and Zhang (2011)
triclosan		AS	800	250	69	Jim *et al.* (2006)
		AS*			85	Singer *et al.* (2002)
benzophenone-3	PPCP	AS*	700–7,800		68–98	Balmer *et al.* (2005)
4-methylbenzylidene camphor	PPCP	AS*	600–6,500		18–82	Balmer *et al.* (2005)
ethylhexyl methoxycinnamate	PPCP	AS*	500–19,000		88–99	Balmer *et al.* (2005)

(*Continued*)

Table 6.1 Summary of hazardous pollutants through full-scale municipal wastewater treatment plants (*Continued*).

Pollutant	Category	Process	Concentration (ng/L)		Removal (%)	Reference
			Influent	Effluent		
octocrylene	PPCP	AS*	100–12,000		97–99	Balmer *et al.* (2005)
chlorfenvinphos	Agricultural	AS*		24		Ormad *et al.* (2011)
erbuthylazine	Agricultural	AS*		1,310		Ormad *et al.* (2011)
Pyrethroid	Agricultural	Wetland			52–94	Budd et al. (2009)
linear alkylbenzene sulfonates	Surfactant	AS*			>99	Brunner *et al.* (1988)
		AS*	15.7×10^6	0.12×10^6		Prats *et al.* (1997)
alcohol ethoxylates	Surfactant	AS*	5.4×10^6	0.38×10^6	95	Prats *et al.* (1997)
DTDMAC	Surfactant	AS*	175,000–334,000	28,000–56,000		We (1984)
SAS	Surfactant	AS	598,000–76,000	<1,000–14,000		Field *et al.* (1995)
PFOS	Surfactant	AS		26 (±2.0)		Boulanger *et al.* (2005)
PFOA	Surfactant	AS		22 (±2.1)		Boulanger *et al.* (2005)
N-EtFOSAA	Surfactant	AS	5.1 (±0.8)	3.6 (±0.2)		Boulanger *et al.* (2005)
TiO_2	Other	AS*	100×10^3–$3{,}000 \times 10^3$	5×10^3–15×10^3		Kiser *et al.* (2009)
		AS*	181×10^3–$1{,}233 \times 10^3$	$<25 \times 10^3$		Westerhoff *et al.* (2011)
benzotriazole	Other	AS	5,760		7–27	Liu *et al.* (2012)
		AS*			30–50	Reemtsma *et al.* (2010)
5-TTri	Other	AS	6,758		60–100	Liu *et al.* (2012)
		AS*			20–70	Reemtsma *et al.* (2010)
CBT	Other	AS	1,130		60–100	Liu *et al.* (2012)
XTri	Other	AS	1,313		60–100	Liu *et al.* (2012)
PBDEs	Other	AS with tertiary			93	Rayne and Ikonomou (2005)
chlorinated OPFRs	Other	AS	570–5,800		Zero	Meyer and Bester (2004)
non-chlorinated OPFRs	Other	AS*	2,400–6,100		57–86	Meyer and Bester (2004)

*Multiple plants were investigated.

WW: Wastewater; PPCP: Pharmaceuticals and personal care products; AS: Activated sludge; DTDMAC: ditallow dimethyl ammonium chloride; SAS: Secondary alkane sulfonates; PFOS: perfluorooctane sulfonates; PFOA perfluorooctanate; N-EtFOSAA: 2-(N-ethyl-perfluorooctanesulfonamidoacetic) acid; 5-TTri: 5-methylbenzotriazole; CBT: 5-chloro-benzotriazole; XTri: 5,6-dimethyl-benzotriazole; PBDEs: polybrominated diphenyl ethers; OPFR: organophosphorus flame retardants.

Previous studies on EDCs suggested that most EDCs were not completely removed by existing WWTPs (Liu *et al.* 2009). Among all of the EDCs that were investigated, three natural estrogens (E1, E2, E3) and one synthetic estrogen (EE2) were intensively studied in the literature and their chemical structures are shown in Figure 6.1. As early as 1999, Ternes and collaborators investigated the occurrence of estrogens in WWTPs in Germany, Canada, and Brazil. Samples were collected and concentrations of estrogens were measured from both influent and effluent via chemical analysis using liquid chromatography and tandem mass spectrometry (LC-MS/MS). The removal efficiencies were 83%, 99.9%, and 78% for E1, E2, and EE2 in WWTPs with AS processes, respectively (Ternes *et al.* 1999). Servos and co-workers found a similar trend by analyzing E1 and E2 in wastewater samples from 18 WWTPs in Canada (Servos *et al.* 2005). The detected concentrations of E1 and E2 were 48 ng/L and 15.6 ng/L in raw sewage, respectively. High removal efficiencies were reported (>75% or as high as 98%) for E2 in most of the conventional WWTPs with AS processes. However, the removal of E1 was more complex, where the removal efficiencies varied from 98% to zero or even negative values (Servos *et al.* 2005). Baronti and collaborators reported similar observations by investigating six WWTPs in Italy, where 95% of E3, 87% of E2, and 61% of E1 were removed. Interestingly, E1 concentrations in the effluent were somehow higher than the influent, which the authors attributed to the conversion from E2 to E1 via hydrolysis (Baronti *et al.* 2000; Servos *et al.* 2005). Anderson and collaborators also observed high removal efficiencies of E1 and E2 (above 98%), while EE2 elimination was slightly lower (Andersen *et al.* 2003).

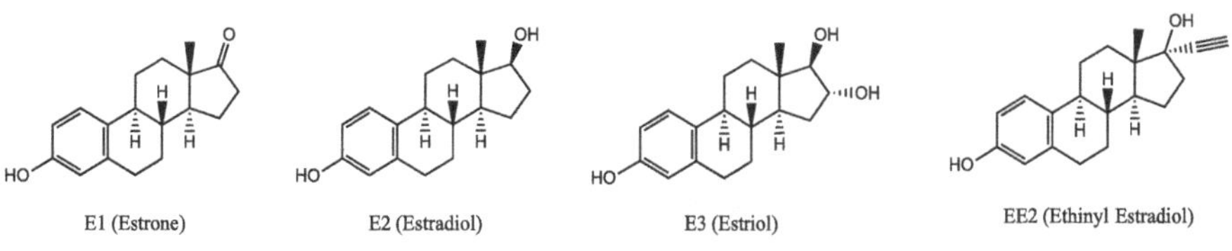

Figure 6.1 Chemical structures of three natural (E1, E2, and E3) and one synthetic (EE2) estrogens.

Various treatment processes were compared to identify the best practice in removing EDCs in full-scale WWTPs. Svenson and collaborators investigated 20 WWTPs with different treatment processes (AS process, trickling filter, and chemical precipitation) in Sweden. Their results suggested that the AS process was more efficient than trickling filters, while chemical precipitation showed little removal by itself (Svenson *et al.* 2003). Similar findings were reported by other groups where the AS processes consistently outperformed biofilm systems (Johnson *et al.* 2004; Chimchirian *et al.* 2007). Joss and co-workers reported over 90% removal of E1, E2, and EE2 in WWTPs with AS processes (Joss *et al.* 2004). Efforts have also been made to identify the key operating parameters of WWTPs in EDCs removal, such as hydraulic residence time (HRT) and solids retention time (SRT). However, due to the inherent complexity among different WWTPs design and wastewaters origins, conflicting results were often reported. Clara and co-workers found that solid retention time was a key parameter in achieving low effluent concentrations in WWTPs (Clara *et al.* 2005). However, Servos' group observed no apparent correlation between EDCs removal and the SRT in 18 WWTPs (Servos *et al.* 2005). It is worth mentioning that the authors found WWTPs with nitrification achieved higher removal of EDCs in comparison to those without nitrification. The authors attributed this to the fact that the cometabolic transformation of estrogens might be favored under nitrifying conditions (Servos *et al.* 2005; Yi *et al.* 2007).

Recently, bioassays have been widely used in measuring EDCs in wastewater samples due to their capability of providing direct information on estrogenic activity, regardless of the chemical structure (Campbell *et al.* 2006). To date, various bioassay methods exist, including E-screen, ER-binding, and the yeast estrogen screen (YES) assay. However, no bioassay is universally accepted due to the significant discrepancies in results obtained from different methods (Liu *et al.* 2009). Drewes and collaborators investigated a total of seven full-scale WWTPs and found that the total estrogenic activity was removed by 96% using E-screen (Drewes *et al.* 2005). However, Servos and co-workers observed removal efficiencies ranging from 0 ~ 100% using the YES assay (Servos *et al.* 2005). In a similar study, Nelson and collaborators compared E-screen and the YES assay and found that the results varied by up to 4-fold on the same split sample (Nelson *et al.* 2007). It is clear that more research is needed to identify a standard bioassay for estrogenic activity measurement. Moreover, the interpretation of bioassay results together with chemical analysis warrants further investigation. In a study of 30 primarily estrogenic organic contaminants in several wastewater samples, Fernandez and collaborators observed weak correlations between chemical concentrations and estrogenic equivalency from bioassays with sometimes unexpected inverse correlations (Fernandez *et al.* 2007).

6.2.2 Antibiotics and antibiotic resistance

There have been increasing concerns over the widespread presence of antibiotics in the aquatic environment and their ecological impacts (Snyder *et al.* 2003; Ternes *et al.* 2003; Verlicchi *et al.* 2012). A study by the U.S. Geological Survey found that 80% of the 139 streams sampled across the U.S. contain trace levels of organic wastewater contaminants and 14 veterinary and human antibiotics were detected at the μg/L range (Kolpin *et al.* 2002). In the European Union, over 3,000 chemical substances were used in medicines such as antibiotics and they can inevitably enter the environment as a combination of intact and metabolized chemical mixtures (Ternes *et al.* 2004). WWTPs could receive most of the antibiotics through domestic wastewaters, hospital discharges, and manufacturers, although many WWTPs were not designed to remove these hazardous pollutants (Kolpin *et al.*

2002). Numerous efforts have been made to understand the efficacy of WWTPs in removing antibiotics and antibiotic genes. The purpose of this section is to summarize existing data on the removal efficiencies of antibiotics from full-scale WWTPs. Five veterinary and human antibiotics (Figure 6.2) with high occurrence frequencies are selected: ciprofloxacin, lincomycin, roxithromycin, sulfamethoxazole, trimethoprim, and erythromycin-H_2O. Moreover, to compare the behavior of these antibiotics in WWTPs and identify the key treatment processes that promote their removal, we focus on literatures that have investigated at least three target antibiotics in full-scale WWTPs with a secondary treatment.

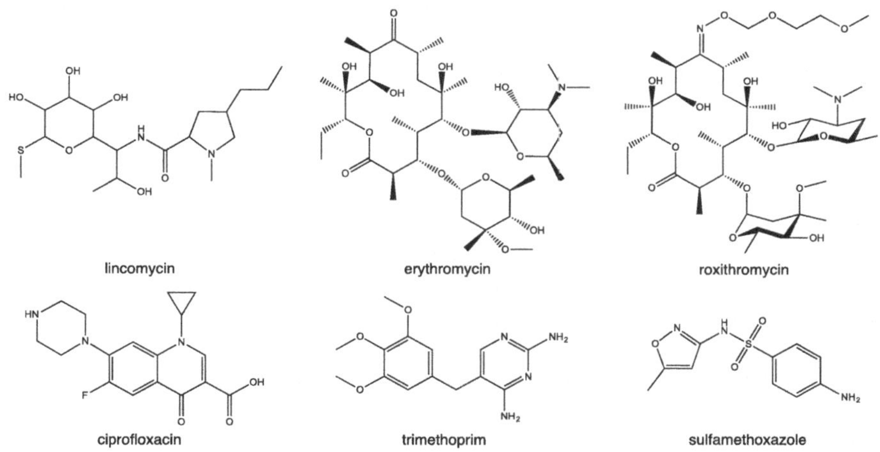

Figure 6.2 Chemical structures of five veterinary and human antibiotics: ciprofloxacin, lincomycin, roxithromycin, sulfamethoxazole, trimethoprim, and erythromycin.

To date, many studies have investigated the removal of antibiotics in full-scale WWTPs with AS processes. Both sorption and biodegradation processes are believed to be important in eliminating antibiotics in WWTPs with a wide range of removal efficiencies from negative to over 90% (McArdell *et al.* 2003; Watkinson *et al.* 2007; Miege *et al.* 2009; Verlicchi *et al.* 2012). For instance, Watkinson and collaborators assessed the removal of antibiotics in a full-scale conventional WWTP with AS process in Australia. All five antibiotics were detected in the influent, with concentrations of 3.8 μg/L, 0.06 μg/L, 0.36 μg/L, and 0.34 μg/L for ciprofloxacin, lincomycin, sulfamethoxazole, and trimethoprim, respectively. Overall, the WWTP investigated was effective in removing the antibiotics but was not able to eliminate their presence, where the measured secondary effluent concentrations were 0.6 μg/L, 0.05 μg/L, 0.27 μg/L and 0.05 μg/L for ciprofloxacin, lincomycin, sulfamethoxazole, and trimethoprim, respectively. The corresponding removal efficiencies were calculated using the concentration differences between the influent and secondary effluent in the aqueous phase, which ranged from 33% for lincomycin to 91% for trimethoprim (Watkinson *et al.* 2007). Other researchers observed similar trends of the removal of antibiotics in full-scale WWTPs. For example, Ghosh and co-workers evaluated the occurrence and removal of antibiotics in full-scale WWTPs in Japan. For ciprofloxacin, lincomycin, roxithromycin, sulfamethoxazole, and trimethoprim, the removal efficiencies varied from –32 to 83% (Ghosh *et al.* 2009). Li and Zhang found the removal efficiencies of antibiotics in WWTPs in China to be 18 ~ 55% for ciprofloxacin, 40 ~ 46% for roxithromycin, 62 ~ 92% for sulfamethoxazole, 13 ~ 42% for trimethoprim, and 15 ~ 26% for erythromycin-H_2O (Li & Zhang, 2011). Although the removal of antibiotics did occur during the AS processes, they were not eliminated completely, indicating that a significant amount of antibiotics will still enter the aquatic environment.

Due to the complicated removal mechanisms of antibiotics in biological treatment processes and the differences among WWTPs, it is extremely difficult to obtain a consistent trend of antibiotics removal from full-scale WWTPs data. Generally, it is believed that sorption to AS is important for more hydrophobic antibiotics, while biodegradation in AS is more important for hydrophilic antibiotics (Verlicchi *et al.* 2012). For instance, Lindberg and co-workers attributed 77% of the observed ciprofloxacin removal in a conventional WWTP to the sorption onto AS and less than 10% of its removal to biodegradation (Lindberg *et al.* 2006). In contrast, sorption was less important for the removal of trimethoprim in WWTPs (<10%), while biodegradation contributed a significant amount (Jelic *et al.* 2011). Operational parameters, SRT and HRT, are also important. Li and Zhang (2011) reported that longer SRT was unfavorable for the removal of antibiotics when sorption was the dominant removal mechanism. However, longer HRT could favor the elimination of antibiotics when biodegradation was important in removing antibiotics. Advanced treatment processes, such as microfiltration/reverse osmosis, were also investigated in some studies. For instance, Watkinson and collaborators found that WWTPs with microfiltration/reverse osmosis were more efficient in eliminating antibiotics, including ciprofloxacin, lincomycin, roxithromycin, sulfamethoxazole, trimethoprim, and erythromycin-H_2O (Watkinson *et al.* 2007).

Antibiotic resistance has been detected in many water bodies downstream of WWTPs and research has suggested that WWTP effluent is a significant contributor (Auerbach *et al.* 2007; Lefkowitz & Duran, 2007; Watkinson *et al.* 2007; LaPAra *et al.* 2011; Harris *et al.* 2012; Pruden *et al.* 2013). Auerbach and co-workers surveyed two WWTPs and two freshwater lakes for the presence of an array of 10 tetracycline resistance determinants. Their findings suggested that both WWTPs contained more types of tetracycline resistance genes in comparison to the lake samples, indicating that antibiotic resistance genes persist in the effluent WWTPs at levels well above typical aquatic environments (Auerbach *et al.* 2007). LaPara and collaborator also observed the presence of antibiotic resistance genes in the effluents of WWTPs that use mixed-media filtration (LaPara *et al.* 2011). In another study, Lefkowitz and Duran investigated the changes in antibiotic resistance patterns in five conventional WWTPs. Briefly, grab samples were collected from raw influent, primary effluent, secondary effluent, and final effluent after disinfection. Resistance of *Escherichia coli* isolates to six antibiotics – amoxicillin, ampicillin, chlorotetracycline, oxytetracycline, streptomycin, and tetracycline – were compared among different stages within samples and no significant difference was observed. Interestingly, the treatment process seemed to promote isolates with multiple antibiotic resistance (Lefkowitz & Duran, 2009).

6.2.3 Antimicrobial agents and sunscreens

Triclosan has gained tremendous popularity in the past 40 years as an antimicrobial agent that is widely used in many consumer care products including soap, toothpaste, mouthwash, and household cleaners. As a consequence, triclosan has often been reported in various water bodies (Singer *et al.* 2002). In the United States, a study by the U.S. Geological Survey found 58% of the 139 streams sampled across the country contain trace levels of triclosan in the range of 0.14 to 2.3 μg/L (Kolpin *et al.* 2002). WWTPs are believed to be one of the major sources that introduces triclosan into surface waters that poses ecological risks (Reiss *et al.* 2002). Therefore, it is critical to understand the removal of triclosan in WWTPs. Many researchers have investigated the occurrence and elimination of triclosan in conventional WWTPs and relatively high removal efficiencies were reported (Singer *et al.* 2002; Verlicchi *et al.* 2012). Yu and collaborators analyzed 24-h influent and effluent composite samples from a conventional WWTP with a SRT between 8 to 10 days. The concentration of triclosan in the influent was 0.8 μg/L, and 0.25 μg/L in the effluent, indicating a 69% incomplete removal (Jim *et al.* 2006). Singer and co-workers reported a higher removal efficiency of 85% on average by investigating seven WWTPs with biological treatment supplemented with an anoxic zone for denitrification. To better understand the removal mechanism of triclosan in WWTPs, the authors also collected grab samples and quantified the amount of sorbed and dissolved triclosan, separately. By applying a mass balance approach over one-week of survey data, the authors concluded that 79% of triclosan was biodegraded and 15% was removed with the excess sludge via sorption (Singer *et al.* 2002).

Another group of consumer care products that draws increasing attention is UV filter material, including organic and/or inorganic compounds that absorb or reflect UV light. They are widely present in many products of daily use, such as sunscreen products, cosmetics, skin creams, body lotions, hair spray, and so forth. These compounds enter the environment either from showering, wash-off via WWTPs, or from recreational activities such as swimming and bathing. Concerns have been raised over their potential impact on the aquatic ecological systems

(Poiger *et al.* 2004). For instance, 4-methylbenzylidene camphor (4-MBC), one organic UV filter, has shown estrogenic activity (Schlumpf *et al.* 2001). Therefore, it is important to understand the occurrence and elimination of UV filter materials in WWTPs. Balmer and collaborators assessed four organic UV filters with high occurrence frequencies (benzophenone-3 (BP-3), 4-methylbenzylidene camphor (4-MBC), octocrylene (OC) and ethylhexyl methoxycinnamate (EHMC)) in wastewater samples from eight municipal WWTPs in Switzerland. Their chemical structures are shown in Figure 6.3. All four contaminants were detected in raw influent from all WWTPs, where the concentrations were 0.6 ~ 6.5 μg/L of 4-MBC, 0.7 ~ 7.8 μg/L of BP-3, 0.1 ~ 12 μg/L of OC, and 0.5 ~ 19 μg/L of EHMC. Although organic UV filters were still detected in the effluent, they were present at much lower concentrations, indicating effective removal but not elimination in WWTPs with conventional AS processes. The removal efficiencies were 18 ~ 82% for 4-MBC, 68 ~ 96% for BP-3, 88 ~ 99% for OC, and 97 ~ 99% for EHMC (Balmer *et al.* 2005).

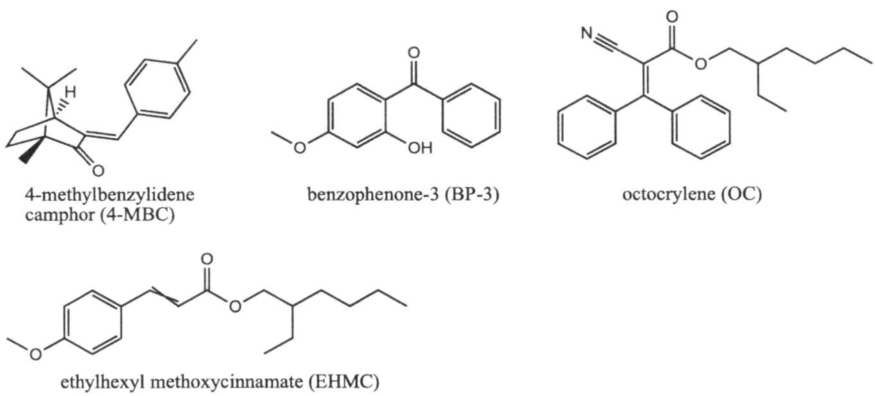

Figure 6.3 Chemical structures of four selected organic UV filters.

6.3 VOLATILE ORGANIC COMPOUNDS

Volatile organic compounds (VOCs) such as trichloroethylene, benzene, toluene, chloroethene (vinyl chloride), and dichloromethane (methylene chloride) fall into the hazardous organic compounds category. The term VOCs is rather vague particularly within the context of indoor and outdoor air quality. However, the definition "organic compounds with boiling point of 100°C or less and/or vapor pressure of greater than 11 mm Hg at 25°C" is generally used for the VOCs of concern in wastewater treatment facilities (Tchobanoglous *et al.* 2014).

The fate of VOCs through WWTPs has been studied extensively, since these compounds are present at various concentrations in sewage and stormwater;

some such as benzene and vinyl chloride are known or suspected carcinogens, and they are ozone precursors. Although VOCs in the effluent from WWTPs are not regulated in the U.S., the Clean Air Act has provisions to mandate inventory and control of VOC emissions if WWTPs are suspected as a source. In fact, the majority of the early research in the U.S. focused on VOC emissions from WWTPs and originated from a potential, stringent regulation of VOC emissions from WWTPs within Clean Air Act Amendments. The book by Tata and others (Tata *et al.* 2003) is an excellent reference on VOCs from WWTPs including a chapter on sampling and analytical methods (Chapter 13). It also summarizes the early research including EPA 40-City Study, Research Triangle Institute Study, and the works of the Association of Metropolitan Sewerage Agencies, East Bay Municipal Utility District (Oakland California), and city of Ontario (Canada). Although these studies were conducted to collect data on emissions of VOCs from WWTPs, they established the baseline of our understanding of the sources and fate of VOCs in WWTPs. Since there is an extensive review of the early work by Tata and colleagues, we will focus on their main findings and the studies since the early 2000s.

The extensive studies covered in the book by Tata and collaborators point out that the main sources of VOCs in wastewater are: chlorination of water and wastewaters; industrial discharges; commercial activities such as dry cleaning; household and consumer products including cleaning products and personal hygiene products; stormwater runoff; and chemical and biological reactions occurring during wastewater treatment (Tata *et al.* 2003). Regardless of the exact sources, WWTPs dealing with negligible industrial wastewater discharge see a few different VOCs with less than 10 μg/L, while plants serving combined sewer systems experience more VOCs and higher concentrations. Chapter 3 of the book by Tata and others (page 43 through 45) summarizes the frequency and maximum concentrations of VOCs in the influent of WTTPs covered by large-scale investigations prior to 2000. It is noteworthy that the largest of those studies reported that the most frequently found VOCs in the influent of 40 WWTPs were toluene, tetrachloroethylene, dichloromethane, chloroform, trichloroethylene, 1,1,1-trichlororethane, ethlybenzene, phenol, trans-1,2-dichloroethene, and benzene (United States Environmental Protection Agency [U.S. EPA], 1982). The results of these extensive early studies showed that there was high variability in the concentrations of VOCs detected in the influent of different WWTPs due to variations in sources and that daily, weekly, and seasonal variations in concentration of VOCs occur.

Although the fate of VOCs through WWTPs could occur via a variety of mechanisms including volatilization, stripping, biodegradation, chemical oxidation/reduction reactions, and sorption, the quantitative contribution of each fate mechanism would vary based on processes and individual plant design. Generally, volatilization and stripping are the two main mechanisms responsible for the fate of VOCs in WWTPs. Stripping increases with inefficient aeration and thus aerobic biological processes with coarse bubble diffusers and surface aerators are expected to have higher VOC emissions than those with fine bubble diffusers. Similarly, aerated grit chambers have been reported to have higher VOC emissions than non-aerated ones. Volatilization increases with turbulence generated by mixing and/or high velocity areas and thus structures such as inlet and weirs where velocity is high tend to experience greater VOC emission. The work by Tansel and Eyma (1999) confirmed that volatilization is the predominant mechanism of VOC fate in wastewater treatment plants. The authors used the influent and effluent liquid phase concentrations of nine VOCs – 1,1,1-trichloroethane, toluene, tetrachloroethene, 1,4-dichlorobenzene, chloroform, trichloroethylene, ethylbenzene, 1,2-dichloroethene, and o-xylene – from a full-scale municipal treatment plant with 227,500 m^3/d average flow to validate a simple, mass balance based model that they developed to study the fate of VOCs going through secondary wastewater treatment. The average influent concentrations of these VOCs were very low, mostly 10 μg/L or less, except for 1,1,1-trichlororethane, which was slightly over 80 μg/L. The effluent concentrations for all of these VOCs, on the other hand, were less than 10 μg/L. The full-scale treatment plant data suggested that volatilization was the primary fate mechanism, most of the VOCs emissions occurring in the aeration tank and secondary clarifier weir.

A more recent study investigated occurrence and fate of 23 VOCs going through five conventional AS processes – capacities ranging from 12,000 to 700,000 population equivalent – in Italy, and three of the plants studied included membrane bioreactors (Fatone *et al.* 2011). The authors found that the most frequently detected VOCs were toluene, ethylbenzene, xylenes, styrene, 1,2,4-trimethylbenzene, and 4-chlorotoluene. Regardless of the detection frequency, the influent concentrations of 23 VOCs were below 1 μg/L, except for toluene, which ranged from 3 to 7 μg/L. Also noted was that volatilization in the headworks accounted for 40–60% removal of the VOCs studied. The fate of these VOCs was similar among the plants with and without membrane bioreactors and across different sludge ages, suggesting that biodegradation and sorption were insignificant in removal of VOCs.

Yang and collaborators collected data to study seasonal changes in concentrations of benzene, toluene, xylenes, chloroform, carbon tetrachloride, trichloroethylene, and tetrachloroethylene in a 325,000 m^3/day capacity treatment plant in northern China and determined their fate through different processes of the plant (Yang *et al.* 2014). It was found that the aromatic hydrocarbon toluene was released to atmosphere more readily than the chlorinated VOCs, with the exception of chloroform, which remained in the solution phase throughout the WWTP. This is contradictory to the earlier findings that chlorinated VOCs tend to be stripped more readily than non-chlorinated ones (Bell *et al.* 1993). Regardless, Yang and co-workers reported that the highest VOC volatilization/stripping occurred during primary treatment in aerated grit chambers and primary clarifiers. The authors found that seasonal temperature variations had a minimal effect on the fate of VOCs in the WWTPs (Yang *et al.* 2014).

In a unique study, Keefe and collaborators studied the fate of VOCs in a constructed wetland and reported that 15 VOCs were found above the detection limit in the influent to a 1.34 ha wetland receiving 3,780 m^3/day wastewater (Keefe *et al.* 2004). The concentrations of the 15 VOCs in the inlet were below 1 μg/L, except for bromodichloromethane and trichloromethane, and their removal efficiencies from inlet to outlet ranged from 63% to 87%.

Toxic effects of VOCs on an anaerobic microbial consortium and their degradation through direct energy metabolism or cometabolic pathways have been studied extensively at bench-scale. Speece (2008) provided an excellent summary of toxicity of VOCs on anaerobic biomass in Chapter 17 and 19 of his book. An earlier review by Zitomer and Speece (1993) is another good summary of VOCs toxicity and biotransformation by sequential anaerobic/aerobic processes. However, there seems to be a lack of literature on effects of VOCs on full-scale anaerobic treatment processes, ostensibly due to the fact that in WWTPs, most VOCs are volatilized and/or stripped before they reach the sludge digestion stages.

6.4 AGRICULTURAL POLLUTANTS

The U.S. EPA defines pesticides as "any substance or mixture of substances intended for preventing, destroying, repelling, or mitigating any pest; any substance or mixture of substances used as a plant regulator, defoliant, or desiccant; and any nitrogen stabilizer" with certain exceptions specified within the pesticide regulation framework 40 CFR 152.6–152.15 (U.S. Government, 2013). Although the European Environment Agency includes pesticides in the biocides category, which is defined as "a diverse group of poisonous substances including preservatives, insecticides, disinfectants, and pesticides used for the control of organisms that are harmful to human or animal health or that cause damage to natural or manufactured products" (European Environment Agency, 2016), this subsection is focused on hazardous substances originating from agricultural activities and it excludes biocides intended as personal care products, such as hand sanitizers, and antimicrobial household cleaners and industrial coating agents.

In the context of this book, pesticides fall under the organic hazardous pollutants category. Although they are considered agricultural pollutants and rarely studied in WWTPs, which are typically urban infrastructures, they do reach the WWTPs via a number of urban uses, including lawn care, vegetation control in parks and golf courses, urban nurseries, and public health protection such as mosquito and other pests control. This subsection reviews not only occurrence and possible transformation of pesticides in full-scale WWTPs, it also summarizes the published literature in terms of their inhibitory effects on biological treatment processes.

An early work investigated the pesticide loads from a rural and an urban WWTP in Germany and reported that isoproturon accounted for three quarters of the total pesticides in the effluent of the rural WWTP (Nitschke & Schüssler, 1998). Diuron on the other hand made up 80% of the effluent pesticides from the urban WWTP. In a similar study from Germany, Muller and collaborators studied the sources of 19 herbicides detected in surface water in a 4,917-ha catchment area with 41% of land used for agriculture (Muller *et al.* 2002). It is reported that 64% of the pesticide loads in surface waters originated from the two WWTPs in the catchment area.

In a recent study from Spain, Köck-Schulmeyer and co-workers monitored 22 selected pesticides in the influent and effluent of three full-scale WWTPs in Catalonia and reported that diazinon and diuron were the most frequently detected – 96 and 88% of 24 influent samples, respectively – while cyanazine, fenitrothion, linuron, metolachlor, and molinate were not detected in any samples at the detection limit of 30 ng/L (Köck-Schulmeyer, 2013). In most cases, the total pesticide concentration was below 1 μg/L. The authors reported that pesticide removal in all three WWTPs was poor and in some cases effluent concentrations were higher than the influent suggesting involvement of sorption and desorption through the WWTPs. An earlier study focused on 40 selected pesticides in the effluent samples from five WWTPs in Comunidad Foral de Navarra, Northern Spain (Ormad *et al.* 2011). Ten pesticides detected, out of the suite of 40 screened for, were chlorfenvinphos, chlorpyrifos, dimethoate, isoproturon, prometon, metholachlor, 3,4-dichloroaniline, terbutryn, terbuthylazine and simazine (quantification limits from 0.015 to 0.1 μg/L). The concentrations of the detected pesticides ranged from a low of 0.024 μg/L (chlorfenvinphos) to a high of 1.31 μg/L (erbuthylazine). Although the authors suggested that the study showed the pesticides were not removed in any of the five WWTPs studies, no influent data were presented making it rather impossible to draw any conclusions about the removal, or lack thereof, of these pesticides during the treatment processes studied.

Cahill and collaborators developed an online solid phase extraction followed by LC-MS/MS based method for detection of 11 pesticides at ng/L concentration range and used the method to analyze effluent samples from five full-scale WWTPs in Ireland. The authors reported that simazine, prometon, diuron, atrazine-desethyl, and atrazine were detected out of the 11 target pesticides and the highest concentrations of the detected pesticides were below 0.2 μg/L (Cahill *et al.* 2011).

Fungicides triazoles and imidazoles are commonly used in agriculture as well as in pharmaceuticals. There had been intensive research efforts on the fate of azole compounds in WWTPs in Switzerland. In an early work, Gerecke and co-workers monitored the concentrations of eight pesticides – three used exclusively in agriculture and the remaining five used both in agriculture and urban applications – in the effluent from 19 WWTPs and in two rivers in the Lake Greifensee catchment area in Switzerland (Gerecke *et al.* 2002). The results showed that 75% of pesticides, used for urban applications and found in the two rivers, resulted from the effluent of the WWTPs, due to the predominantly combined sewer systems in the urban areas. Later, the occurrence and fate of nine agricultural azole fungicides in WWTPs and lakes throughout Switzerland were investigated by Kahle and collaborators (Kahle *et al.* 2008). The authors indicated that there are

17 azole fungicides registered for agricultural use in Switzerland and among those prochloraz, tebuconazole, cyproconazole, difenconazole, flusilazole, propiconazole, and epoxiconazole are most frequently sold in the total annual consumption range of 1 to 6 tons. Propiconazole and tebuconazole were consistently detected in the effluent of WWTPs investigated in the range of 1–30 ng/L concentration. Although the influent and effluent load measurements indicated that both of these pesticides moved through the WWTPs unchanged, incubation with AS biomass showed they are slowly degraded with degradation half-lives of 2 to 3 days. These findings suggest that these two pesticides might be degraded in AS processes with high HRTs. In another study from Switzerland, Morasch and collaborators screened the influent and effluent samples from WWTPs for 58 micropollutants including biocides and pesticides (Morasch *et al.* 2010). The influent and effluent concentrations of the screened pesticides were always below 1 μg/L. While atrazine, mecoprop and propiconazole were not removed, carbendazim, diazinon and terbutryn showed slight removal. On the other hand, 70 to 98% removal was observed for chloridazon, tebufenozide, IPBC (3-Iodo-2-propynyl butylcarbamate), and irgarol. The removal mechanisms were not studied.

In a rare study on pesticides in the U.S., Bud and collaborators looked at the efficacy of two full-scale constructed wetlands in Central Valley, California in removal of pyrethroid and organophosphate insecticides including chlorpyrifos and diazinon (Budd *et al.* 2009). Both treatment systems were effective in pyrethroid removal achieving 52 to 94% reduction in concentration. The authors reported that the seasonal average mass reduction of pyrethroid reached 95 to 100%. Sorption to SS and subsequent settling was found to be the primary pyrethroid removal mechanism. The wetlands were less effective in removal of chlorpyrifos with 52–61% seasonal average reduction in mass and the reduction in mass of diazinon was negligible in both wetlands.

6.5 SURFACTANTS

Surfactants, categorized under "organic hazardous compounds" in this book, are a group of usually organic chemicals that lower surface tension between two liquid phases or a liquid and solid surface. Also called surface active agents, they are the main active component of many consumer products including detergents, fabric treatment solutions, and fire-fighting applications. They are also used as dispersing agents during oil spills cleaning, and soil and groundwater remediation. Surfactants are commonly classified according to the polar (hydrophilic) head they contain (anionic vs. cationic) and/or the hydrophobic hydrocarbon in their tails (alkyl based surfactants vs. silicone based surfactants). Approximately 60% of surfactants are used in detergents and household cleaning products while the remainder is used for industrial purposes including textiles and pulp and paper (Karsa, 2008).

It is important to note that there is a wealth of literature on surfactants covering their chemistry and fate in natural environments. The aim of this section is to focus on the studies that specifically address the surfactants in full-scale WWTPs. Similarly, there was a surge of research in the earlier 1980s on biodegradation of surfactants in treatment processes and those studies have been reviewed by Swisher in his seminal book "Surfactant Biodegradation" (Swisher, 1987). In order the avoid repetition, only studies after the mid-1980s that have focused on the fate of surfactants in full-scale WWTPs are reviewed in this subsection.

In a comprehensive study, Brunner and collaborators quantified mass flow of linear alkylbenzene sulfonate (LAS), nonylphenol (NP), nonylphenol monoethoxylate (NP1EO), and nonylphenol diethoxylate (NP2EO) through 29 WWTPs in Switzerland (Brunner *et al.* 1988). The authors quantified these surfactants in raw sewage, primary and secondary effluents, and sewage sludge. Although there was significant variation in mass flows among the different treatment plants, LAS was shown to be removed almost completely (greater than 99%) while 15 to 20% of it was transferred to the digested sludge, indicating that main mechanism of LAS removal in WWTPs is aerobic biodegradation in the AS process. Approximately 50% of NP2EO on the other hand was transformed into NP, a metabolic product of NP2EO, and NP was almost entirely transferred to the digested sludge. The same research group reported similar results on the fate of LAS through WWTPs in another study (Giger, 1989).

Adding to his earlier work on the fate of surfactants in natural environments, Federle focused on mineralization of surfactants in treatment processes during the late 1980s and early 1990s. In one case, he studied the biodegradation of three surfactants – LAS, stearyltrimethylammonium chloride (STAC), and linear alcohol ethoxylate (LAE) – in the sediments of a wastewater pond receiving discharge from a laundromat (Federle & Schwab, 1992). Using ^{14}C labeled surfactants, the authors demonstrated that LAE was rapidly biodegraded by anaerobic consortia acclimated to laundromat wastewater over 25 years. On the other hand, neither LAS nor STAC was biodegraded under anaerobic conditions, ostensibly due to lack of monooxygenase activity. It was suggested that biodegradation of LAE may involve sulfate reducers and methanogens under anaerobic conditions.

In another comprehensive study on LAS, Prats and collaborators investigated the fate of LAS, an ionic surfactant, and compared it to that of non-ionic linear alcohol ethoxylates (AE) through a typical WWTP with anaerobic digestion in Alicante, Spain (Prats *et al.* 1997). The total LAS concentration in raw wastewater was 15.7 mg/L, 6.6 mg/L dissolved and the rest sorbed on suspended solids (SS). The aeration process reduced the dissolved LAS to 0.12 mg/L in the effluent of the secondary settler. The LAS mass flow to the AD was 166 kg/d and it was reduced to 135 kg/d after the AD. The nonionic AE on the other hand was much lower in the raw wastewater, 5.4 mg/L with 2.5 mg/L being in dissolved form, and 95.3% removal was observed in the aeration tank. The observation that AE was mostly removed during aerobic treatment and no significant sorption to AS biomass occurred was later verified by Szymanski and co-workers (Szymanski *et al.* 2003). It was reported that 63 kg/d AE from the primary settlers – AE sorbed on the SS in raw wastewater – in the Alicante wastewater treatment was effectively reduced to 7 kg/d during the anaerobic digestion process (Prats *et al.* 1997).

Dialkyl dimethyl ammonium chlorides were popular surfactants in anti-static and fabric softener products during the 1970s and 1980s until the introduction of silicone based surfactants. In an effort to compare three analytical methods for quantification of ditallow dimethyl ammonium chloride (DTDMAC), Wee collected influent and effluent grab samples from a secondary WWTP (Wee, 1984). The influent DTDMAC concentration was in the range of 175 to 334 μg/L (five samples) and it was reduced to 28 to 56 μg/L in the final effluent. The author did not attempt to study the DTDMAC removal mechanisms through the WWTP.

Secondary alkane sulfonates (SAS) attracted attention in the 1990s as an alternative to LAS in cleaning products due to their faster biodegradation rates, mainly because of the absence of a benzene moiety in SAS. Field and co-workers built upon their earlier analytical work on developing an extraction/derivatization followed by gas chromatography mass spectrometry (GC-MS) quantification of SAS to investigate SAS fate through a municipal WWTP in Zurich, Switzerland (Field *et al.* 1995). The study provided quantitative estimates of SAS mass flow through different stages of the WWTP including sludge digestion. During the 10-day sampling event, SAS in the influent raw wastewater ranged from 598 to 976 μg/L. The preliminary treatment was not effective in removal of SAS in the liquid phase and thus the influent to the AS process contained SAS in the range between 545 and 890 μg/L. The low secondary effluent SAS concentration from below the detection limit (1 μg/L) to 14 μg/L demonstrated that SAS are removed effectively in the AS process. However, the mass flow analysis showed that these surfactants are not removed during the anaerobic digestion process. The raw sludge, 40 and 60 percent by volume combination of primary and waste activated sludge, containing 463 to 593 mg SAS per kg dry matter (dm) was fed to a two-stage anaerobic digestion process, each stage with 20 d retention time. The SAS in anaerobically digested sludge increased to a range from 649 to 738 mg/kg dm. SAS mass flow analysis demonstrated that 84.1% of SAS coming to the WWTP goes through AS processes and is removed almost entirely. Of the remaining SAS mass, 16.3% is accumulated in the anaerobically digested biosolids indicating the tendency of SAS to be sorbed on primary sludge and to resist biodegradation during anaerobic digestion.

Due to their persistence in the environment, perfluoroalkyl surfactants (PAS) have been studied extensively. The persistence of PAS in the environment comes from the strong carbon-fluorine bonds that make these compounds difficult to metabolize or physically degrade (Sinclair & Kannan, 2006). In one of the earliest studies on surfactants in WWTPs, 3M Company monitored perfluorooctane sulfonate (PFOS) and perfluorooctanate (PFOA) in the effluent of six WWTPs (3M, 2000). Both compounds showed significant variability with average concentrations of 1,140 (±1,823) and 549 (±840) ng/L, respectively. PFOS was detected in biosolids from five of the six plants. Similarly, PFOA was present in biosolids from four WWTPs. Later, Boulanger and collaborators built upon the 3M study and evaluated the fate of eight perfluorooctane surfactants in the influent and effluent of a 22,750 m^3/d WWTP (Boulanger *et al.* 2005). The authors also monitored these eight surfactants in the river water at the point of discharge from the plant. The reported average 2-(N-ethyl-perfluorooctanesulfonamidoacetic) acid (N-EtFOSAA) concentrations were 5.1 (±0.8) ng/L in the influent, 3.6 (±0.2) ng/L in the effluent, and 1.2 (±0.3) ng/L in the river water. Similarly, PFOS was 26 (±2.0) ng/L in the effluent and 23 (±1.5) ng/L in the river water. PFOA was also observed at 22 (±2.1) ng/L in the effluent and 8.7 (±0.8) ng/L in the river water. The authors noted that both PFOS and PFOA signals were observed in the influent samples, however exact quantification was not possible due to low recovery. N-Ethyl Perfluorooctane Sulfonamidoethanol (N-EtFOSE) biodegradation study under both aerobic and anaerobic conditions used biomass from the WWTP and 96 h incubation. Although no degradation occurred under anaerobic conditions, perfluorooctane sulfonate was formed as the metabolites under aerobic conditions.

6.6 OTHERS

Literature on the occurrence and elimination of nanomaterials in full-scale WWTPs is scarce. To date, the most investigated nanomaterial is titanium dioxide (TiO_2), which is widely used in nanotechnology-based consumer products such as food additives, pharmaceuticals, clothing, sunscreens and paints. Inevitably, these nanomaterials can enter the aquatic environment and cause potential ecological risks and human health concern. However, little is known regarding their behavior in WWTPs. Kiser and co-workers investigated the occurrence and removal of nano- and larger sized Ti at a full-scale wastewater reclamation facility in central Arizona. The concentration of Ti was 100 ~ 3,000 μg/L in the raw sewage, where particle size was larger than 0.7 μm for the majority, and this was effectively removed during the treatment. The concentration of Ti in the effluent was in the range of 5 ~ 15 μg/L and nearly all particles were below 0.7 μm size fraction (Kiser *et al.* 2009). Westerhoff and collaborators later assessed the concentrations of Ti in raw sewage and treated effluent from 10 WWTPs. Raw sewage concentrations ranged from 181 to 1,233 μg/L. Effluent concentrations were less than 25 μg/L with diameters in the range of 4 to 30 nm (Westerhoff *et al.* 2011).

Benzotriazoles are a group of heterocyclic chemicals that are widely used in a variety of applications, such as antifogging agents, UV light stabilizers in plastics, and corrosion inhibitors in detergents and de-icing fluids (McNeil & Cancilla, 2009). As a consequence, these chemicals are often detected in surface water, groundwater, sediment, and sewage sludge in the range of ng/L to μg/L. Several studies evaluated the removal efficiencies of these contaminants in full-scale WWTPs. For example, Liu and collaborators evaluated four benzotriazoles in a full-scale WWTP in South Australia: benzotriazole (BT), 5-methylbenzotriazole (5-TTri), 5-chloro-benzotriazole (CBT), and 5,6-dimethyl-benzotriazole (XTri). The influent concentrations were

5,706 ± 928 ng/L, 6,758 ± 1,438 ng/L, 1,130 ± 480 ng/L, and 1,313 ± 370 ng/L for BT, 5-TTri, CBT and XTri, respectively. Most contaminants were effectively removed during the AS process (60% to 100%). However, BT was very resistant to microbial degradation and the biological treatment process was found to only contribute 7% to 27% of the removal (Liu *et al.* 2012). Similar results were reported by Weiss and co-workers (Weiss *et al.* 2006). Reemtsma and collaborators investigated the removal of BT and 5-TTri in four wastewater treatment plants in Germany and found that both BT and 5-TTri were recalcitrant to the biological treatment process, where the removal for BT and 5-TTri was 30 to 55% and 20 to 70%, respectively (Reemtsma *et al.* 2010). Overall, the conventional AS process may not be suitable for removal of these emerging contaminants and other post-treatment processes might be needed. However, due to the complex nature of the biological process and the significant variations in operating parameters among WWTPs, results vary from one to another and systematic studies are needed to provide more insights.

Flame retardants are another important group of hazardous pollutants, including polybrominated diphenyl ethers (PBDEs) and organophosphorus flame retardants. Rayne and Ikonomou evaluated the removal of PBDEs in a tertiary WWTP with a post-filtration ultraviolet light disinfection (Rayne & Ikonomou, 2005). The authors reported no substantial biological degradation of PBDEs. However, 93% overall PBDEs removal occurred through sorption to sludge, which was in line with the high concentrations observed in WWTP biosolids (approximately 2.4 mg/kg dry weight). Meyer and Bester investigating the removal of both chlorinated and non-chlorinated organophosphorus flame retardants in two WWTPs reported that the removal efficiency ranged from 57–86% for non-chlorinated flame retardants with an influent concentration of 2.4–6.1 μg/L. In contrast, no elimination of the chlorinated organophosphorus flame retardants was observed in any of the treatment plants with an influent concentration of 0.57–5.8 μg/L (Meyer & Bester, 2004).

Can and collaborators investigated the removal efficiencies of three alkyl phenols, 4-t-octylphenol (4-tOP), nonylphenol (NP) and 4-n-octylphenol (4-OP), in three WWTPs with different treatment processes, including advanced biological treatment process (A2/O process), AS process, and just primary treatment (Can *et al.* 2014). The removal efficiencies of the three alkyl phenols were in the range of 49–82% for all three WWTPs. The WWTP with A2/O process consistently achieved the highest removal efficiency for all three contaminants. However, the removal mechanisms were not discussed.

6.7 REFERENCES

3M (2000). Docket AR-226. U.S. Environmental Protection Agency, Office of Pollution Prevention and Toxic Substances. Washington DC.

Andersen H., Siegrist H., Halling-Sørensen B. and Ternes T. A. (2003). Fate of estrogens in a municipal sewage treatment plant. *Environmental Science and Technology*, **37**(18), 4021–4026.

Auerbach E. A., Seyfried E. E. and McMahon K. D. (2007). Tetracycline resistance genes in activated sludge wastewater treatment plants. *Water Research*, **41**(5), 1143–1151.

Balmer M. E., Buser H. R., Müller M. D. and Poiger T. (2005). Occurrence of some organic UV filters in wastewater, in surface waters, and in fish from Swiss lakes. *Environmental Science and Technology*, **39**(4), 953–962.

Baronti C., Curini R., D'Ascenzo G., Di Corcia A., Gentili A. and Samperi R. (2000). Monitoring natural and synthetic estrogens at activated sludge sewage treatment plants and in a receiving river water. *Environmental Science and Technology*, **34**(24), 5059–5066.

Bell J., Melcer H., Monteith H., Osinga I. and Steel P. (1993). Stripping of volatile organic compounds at full-scale municipal wastewater treatment plants. *Water Environment Research*, **65**(6), 708–716.

Blair R. M., Fang H., Branham W. S., Hass B. S., Dial S. L., Moland C. L., Tong W., Shi L., Perkins R. and Sheehan D. M. (2000). The estrogen receptor relative binding affinities of 188 natural and xenochemicals: structural diversity of ligands. *Toxicological Sciences*, **54**(1), 138–153.

Boulanger B., Vargo J. D., Schnoor J. L. and Hornbuckle K. C. (2005). Evaluation of perfluorooctane surfactants in a wastewater treatment system and in a commercial surface protection product. *Environmental Science and Technology*, **39**(15), 5524–5530.

Brunner P. H., Capri S., Marcomini A. and Giger W. (1988). Occurrence and behavior of linear alkylbenzenesulphonates, nonylphenol, nonylphenol mono- and nonylphenol diethoxylates in sewage and sewage sludge treatment. *Water Research*, **22**(12), 1465–1472.

Budd R., O'Geen A., Goh K. S., Bondarenko S. and Gan J. (2009). Efficacy of constructed wetlands in pesticide removal from tailwaters in the Central Valley, California. *Environmental Science and Technology*, **43**(8), 2925–2930.

Cahill M. G., Caprioli G., Stack M., Vittori S. and James K. J. (2011). Semi-automated liquid chromatography–mass spectrometry (LC–MS/MS) method for basic pesticides in wastewater effluents. *Analytical and Bioanalytical Chemistry*, **400**(2), 587–594.

Campbell C. G., Borglin S. E., Green F. B., Grayson A., Wozei E. and Stringfellow W. T. (2006). Biologically directed environmental monitoring, fate, and transport of estrogenic endocrine disrupting compounds in water: a review. *Chemosphere*, **65**(8), 1265–1280.

Can Z. S., Firlak M., Kerc A. and Evcimen S. (2014). Evaluation of different wastewater treatment techniques in three WWTPs in Istanbul for the removal of selected EDCs in liquid phase. *Environmental Monitoring and Assessment*, **18**, 525–539.

Chimchirian R. F., Suri R. P. and Fu H. (2007). Free synthetic and natural estrogen hormones in influent and effluent of three municipal wastewater treatment plants. *Water Environment Research*, **79**(9), 969–974.

Clara M., Kreuzinger N., Strenn B., Gans O. and Kroiss, H. (2005). The solids retention time – a suitable design parameter to evaluate the capacity of wastewater treatment plants to remove micropollutants. *Water Research*, **39**(1), 97–106.

Crisp T. M., Clegg E. D., Cooper R. L., Anderson D. G., Baetcke K., Hoffmann J. L., Morrow M. S., Rodier D. J., Schaeffer J. E. and Touart L. W. (1997). Special Report on Environmental Endocrine Disruption: an Effects Assessment and Analysis. United States Environmental Protection Agency, Washington DC.

Drewes J. E., Hemming J., Ladenburger S. J., Schauer J. and Sonzogni W. (2005). An assessment of endocrine disrupting activity changes during wastewater treatment through the use of bioassays and chemical measurements. *Water Environment Research*, **77**(1), 12–23.

European Environment Agency (2016). Environmental Terminology and Discovery Service. Online document from http://glossary.eea.europa.eu/terminology/concept_html?term=biocide (accessed October 6, 2016).

Fatone F., Di Fabio S., Bolzonella D. and Cecchi F. (2011). Fate of aromatic hydrocarbons in Italian municipal wastewater systems: An overview of wastewater treatment using conventional activated-sludge processes (CASP) and membrane bioreactors (MBRs). *Water Research*, **45**(1), 93–104.

Federle T. W. and Schwab B. S. (1992). Mineralization of surfactants in anaerobic sediments of a laundromat wastewater pond. *Water Research*, **26**(1), 123–127.

Fernandez M. P., Ikonomou M. G. and Buchanan I. (2007). An assessment of estrogenic organic contaminants in Canadian wastewaters. *Science of the Total Environment*, **373**(1), 250–269.

Field J. A., Field T. M., Poiger T., Siegrist H. and Giger W. (1995). Fate of secondary alkane sulfonate surfactants during municipal wastewater treatment. *Water Research*, **29**(5), 1301–1307.

Gerecke A. C., Schärer M., Singer H. P., Müller S. R., Schwarzenbach R. P., Sägesser M., Ochsenbein U. and Popow G. (2002). Sources of pesticides in surface waters in Switzerland: pesticide load through waste water treatment plants – current situation and reduction potential. *Chemosphere*, **48**(3), 307–315.

Ghosh G. C., Okuda T., Yamashita N. and Tanaka H. (2009). Occurrence and elimination of antibiotics at four sewage treatment plants in Japan and their effects on bacterial ammonia oxidation. *Water Science and Technology*, **59**(4), 779–786.

Giger W., Alder A. C., Brummer P. H., Marcomimi A. and Siegrist H. (1989). Behavior of LAS in sewage sludge treatment and in sludge-treated soil. *Tenside Surfactants Detergents*, **26**, 95–100.

Government U.S. (2013). 40 CFR 152.6–152.15. US Code Title 7, Chapter 6. Retrieved July 1, 2016, from http://uscode.house.gov/browse/prelim@title7/chapter6&edition=prelim.

Harris S., Cormican M. and Cummins E. (2012). The effect of conventional wastewater treatment on the levels of antimicrobial-resistant bacteria in effluent: a meta-analysis of current studies. *Environmental Geochemistry and Health*, **34**(6), 749–762.

Jelic A., Gros M., Ginebreda A., Cespedes-Sanchez R., Ventura F., Petrovic M. and Barcelo D. (2011). Occurrence, partition and removal of pharmaceuticals in sewage water and sludge during wastewater treatment. *Water Research*, **45**(3), 1165–1176.

Jim T. Y., Bouwer E. J. and Coelhan M. (2006). Occurrence and biodegradability studies of selected pharmaceuticals and personal care products in sewage effluent. *Agricultural Water Management*, **86**(1), 72–80.

Johnson A. C., Williams R. J., Simpson P. and Kanda R. (2007). What difference might sewage treatment performance make to endocrine disruption in rivers? *Environmental Pollution*, **147**(1), 194–202.

Joss A., Andersen H., Ternes T., Richle P. R. and Siegrist H. (2004). Removal of estrogens in municipal wastewater treatment under aerobic and anaerobic conditions: consequences for plant optimization. *Environmental Science and Technology*, **38**(11), 3047–3055.

Kahle M., Buerge I. J., Hauser A., Müller M. D. and Poiger T. (2008). Azole fungicides: Occurrence and fate in wastewater and surface waters. *Environmental Science and Technology*, **42**(19), 7193–7200.

Karsa D. R. (2008). History and applications of surfactants. Chemistry and technology of surfactants. In: Chemistry and Technology of Surfactants. Farn R. J. (ed.), Wiley, Hoboken, New Jersey, USA, pp. 1–23.

Keefe S. H., Barber L. B., Runkel R. L. and Ryan J. N. (2004). Fate of volatile organic compounds in constructed wastewater treatment wetlands. *Environmental Science and Technology*, **38**(7), 2209–2216.

Kiser M. A., Westerhoff P., Benn T., Wang Y., Perez-Rivera J. and Hristovski K. (2009). Titanium nanomaterial removal and release from wastewater treatment plants. *Environmental Science and Technology*, **43**(17), 6757–6763.

Knobil E., Bern H., Burger J., Fry D., Giesy J., Gorski J., Grossman C., Guillette Jr. L., Hulka B. and Lamb J. (1999). Hormonally Active Agents in The Environment. National Academy Press, Washington DC.

Köck-Schulmeyer M., Villagrasa M., López de Alda M., Céspedes-Sánchez R., Ventura F. and Barceló D. (2013). Occurrence and behavior of pesticides in wastewater treatment plants and their environmental impact. *Science of the Total Environment*, **458–460**, 466–476.

Kolpin D. W., Furlong E. T., Meyer M. T., Thurman E. M., Zaugg S. D., Barber L. B. and Buxton T. (2002). Pharmaceuticals, hormones, and other organic wastewater contaminants in US streams, 1999–2000: a national reconnaissance. *Environmental Science and Technology*, **36**(6), 1202–1211.

LaPara T. M., Burch T. R., McNamara P. J., Tan D. T., Yan M. and Eichmiller J. J. (2011). Tertiary-treated municipal wastewater is a significant point source of antibiotic resistance genes into Duluth-Superior Harbor. *Environmental Science and Technology*, **45**(22), 9543–9549.

Lefkowitz J. R. and Duran M. (2009). Changes in antibiotic resistance patterns of *Escherichia coli* during domestic wastewater treatment. *Water Environment Research*, **81**(9), 878–885.

Li B. and Zhang T. (2011). Mass flows and removal of antibiotics in two municipal wastewater treatment plants. *Chemosphere*, **83**(9), 1284–1289.

Lindberg R. H., Olofsson U., Rendahl P., Johansson M. I., Tysklind M. and Andersson B. A. (2006). Behavior of fluoroquinolones and trimethoprim during mechanical, chemical, and active sludge treatment of sewage water and digestion of sludge. *Environmental Science and Technology*, **40**(3), 1042–1048.

Liu Y. S., Ying G. G., Shareef A. and Kookana R. S. (2012). Occurrence and removal of benzotriazoles and ultraviolet filters in a municipal wastewater treatment plant. *Environmental Pollution*, **165**, 225–232.

Liu Z. H., Kanjo Y. and Mizutani S. (2009). Removal mechanisms for endocrine disrupting compounds (EDCs) in wastewater treatment – physical means, biodegradation, and chemical advanced oxidation: a review. *Science of the Total Environment*, **407**(2), 731–748.

María P., Ormad N. M., Mosteo R., Rodríguez J. and Ovelleiro J. L. (2011). Study of the presence of pesticides in treated urban wastewaters. In: Pesticides in the Modern World-Risks and Benefits, Stoytcheva M. (ed.), Intech, Rijeka, Croatia, pp. 453–470.

McArdell C. S., Molnar E., Suter M. J. F. and Giger W. (2003). Occurrence and fate of macrolide antibiotics in wastewater treatment plants and in the Glatt Valley Watershed, Switzerland. *Environmental Science and Technology*, **37**(24), 5479–5486.

McNeill K. S. and Cancilla D. A. (2009). Detection of triazole deicing additives in soil samples from airports with low, mid, and large volume aircraft deicing activities. *B Environmental Contamination and Toxicology*, **82**(3), 265–269.

Meyer J. and Bester K. (2004). Organophosphate flame retardants and plasticisers in wastewater treatment plants. *Journal of Environmental Monitoring*, **6**, 599–605.

Miege C., Choubert J. M., Ribeiro L., Eusebe M. and Coquery M. (2009). Fate of pharmaceuticals and personal care products in wastewater treatment plants – Conception of a database and first results. *Environmental Pollution*, **157**(5), 1721–1726.

Morasch B., Bonvin F., Reiser H., Grandjean D., de Alencastro L. F., Perazzolo C., Chevre N. and Kohn T. (2010). Occurrence and fate of micropollutants in the Vidy Bay of Lake Geneva, Switzerland. Part II: micropollutant removal between wastewater and raw drinking water. *Environmental Toxicology and Chemistry*, **29**(8), 1658–1668.

Muller K., Bach M., Hartmann H., Spiteller M. and Frede H. G. (2002). Point- and nonpoint-source pesticide contamination in the Zwester Ohm catchment, Germany. *Journal of Environmental Quality*, **31**(1), 309–318.

Nelson J., Bishay F., van Roodselaar A., Ikonomou M. and Law F. C. (2007). The use of in vitro bioassays to quantify endocrine disrupting chemicals in municipal wastewater treatment plant effluents. *Science of the Total Environment*, **374**(1), 80–90.

Nitschke L. and Schüssler W. (1998). Surface water pollution by herbicides from effluents of waste water treatment plants. *Chemosphere*, **36**(1), 35–41.

North D. (2004). Tracking polybrominated diphenyl ether releases in a wastewater treatment plant effluent, Palo Alto, California. *Environmental Science and Technology*, **38**, 4484–4488.

Poiger T., Buser H. R., Balmer M. E., Bergqvist P. A. and Müller M. D. (2004). Occurrence of UV filter compounds from sunscreens in surface waters: regional mass balance in two Swiss lakes. *Chemosphere*, **55**(7), 951–963.

Prats D., Ruiz F., Vázquez B. and Rodriguez-Pastor M. (1997). Removal of anionic and nonionic surfactants in a wastewater treatment plant with anaerobic digestion. A comparative study. *Water Research*, **31**(8), 1925–1930.

Pruden A., Larsson D. J., Amézquita A., Collignon P., Brandt K. K., Graham D. W., Lazorchak J. M., Suzuki S., Silley P. and Snape J. R. (2013). Management options for reducing the release of antibiotics and antibiotic resistance genes to the environment. *Environmental Health Perspectives*, **121**(8), 878–785.

Rayne S. and Ikonomou M. G. (2005). Polybrominated diphenyl ethers in an advanced wastewater treatment plant. Part 1: Concentrations, patterns, and influence of treatment processes. *Journal of Environmental Engineering Science*, **4**(5), 353–367.

Reemtsma T., Miehe U., Duennbier U. and Jekel M. (2010). Polar pollutants in municipal wastewater and the water cycle: occurrence and removal of benzotriazoles. *Water Research*, **44**, 596–604.

Reiss R., Mackay N., Habig C. and Griffin J. (2002). An ecological risk assessment for triclosan in lotic systems following discharge from wastewater treatment plants in the United States. *Environmental Toxicology and Chemistry*, **21**(11), 2483–2492.

Schlumpf M., Cotton B., Conscience M., Haller V., Steinmann B. and Lichtensteiger W. (2001). In vitro and in vivo estrogenicity of UV screens. *Environmental Health Perspectives*, **109**(3), 239–244.

Servos M., Bennie D., Burnison B., Jurkovic A., McInnis R., Neheli T., Schnell A., Seto P., Smyth S. and Ternes T. (2005). Distribution of estrogens, 17β-estradiol and estrone, in

Canadian municipal wastewater treatment plants. *Science of the Total Environment*, **336**(1), 155–170.

Sinclair E. and Kannan K. (2006). Mass loading and fate of perfluoroalkyl surfactants in wastewater treatment plants. *Environmental Science and Technology*, **40**(5), 1408–1414.

Singer H., Müller S., Tixier C. and Pillonel L. (2002). Triclosan: occurrence and fate of a widely used biocide in the aquatic environment: field measurements in wastewater treatment plants, surface waters, and lake sediments. *Environmental Science and Technology*, **36**(23), 4998–5004.

Snyder S. A., Westerhoff P., Yoon Y. and Sedlak D. L. (2003). Pharmaceuticals, personal care products, and endocrine disruptors in water: implications for the water industry. *Environmental Engineering Science*, **20**(5), 449–469.

Speece R. E. (2008). Anaerobic Biotechnology and Odor/Corrosion Control for Municipalities and Industries. Archae Nashville, TN USA.

Svenson A., Allard A. S. and Ek, M. (2003). Removal of estrogenicity in Swedish municipal sewage treatment plants. *Water Research*, **37**(18), 4433–4443.

Swisher R. D. (1987). Surfactant Biodegradation, 2nd edn, Marcel Dekker, Inc. New York, N.Y. USA.

Szymanski A., Wyrwas B. and Lukaszewski Z. (2003). Determination of non-ionic surfactants and their biotransformation by-products adsorbed on alive activated sludge. *Water Research*, **37**(2), 281–288.

Tansel B. and Eyma R. R. (1999). Volatile organic contaminant emissions from wastewater treatment plants during secondary treatment. *Water Air Soil Pollution*, **112**(3), 315–325.

Tata P., Witherspoon J. and Lue-Hing C. (2003). VOC Emissions from Wastewater Treatment Plants: Characterization, Control, and Compliance. Lewis Publishers, Boca Raton, Florida, USA.

Tchobanoglous G., Stensel H. D., Tsuchihashi R., Burton F. L., Abu-Orf M., Bowden G. and Pfrang W. (2014). Wastewater Engineering: Treatment and Resource Recovery. McGraw-Hill Education, New York, NY.

Ternes T. A., Stumpf M., Mueller J., Haberer K., Wilken R. D. and Servos M. (1999). Behavior and occurrence of estrogens in municipal sewage treatment plants – I. Investigations in Germany, Canada and Brazil. *Science of the Total Environment*, **225**(1), 81–90.

Ternes T. A., Joss A. and Siegrist H. (2004). Peer reviewed: scrutinizing pharmaceuticals and personal care products in wastewater treatment. *Environmental Science and Technology*, **38**(20), 392A–399A.

U.S. EPA (1982). Fate of Priority Pollutants in Publicly Owned Treatment Works. United States Environmental Protection Agency, Washington DC.

Verlicchi P., Al Aukidy M. and Zambello E. (2012). Occurrence of pharmaceutical compounds in urban wastewater: removal, mass load and environmental risk after a secondary treatment-A review. *Science of the Total Environment*, **429**, 123–155.

Watkinson A., Murby E. and Costanzo S. (2007). Removal of antibiotics in conventional and advanced wastewater treatment: implications for environmental discharge and wastewater recycling. *Water Research*, **41**(18), 4164–4176.

Wee V. T. (1984). Determination of cationic surfactants in waste- and river waters. *Water Research*, **18**(2): 223–225.

Weiss S., Jakobs J. and Reemtsma T. (2006). Discharge of three benzotriazole corrosion inhibitors with municipal wastewater and improvements by membrane bioreactor treatment and ozonation. *Environmental Science and Technology*, **40**(23), 7193–7199.

Westerhoff P., Song G. X., Hristovski K. and Kiser M. A. (2011). Occurrence and removal of titanium at full scale wastewater treatment plants: implications for TiO_2 nanomaterials. *Journal of Environmental Monitoring*, **13**(5), 1195–1203.

Yang J., Wang K., Zhao Q., Huang L., Yuan C. S., Chen W. H. and Yang W. B. (2014). Underestimated public health risks caused by overestimated VOC removal in wastewater treatment processes. *Environmental Science: Processes Impacts*, **16**(2), 271–279.

Yi, T. and Harper W. F. (2007). The link between nitrification and biotransformation of 17α-ethinylestradiol. *Environmental Science and Technology*, **41**(12), 4311–4316.

Zitomer D. H. and Speece R. E. (1993). Sequential environments for enhanced biotransformation of aqueous contaminants. *Environmental Science and Technology*, **27**(2), 226–244.

Chapter 7

Integrating microbial and molecular tools to determine the fate and impact of hazardous pollutants

Berat Z. Haznedaroglu[1] *and Ulaş Tezel*[1]

[1]*Boğaziçi University, Institute of Environmental Sciences, 34342 Bebek, Istanbul, Turkey; E-mail: berat.haznedaroglu@boun.edu.tr; ulas.tezel@boun.edu.tr.*

7.1 INTRODUCTION

As covered in earlier chapters, our progress in the biological removal of hazardous pollutants in water and wastewater systems does not always achieve the same momentum with the accumulation of "hard to treat" and/or "hard to detect" chemicals in aquatic environments. Similarly, recent developments in microbial biotechnology fields are not necessarily catching up with the optimization of biotransformation protocols for the removal of hazardous pollutants.

Many new chemicals are continuously introduced into the market and most of them first past through wastewater treatment plants. Therefore, wastewater treatment plants are the hotspots of evolution of genes that involve the biotransformation of those new chemicals but are different from each other with respect to composition and structure depending on the characteristics of wastewater, geographical and operational conditions. From that point of view, answers to the questions that every regulatory agency in the U.S. and Europe asks such as "Is this chemical biodegradable?" may vary. These variations are mainly due to facts such as "microorganisms evolve fast", "not every microorganism can degrade every chemical", and "genes are mobile".

Atrazine is an s-triazine class herbicide having a diazine structure containing a third N within a ring and one Cl and two N-alkyl functional groups attached to the

 Hazardous Pollutants in Biological Treatment Systems: Fundamentals and a Guide to Experimental Research
Ferhan Çeçen, Ulaş Tezel
doi: 10.2166/9781780407708_265

ring. Atrazine is one of the water pollutants listed under the priority pollutant list of the European Commission with a maximum allowable surface water concentration of 2 μg/L. Atrazine was introduced into the market in the 1960s. Until the 1990s, there was no evidence of biodegradation of atrazine by bacteria. After 1993, different species of bacteria that can utilize atrazine as a carbon and nitrogen source have been isolated. Detailed analysis of the genes involved in the atrazine biodegradation revealed that the metabolic pathway converts atrazine to cyanuric acid in three steps which involve the *atzABC* gene cluster (Figure 7.1). All three enzymes that were coded by the *atzABC* cluster belong to the aminohydrolase family of enzymes closely related to *trzA* that catalyzes the melamine biotransformation. Few bacteria catabolize atrazine, while many catabolize cyanuric acid. One atrazine-degrading bacterium, *Pseudomonas* sp. ADP, was shown to contain all six genes, *atzABCDEF*, on a broad-host-range 96-kb plasmid. Several genera of bacteria such as *Ralstonia*, *Alcaligenes*, *Agrobacterium*, and *Clavibacter*, which contain virtually identical *atzABC* genes and grow on atrazine as their sole nitrogen and carbon source, have been isolated in North America, Europe, Australia, and Asia. Although the genes have been found on plasmids, no single type of plasmid is common to all atrazine-degrading bacteria, suggesting that they are mobile and easily transferred among bacteria (Singh & Jauhari, 2017).

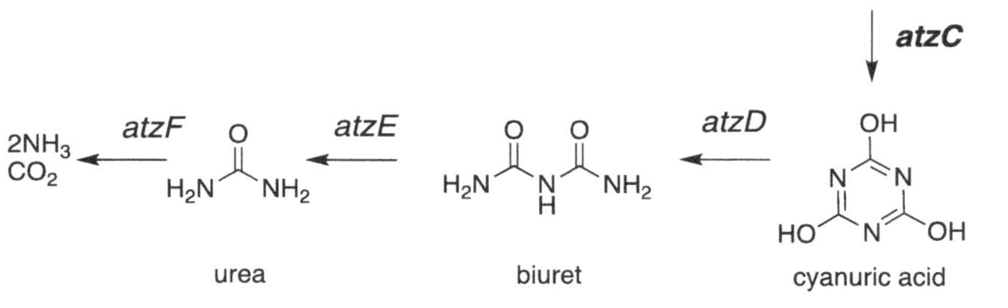

Figure 7.1 Biodegradation pathway of atrazine.

Biotransformation of atrazine is a very famous example of pathway evolution. In this example, we can understand that genes responsible for biodegradation of a compound may evolve to code enzymes that can transform structurally similar compounds which were previously nonbiodegradable within time as microorganisms are exposed to that compound. These genes may be integrated into mobile genetic elements and transferred among different bacteria. That is, a bacterium which cannot degrade a certain pollutant can acquire a mobile genetic element that contains a gene and become a degrader.

Wastewater contains hundreds and thousands of pollutants. Microorganisms inhabiting biological treatment systems are continuously exposed to those chemicals. Given the fact that microbial diversity in the treatment systems is high, wastewater treatment systems are one of the hotspots of evolution of pollutant biotransformation pathways and dissemination of the genes involved in the pathways. For that reason, wastewater treatment is one of the fields in which molecular biology tools are crucial. The molecular biology tools are used to address three crucial questions of microbial ecology. These are "Who is there?", "How many are there?" and "What are they doing?" Similarly, these tools can be used to answer the same questions in the removal of hazardous pollutants: "Who are the degraders?", "How are they doing the degradation?" and "How many of them are present in my system?" Although molecular biology is more than a tool to assess the biodegradation of hazardous pollutants, it has been always overruled by engineers, plant operators and regulation agencies. Therefore, there is no systematic roadmap to guide people of environmental biotechnology on how to use molecular biology to answer the crucial questions given above.

In an effort to address the gap of knowledge, or window of opportunity as we would like to call it, a roadmap of identification of microorganisms that play a role in the fate of hazardous chemicals in wastewater treatment systems, and determination of their numbers, as well as elicitation of mechanisms and genetics of biodegradation pathways are given in this chapter.

7.2 IDENTIFICATION OF THE "DEGRADER"

While assessing the biodegradability of a hazardous pollutant or the potential of a biological treatment system to remove the pollutant, identification of the microorganisms that can degrade the pollutant ("degraders") is important. There are culture-dependent and independent methods as well as in situ methods to identify the "degrader". Culture-dependent methods involve development of a microbial community that contains mostly degraders by feeding the inoculum taken from the source, i.e. activated sludge with a pollutant. This is followed by the isolation and identification of the degrader from the enrichment culture (Figure 7.2). On the other hand, culture-independent methods aim to identify degraders in their natural environment without culturing them in the laboratory, thus avoiding the enrichment bias. In culture-independent methods, mainly the pollutant is labeled so that the microorganism that can degrade the pollutant can be distinguished from the other microorganisms in the complex microbial community (Figure 7.2). Recent developments in culture-independent methods now enable us to identify degraders in the environment that are not cultivable.

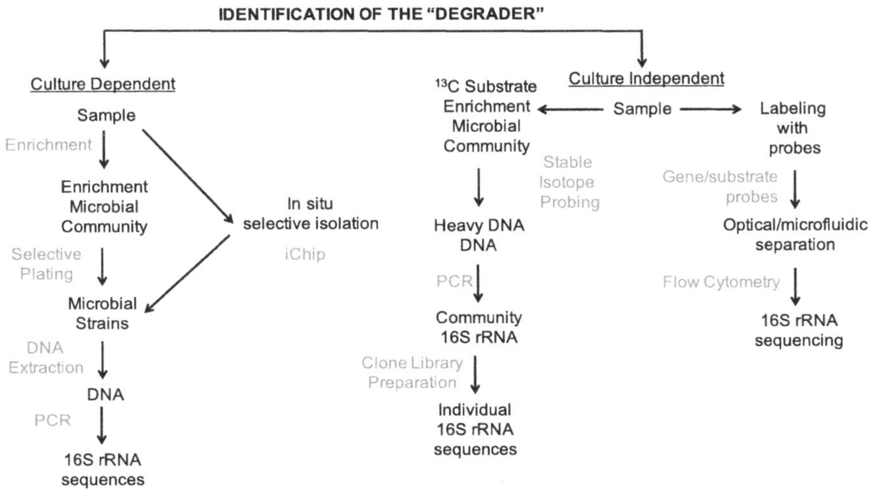

Figure 7.2 Flow chart of culture-dependent and independent methods to identify microbial strains that can degrade a hazardous pollutant in a microbial community.

7.2.1 Culture-dependent methods

The main objective of culture-dependent methods is to isolate the microbial strains that can degrade the target hazardous pollutant from a complex microbial community. Culture-dependent methods to identify the degraders in a microbial community involve enrichment, isolation and identification of the isolates based on their 16S rRNA gene sequences (Figure 7.3).

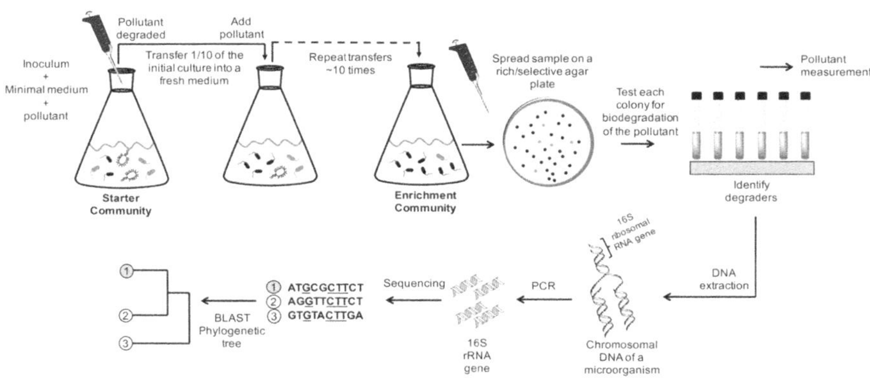

Figure 7.3 Pipeline of isolation and identification of "degraders" in a microbial community.

7.2.1.1 Enrichment and isolation

Most of the conventional wastewater treatment systems use activated sludge as the main process to remove carbon and nutrients from wastewater. Activated sludge is a biological process which utilizes mixed-culture microbial communities. About 70% of microorganisms present in activated sludge are common in most of the treatment plants (Xia *et al.* 2010; Saunders *et al.* 2016). This core community composed of microorganisms belong to about 70 different genera (Xia *et al.* 2010; Saunders *et al.* 2016). The rest of the community is diverse and changing based on geography, wastewater characteristics, plant operation etc. (Griffin & Wells, 2017). Besides the readily biodegradable organic matter in a wastewater which shapes the core community, there is a minor portion of organics which is composed of hazardous pollutants or other micropollutants. Microorganisms that are tolerant and capable of degrading hazardous pollutants make up the variable portion of the activated sludge community. Biodiversity in this variable portion is high. Therefore, identification of microorganisms that can degrade a specific pollutant is challenging, not only due to high diversity in the microbial community but also the low abundance of the degrader which subsists on trace amounts of organics in that community.

One way of circumventing this challenge is isolation of the degrader from the microbial community. As a result, not only can the identity of the degrader be determined, but also the biochemical pathways by which a specific hazardous pollutant is biodegraded can be elucidated. Isolation involves pre- and post-processes. Before isolation, enrichment is crucial since a typically activated sludge microbial community contains about 10^9 cells per mL, most of which are involved in degradation of readily biodegradable organics (Manti *et al.* 2008). If the compound being studied is an emerging contaminant and present in the wastewater at very low concentrations, there is either no or a very small number of bacteria capable of metabolizing it. Even if some are present, and they are plated at a typical dilution of 10^4 to 10^6 on a selective plate, it is likely that no colonies of bacteria that can degrade the pollutant will be obtained. Moreover, if a sample taken from activated sludge is plated on a selective agar containing only the pollutant of interest without dilution, it is unlikely to obtain colonies that degrade the substance, perhaps because of antibiosis or other competing effects derived from some members of the large non-degrading population. To overcome the problems associated with isolation of the degraders directly from the original sample, enrichment is performed.

The aim of enrichment is to increase the abundance of the degraders in the original community. The compound to be degraded is supplied as the growth-limiting, and usually sole source of an essential nutrient in a culture-medium. Only the organism(s) with the necessary degradative ability will grow significantly under these conditions and these organisms will outgrow the very large number of other organisms also added at the start of the experiment. The community is diluted several times to obtain a population composed of microorganisms degrading the pollutant. The transfers are typically performed with a dilution factor of 10, that is

1/10 of the initial culture is transferred into a new medium containing the pollutant and this is repeated at least 10 times after all the pollutants are degraded.

Before enrichment cultures are started, growth medium, growth conditions, inoculum size and the physicochemical properties of the pollutant, as well as the analytical methods to monitor the pollutant, should be determined.

Different minimal media have been developed to isolate different genera of bacteria. However, the type of bacteria that is actively degrading the pollutant in an activated sludge is not known. In the absence of such information, it is often prudent to use a relatively rich minimal medium containing many different trace metals to avoid limitations of inorganic compounds. It is generally wise to use a high buffer content. The pH chosen depends on the organisms that have to be enriched; typically, a pH around neutrality is used. Microorganisms consist of the elements C, O, N, H, P and S, varying in that order from near 50% to near 1% of dry weight, as well as ions (Mg^{2+}, K^{+}) and trace elements (metals), and these components at least must be present in the growth medium. From this point of view, the medium used in Organisation for Economic Co-operation and Development (OECD) biodegradation experiments (OECD medium, **Chapter 5**) is preferred to start enrichments. The M9 minimal medium which has all the constituents in the OECD medium but with a higher ionic strength and buffering capacity at pH 7, is preferred by microbiologists (Table 7.1). Both media have about neutral pH values. The M9 medium should be preferred when enrichment is done at a high pollutant concentration to overcome nutrient limitations. The ionic strength of M9 medium can be adjusted to a desired value by diluting it with water if needed. Both media should be supplied with trace metals and vitamins since they are constituents of many enzymes or serve as cofactors or coenzymes. For instance, iron which is present in the catalytic site of many oxidoreductases should be added into the medium. In addition, Ni is essential for methanogens. Vitamins act as cofactors and the absence of some vitamins may prevent biodegradation although a degrader is present in the community. For instance, vitamin B12 enhances the dechlorination of chlorinated solvents by *Dehalococcoides* species, since the strains of this species are corrinoid-auxotrophic and axenic which require B12 to conserve energy in halorespiration (He *et al.* 2007; Yan *et al.* 2013).

Growth conditions are important and must mimic the conditions of the environment from where the sample is taken for enrichment. Activated sludge microorganisms are mesophilic, aerobic and neutrophilic organisms (Figure 7.4). Therefore, enrichment performed in a medium with a pH around 7.0 at room temperature in a flask loosely capped to avoid contamination is successful. The doubling time of the degrader is important since it is related to the specific growth rate. As known, activated sludge supports the retention of those microorganisms with a doubling time less than the solids retention time (SRT) applied to the bioreactor. The preferred SRT for enhanced micropollutant degradation in activated sludge was determined as 10 days (Clara *et al.* 2005). This information is important since it limits the optimum transfer frequency. After starting the enrichment process, the typical time of transferring culture to a new medium is when the whole pollutant is degraded. Given the fact that the culture is going to be diluted 10-fold, the degrader population after the first amendment of the pollutant should reach a level such that the transfer of the degrader to the next culture is ensured. Otherwise, multiple feeding of the start-up culture is suggested. If the pollutant is not degraded fast, 10 days is a reasonable time to wait for an indication of biodegradation, otherwise this suggests that the sample did not contain any degraders.

Table 7.1 The compositions of two commonly used media for enrichment of degraders in activated sludge.

OECD Medium		M9 Minimal Medium	
Ingredient	**Amount (mg/L)**	**Ingredient**	**Amount (mg/L)**
KH_2PO_4	85	KH_2PO_4	3,000
K_2HPO_4	217.5	$Na_2HPO_4 \cdot 2H_2O$	7,520
$Na_2HPO_4 \cdot 2H_2O$	334.0	NaCl	500.0
NH_4Cl	5.0	NH_4Cl	500.0
$CaCl_2 \cdot 2H_2O$	36.4	$CaCl_2 \cdot 2H_2O$	147.0
$MgSO_4 \cdot 7H_2O$	22.5	$MgSO_4 \cdot 7H_2O$	246.5
Trace metals[a]	see Table 7.2	Trace metals[a]	see Table 7.2
Vitamins[a]	see Table 7.2	Vitamins[a]	See Table 7.2

[a]Optional, not present in the original recipe.

Table 7.2 Suggested concentrations of typical trace metals and vitamins in an enrichment medium.

Trace Metals		Vitamins	
Ingredient	**Amount (mg/L)**	**Ingredient**	**Amount (mg/L)**
$ZnCl_2$	0.5	Biotin	0.2
$MnCl_2 \cdot 4H_2O$	0.3	Folic acid	0.2
H_3BO_3	3.0	Pyridoxine hydrochloride	1.0
$CoCl_2 \cdot 6H_2O$	2.0	Riboflavin	0.5
$CuCl_2 \cdot 2H_2O$	0.1	Thiamine	0.5
$NiSO_4 \cdot 6H_2O$	0.2	Nicotinic acid	0.5
$Na_2MoO_4 \cdot 2H_2O$	0.3	Pantothenic acid	0.5
$FeCl_2 \cdot 6H_2O$	0.25	*Vitamin* B12	0.01
		p-Aminobenzoic acid	0.5
		Thioctic acid	0.5

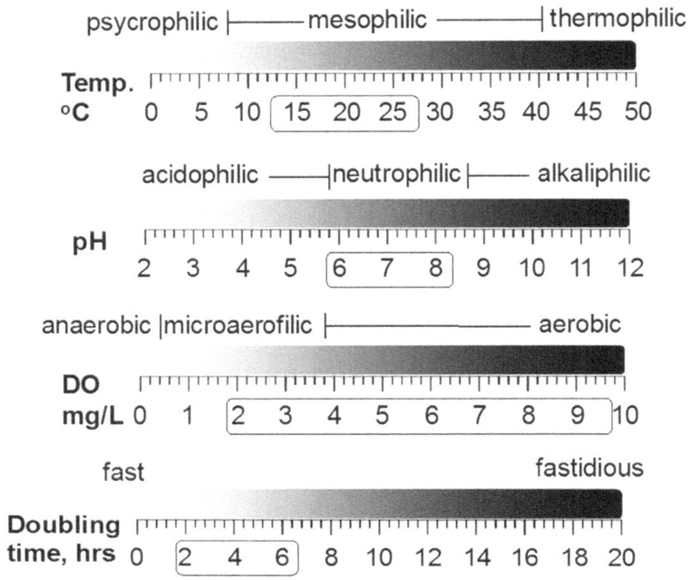

Figure 7.4 Important conditions influencing microorganisms and the range of typical values (shown in grey boxes) that can be used while performing enrichment from activated sludge.

If the terminal electron accepting process (TEAP) under which the pollutant is getting degraded is known, the electron acceptor has to be selected accordingly before enrichment. The TEAP under which the enrichment is performed often affects the choice of the medium, whether reducing agents and redox indicators are added, and the types of reactor in which the enrichment is performed. For instance, if the pollutant is known to degrade in the anoxic compartment of a biological nutrient removal system, the sample taken from that compartment is transferred into a bottle along with the medium and the bottle is closed with a crimp-sealed septum. Intrusion of oxygen must be eliminated. In addition, nitrate is added as the electron acceptor into the bottle to facilitate pollutant degradation.

The size of the inoculum to start up an enrichment culture is another factor that should be considered. Best practice is to start the culture without diluting it, to avoid the loss of less abundant species that may be degraders. However, starting a culture with the original sample may be problematic due to particles, residual organics or high biomass concentration. The presence of such solid material may interfere in the analysis of the pollutant or lead to adsorption of the pollutant. Therefore, multiple start-up cultures should be prepared which are diluted with the growth medium to different extents.

While preparing an enrichment culture, the target pollutant is used as the only source of carbon in the medium. Sometimes a pollutant cannot be obtained in its pure form, or it is expensive. Therefore, it is not economically feasible to use it at the very high amounts which the enrichment process requires. Since structurally similar compounds are degraded similarly, model chemicals are used in enrichment to isolate microorganisms that can degrade pollutants in the same class (Table 7.3). However, before starting enrichment, physicochemical properties of the pollutant such as volatility, hydrophobicity, solubility, etc. as well as its toxicity to microorganisms should be identified, so that the chemical can be used correctly in the enrichment medium. For example, it is often a good practice to add pollutants with low water solubility and supplied in powder form at concentrations well above their solubility in the enrichment medium. The compound will continuously dissolve as it is being degraded in the liquid phase, and it is rarely toxic to bacteria. On the other hand, pollutants that have low solubility and present in liquid form at room temperature, such as hydrocarbons and halogenated solvents, must not be added directly into the medium since formation of two liquid phases in the enrichment medium is generally very toxic to many bacteria. This problem can be overcome neatly with volatile, water-insoluble pollutants by providing the compound as a vapor, with the liquid suspended in a glass bulb above the growth medium in an enrichment flask. This provides the bacteria with a continuous supply of substrate without generating a two-phase system (Wackett & Hershberger, 2001a). In addition, highly water soluble but toxic compounds such as biocides, like quaternary ammonium compounds or linear alkyl sulfonates, should be supplied cautiously. Addition of those pollutants adsorbed on inorganic solid particles such as silica into the growth medium is suggested to sustain high growth but low toxicity (van Ginkel *et al.* 2008). The initial concentration of the pollutant at the start-up as well as in the following generations has to be as high as possible to facilitate the growth of the degrader. This would then increase the abundance of the degrader over other microorganisms that cannot degrade the pollutant. However, toxicity of the pollutant to the degrader should be considered. Therefore, the preferred initial concentration should be the maximum concentration that does not exert toxicity to the degrader, in the case where the toxicity of the pollutant is known. Otherwise, the pollutant should be supplied as adsorbed on a particle or different enrichments should be prepared at different initial concentrations.

After transfers, the resulting microbial community should ideally contain only the degraders which may be different species or different strains of the same species. Unfortunately, this is not true all the time. Enrichment cultures generally contain other microorganisms that can degrade the chemical products produced during the transformation of the pollutant if the pollutant follows a multistep pathway. That is, transformation by-products of a pollutant may serve as a carbon source for other microorganisms and support their survival and proliferation in the culture. In order to differentiate degraders from non-degraders and identify the diversity of the degraders in an enrichment culture, isolation is applied.

Table 7.3 Representative substrates utilized for enrichment methods for culturing bacteria.

Chemical Group	Chemical	Reference
Alkane hydrocarbons	CH_4 Methane	(Sukesan & Watwood, 1998)
	 Pentane	(Garnier *et al.* 1999)
	 Propane	(Sukesan & Watwood, 1998)
Alkene hydrocarbons	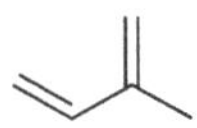 Isoprene	(van Hylckama Vlieg *et al.* 1998)
Alkyl acids (tertiary)	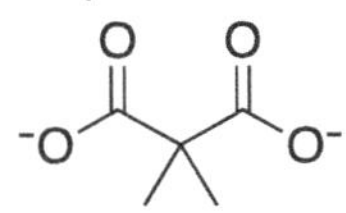 Dimethylmalonate	(Kniemeyer *et al.* 1999)
Alkyl amines	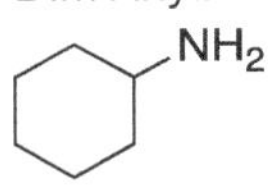 Cyclohexylamine	(Iwaki *et al.* 1999)
	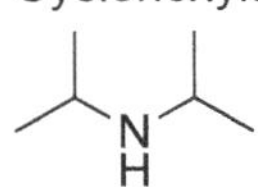 Diisopropylamine	(Gieg *et al.* 1999)
	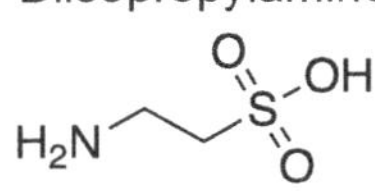 Taurine	(Lie *et al.* 1999)
Alkyl sulfides	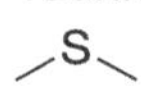 Dimethyl sulfide	(Lomans *et al.* 1999)
Aromatic acids	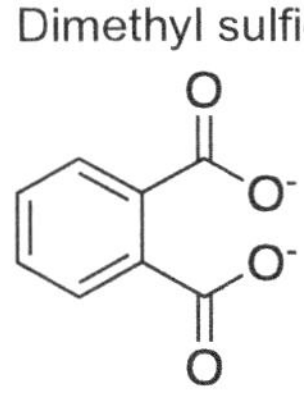 Phthalates	(Kleerebezem *et al.* 1999)
Aromatic amines	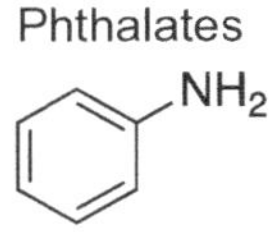 Aniline	(Tan *et al.* 2015)

Table 7.3 Representative substrates utilized for enrichment methods for culturing bacteria (*Continued*).

Chemical Group	Chemical	Reference
Aromatic hydrocarbons	 Benzene	(Burland & Edwards, 1999; Veiga *et al.* 1999)
	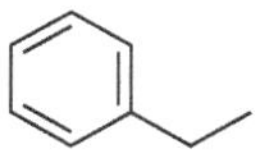 Ethylbenzene	(Phelps & Young, 1999; Veiga *et al.* 1999)
	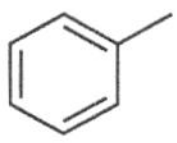 Toluene	(Hubert *et al.* 1999; Veiga *et al.* 1999)
	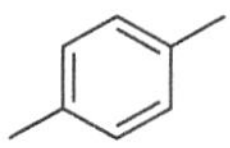 Xylene	(Harms *et al.* 1999)
Chlorinated alkenes	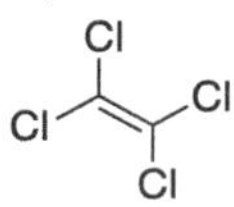 Tetrachloroethene	(Kengen *et al.* 1999)
	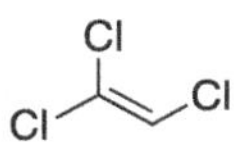 Trichloroethene	(Bielefeldt & Stensel, 1999)
Chlorobiphenyls	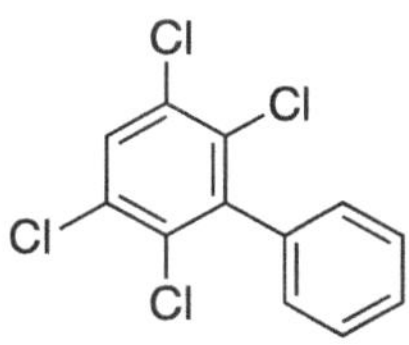 2,3,5,6-Tetrachlorobiphenyl	(Cutter *et al.* 1998; Holoman *et al.* 1998)
Halobenzoates	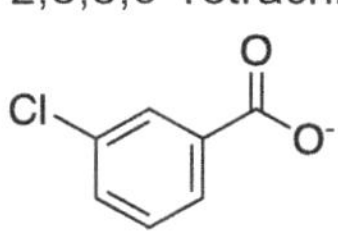 3-Chlorobenzoate	(Häggblom & Young, 1999)
N-Heterocyclic rings	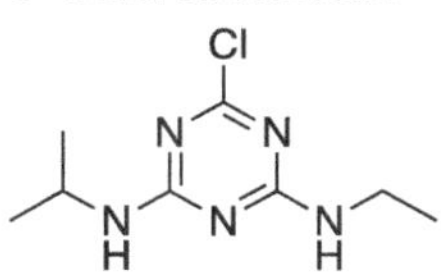 Atrazine	(Kontchou & Gschwind, 1999)

(*Continued*)

Table 7.3 Representative substrates utilized for enrichment methods for culturing bacteria (*Continued*).

Chemical Group	Chemical	Reference
	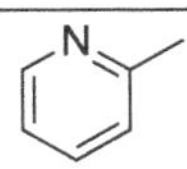 2-Methylpyridine	(O'loughlin *et al.* 1999)
	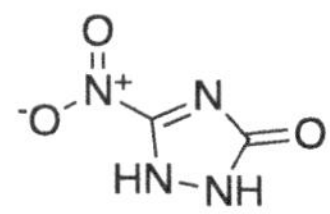 5-Nitro-1,2-dihydro-1,2,4-triazol-3-one	(Le Campion *et al.* 1999)
O-Heterocyclic rings	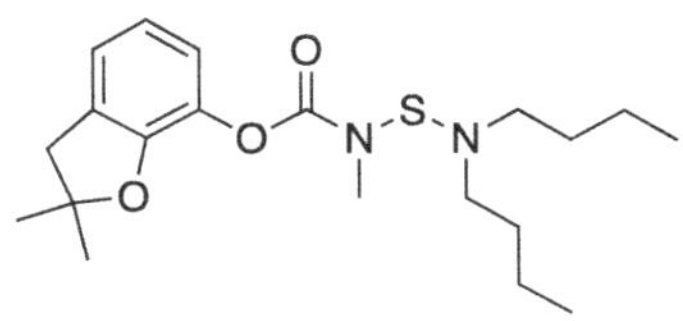 Carbosulfan	(Sahoo *et al.* 1998)
S-Heterocyclic rings	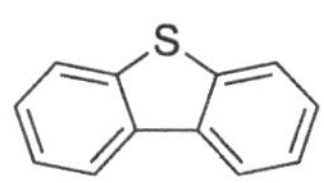 Dibenzothiophene	(Chang *et al.* 1998)
Nitriles	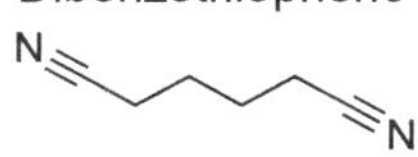 Adiponitrile	(Dhillon & Shivaraman, 1999)
Nitroaliphatics	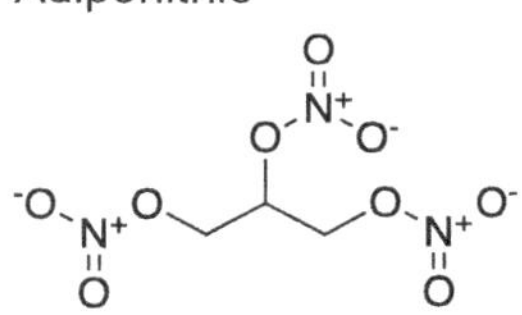 Nitroglycerin	(Accashian *et al.* 1998)
Nitroaromatics	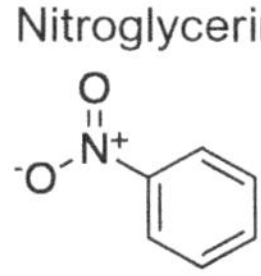 Nitrobenzene	(Zhao & Ward, 1999)
Organophosphorus pesticides	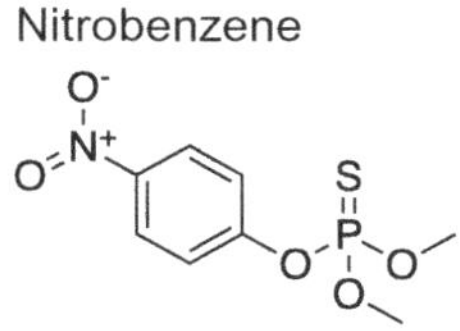 Methylparathion	(Ramanathan & Lalithakumari, 1999)

Table 7.3 Representative substrates utilized for enrichment methods for culturing bacteria (*Continued*).

Chemical Group	Chemical	Reference
Phenols	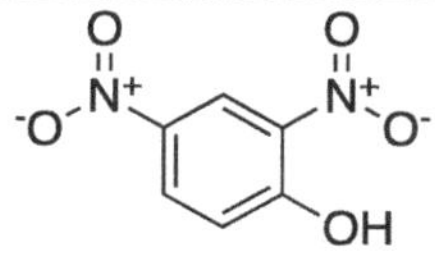 2,4-Dinitrophenol	(Blasco *et al.* 1999)
	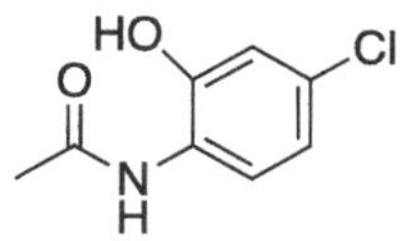 4-Chloro-2-hydroxyacetanilide	(Park *et al.* 1999)
	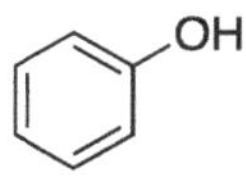 Phenol	(van Schie & Young, 1998; Watanabe *et al.* 1998)
	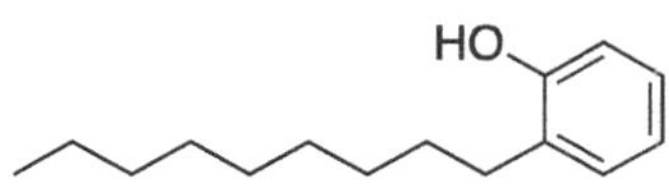 Nonylphenol	(Tanghe *et al.* 1999)
Phenylureas	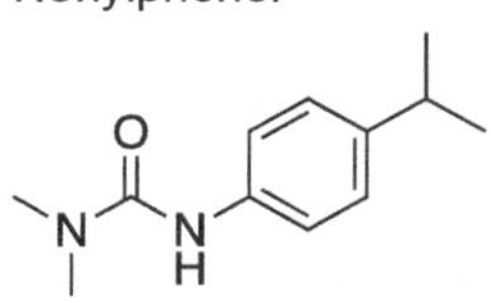 Isoproturon	(Roberts *et al.* 1998)
Polycyclic aromatic hydrocarbons	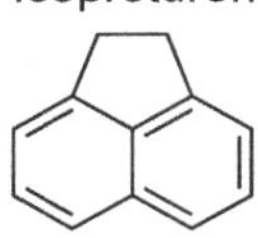 Acenaphthene	(Selifonov *et al.* 1998)
	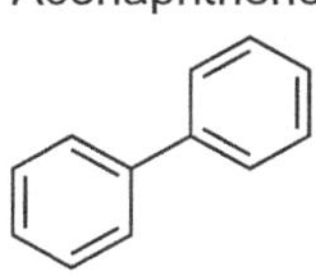 Biphenyl	(Wagner-Döbler *et al.* 1998)
	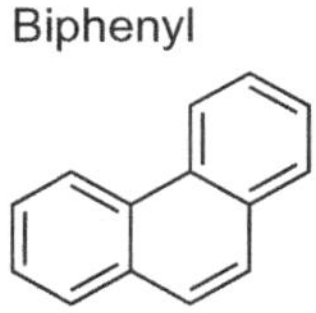 Phenanthrene	(Aitken *et al.* 1998)

(*Continued*)

Table 7.3 Representative substrates utilized for enrichment methods for culturing bacteria (*Continued*).

Chemical Group	Chemical	Reference
	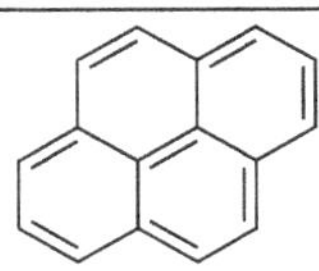 Pyrene	(Molina *et al.* 1999)
Polymers	Poly-hexamethylene carbonate	(Suyama *et al.* 1998)
Sulfonates	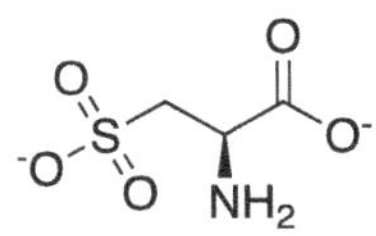 Cysteate	(Mikosch *et al.* 1999)
Terpenes	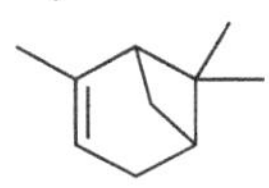 *alpha*-Pinene	(Kleinheinz *et al.* 1999)
	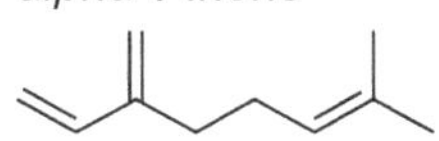 *beta*-Myrcene	(Lurescia *et al.* 1999)

Source: Adapted from Wackett and Hershberger (2001a).

Isolation is performed with a sample taken from the enrichment culture. A certain amount of sample is diluted, till it contains about 100 microorganisms, and spread onto a solid growth medium. This technique is called spread plating. The solid growth medium contains the ingredient of the enrichment medium plus the pollutant as the carbon source and 1.5% agar as the solidifying agent. The solid medium described above is called the "defined selective medium" since the concentration of every ingredient in the medium is known and it contains a single carbon source that supports the growth of microorganisms that can degrade it. Ideally every colony formed on the solid medium is produced from a single microorganism that can degrade the pollutant. However, spread plating on a selective medium should follow streaking the colonies formed on the plate to a fresh selective plate. After several successive streaking of well separated colonies, a pure culture of a degrader can usually be obtained (Chess, 2015). On the other hand, the pollutant degradation capability of the isolates should be confirmed in liquid medium after isolation. Sometimes organic impurities in agar promote the growth of non-degraders and they form colonies on solid medium although they cannot degrade the pollutant.

The pollutant concentration in the solid medium must be high in order to promote growth of the degrader and formation of visible colonies. However, very high pollutant concentrations may be toxic to microorganisms. In that case, a solid medium rich in organics such as Luria Bertani, Mueller Hinton or M2 agars containing yeast extract, dextrose, tryptone, casein, meat infusion, etc. is used. If isolation is done using a rich medium, biodegradation activity of each isolate must be confirmed in a liquid medium.

Phylogenetic classification of each isolate is performed as described in Section 7.2.1.3.

7.2.1.2 iChip

A defined medium is used while developing enrichment cultures. The limited number of salts, trace metals and vitamins used in an enrichment medium do not fully represent the original matrix from where the inoculum is taken. The media described are designed to isolate aerobic organisms with simple nutritional requirements. Thus, strains of *Pseudomonas*, *Alcaligenes*, *Klebsiella, Rhodococcus, Arthrobacter, Sphingomonas*, and *Burkholderia* are commonly isolated (Wackett & Hershberger, 2001b). Those species adopt better and take over desired communities under laboratory conditions while having less of a role in natural environments. In addition, microorganisms isolated by enrichment might not behave in the same way in their native environment compared to the laboratory.

It has been long known that enrichment-based culturing methods are limited, and only less than 1% of microorganisms on earth are cultivable due to unsuitable growth media and conditions used during enrichment. As the culture-independent methods emerge, it is shown that diversity of microorganisms that have biodegradation capacities is higher than expected. Although we can identify degraders in the environment with culture-independent methods as described below, isolation of degraders is still crucial to understand the biochemistry and genetics of pollutant degradation.

Relatively recently an in situ isolation technique called iChip has been developed. iChip circumvents the limitations of enrichment, mainly by eliminating the use of a growth medium. iChip apparatus is designed to capture the microorganisms while they are growing in their natural environment at native conditions (Nichols *et al.* 2010). It was shown that iChip technology can recover about 30% of the microorganisms in seawater and soil which was ten times higher than the amount recovered by standard plating. Most of the microorganisms isolated by iChip were novel, suggesting that uncultivable microorganisms can be isolated using iChip technology (Nichols *et al.* 2010). Recently, iChip was also used to isolate another novel bacterium which produces a new type of antibiotic (Ling *et al.* 2015).

The iChip is an assembly of flat plates made of hydrophobic plastic polyoxymethylene containing an array of through-holes (Figure 7.5). It is composed of one central plate and two cover plates. The assembly is 72 mm × 19 mm (length × width) in size. The central plate is 1 mm thick whereas the cover plates are 6.5 mm thick. The plates have multiple through-holes 1 mm in diameter, arranged in two arrays with 192 through-holes per array. The size of the array is such that it can be completely covered by standard 25- or 47-mm-diameter membranes (Nichols *et al.* 2010).

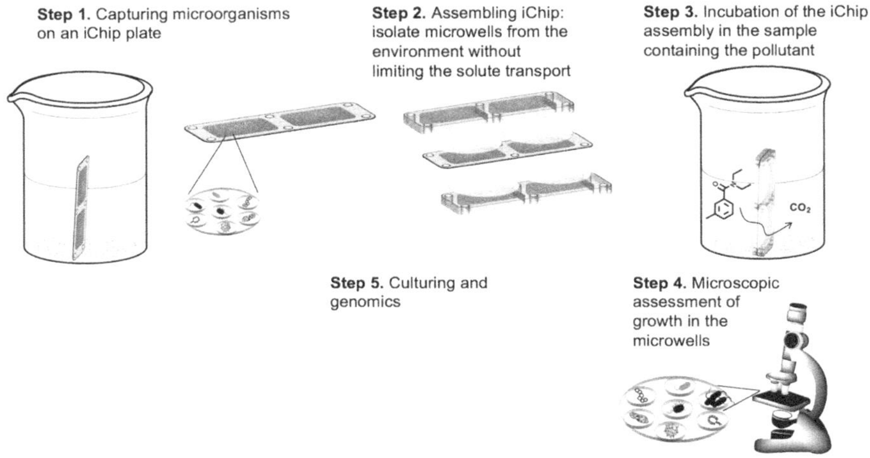

Figure 7.5 In situ isolation of degraders by iChip technology.

The pipeline of isolation of degraders by iChip starts with sterilization of plates with ethanol and rinsing them with particle-free DNA-grade water. The central plate is then dipped into a sample which is the source of microorganisms targeted for cultivation (Figure 7.5, Step 1). As the central plate is taken out of the sample, each through-hole captures a drop of suspension containing a certain number of microorganisms. The number of cells in the holes depends on the degree of dilution applied to the sample. Ideally, the target is to capture one cell into each hole (Figure 7.5, Step 1). Addition of solidifying agents with low melting point into the sample suspension is recommended for better immobilization of the cells entrapped in the holes. In the next step, the top and bottom of the central plate are covered with 47 mm polycarbonate membranes with an average pore size of 0.03 μm. Then, the cover plates and the central plate are attached to each other (Figure 7.5, Step 2). Afterwards, the iChip assembly is incubated in the original sample where the isolation is performed. The assembly allows nutrients in the original environment of the microorganisms to diffuse into the holes while preventing the intrusion of microorganisms in the environment as well as escape of the microorganisms entrapped in the holes. Therefore, subsequent in situ incubation of the microorganisms in the original environmental niche provides the immobilized microorganisms with their naturally occurring nutrients and growth factors (Figure 7.5, Step 3). Adding the pollutant into the original sample may facilitate the growth of the degraders in the holes (Figure 7.5, Step 3). After incubation, iChip is washed with sterile DNA-free water and disassembled. The central plate is then examined under a compound or high-power dissecting microscope and growth in the holes can be measured (Figure 7.5, Step 4). Agar plugs containing the microorganisms are extracted for phylogenetic identification as well as further culturing of the isolates (Figure 7.5, Step 5).

7.2.1.3 Phylogenetic classification of the degrader isolates

The next step in the culture-based identification of the degraders in a community is phylogenetic classification of the isolates. Each isolate obtained via enrichment or iChip with a confirmed pollutant biodegradation capability has to be classified based on 16S rRNA gene sequence. The procedure involves DNA extraction, polymerase chain reaction (PCR) amplification of the 16S rRNA gene, amplicon clean-up, sequencing and bioinformatics analysis.

Phylogenetic classification of a microorganism is done based on its 16S rRNA gene sequence since the 16S rRNA gene is a part of a gene cluster that encodes ribosome which is present in all prokaryotic microorganisms. This gene has both evolutionary highly conserved positions as well as highly variable positions in its nucleotide sequence. In addition, this gene has not been subjected to lateral gene transfer which makes it an ideal marker for phylogenetic classification of prokaryotes. This gene is present in the chromosome of every microorganism. In order to get the sequence of this gene, first the chromosome, i.e. the DNA, of the microorganism must be obtained.

For isolated microorganisms, DNA extraction is as simple as boiling a well grown colony in water. However, depending on the species, DNA yield and quality may vary. For reproducible results, a DNA extraction should involve cell disruption and lysis, protein removal, chemical removal and DNA purification. These steps can be modified and combined in different ways depending on the target organisms, type of sample, and the purpose for the extracted DNA. Today, DNA extraction is performed with commercial kits. Various studies confirmed that the FastDNA Spin kit for soil (MP Biomedicals, USA) is the most efficient kit on the market that yields a high amount of high quality DNA from a broad spectrum of microorganisms that can be used not only for PCR but also for next generation sequencing (Guo & Zhang, 2013; Aziz *et al.* 2015).

Depending on the downstream processing, it is important to quantify the DNA, and check its purity and quality. Quantification is performed either by fluorimetric or spectrophotometric methods. Among them, the most popular quantification method is the Nanodrop photometric method. The absorbance is a natural property of DNA, RNA, free nucleotides, proteins and some amino acids, and many other compounds. Photometric methods quantify DNA based on its absorbance at 260 nm. On the other hand, many other molecules absorb light at 260 nm, and this measurement is subject to inaccuracy due to potential contamination of the sample with other molecules such as RNA. Therefore, it is not possible to distinguish between DNA, RNA, protein or free nucleotides, or amino acids in the sample using photometric methods which leads potentially to highly inaccurate measurements. However, extraction of DNA with a decent kit generally yields a very good quality DNA. The quantity measured with photometric methods can be verified using the quality criterion which is the ratio of sample absorbance at 260 and 280 (A_{260}/A_{280}). This value should be around 1.8 in order to trust the quantification reported by a nanophotometer. In addition, nanophotometers are generally accurate at measuring DNA concentrations above 20 ng/μL. Fluorimetric methods are more sensitive and accurate. For instance, the Qubit DNA assay kit uses fluorescence dyes specific for DNA, that is why quantification is not affected by the RNAs, proteins or other molecules present in the sample. This kit can quantify DNA accurately at the very low concentrations needed for most sequencing based methods. The size distribution of the DNA can be determined using classical gel electrophoresis or via newer, more sensitive instruments such as the 2100 Bioanalyzer and TapeStation 4200 (Agilent Genomics).

PCR is used to selectively amplify the 16S RNA gene in the genomic DNA of the microorganism to bring its concentration to a level such that it can be sequenced by the current Sanger sequencing technology. PCR uses a thermostable polymerase enzyme to duplicate a certain fragment of DNA. Duplication of this particular fragment, i.e. the 16S rRNA gene, is guided by small DNA sequences called primers which are externally added to the reaction mixture. If the 16S rRNA gene is targeted for amplification, two primers complement to the start (Forward Primer) and the end (Reverse Primer) of the fragment that is amplified are designed. There are universal primers that are used to amplify the 16S rRNA gene of almost all bacterial and archaeal species. The consensus is to use 27F and 1492R as the bacterial primers (Mao *et al.* 2012), and 340F and 1000R as the archaeal primers (Gantner *et al.* 2011) (Table 7.4). The sequence length of the fragments amplified with these primers are sufficient for phylogenetic classification of microorganisms at species level.

Table 7.4 Universal primers used for amplification of 16S rRNA gene of bacteria and archaea.

Primer	Sequence
Bacteria	
27F	5′-AGA GTT TGA TYM TGG CTC AG- 3′
1492R	5′-TAC CTT GTT AYG ACT T- 3′
Archaea	
340F	5′-CCC TAY GGG GYG CAS CAG- 3′
1000R	5′-GGC CAT GCA CYW CYT CTC- 3′

During the PCR, cycles of heating and cooling are used to enable the amplification. First, the double stranded DNA is denatured by heating, which separates the two strands from each other. Secondly, the reaction is cooled to enable the primers to anneal to their target sites on the DNA. Thirdly, the temperature is increased again to provide an optimal temperature for the polymerase activity, which starts duplicating the 16S rRNA gene by extending the primers. One cycle creates two copies of the 16S rRNA gene from the template DNA. The cycle is repeated

25–35 times during the PCR, which results in an exponential amplification of the 16S rRNA gene. Other than the isolate DNA, primers and polymerase enzyme, the reaction contains the mixture of nucleotides which is called dNTPs, buffers and other additives such as Mg^{2+} salts that provide the optimal conditions for primer annealing and polymerase activity. PCR reaction may need to be optimized depending on the species. The optimization parameters are generally denaturation temperature, salt concentration and the number of cycles. PCR amplicons are purified and then sequenced.

Usually it is possible to determine phylogenetic classification of the isolate at species level by its 16S rRNA gene sequences using the Basic Local Alignment Search Tool (BLAST) algorithm and online databases such as GenBank (Benson *et al.* 2017), the European Molecular Biology Laboratory (EMBL) Nucleotide Sequence Database (Stoesser *et al.* 1997), Ribosomal Database Project (Maidak *et al.* 1997) or SILVA (Quast *et al.* 2013).

7.2.2 Culture-independent methods

There are various methods to identify microorganisms that can degrade pollutants in the environment without culturing and isolation. These methods generally use labeled pollutants with stable isotopes. Uptake as well as assimilation of the pollutant as a carbon or nitrogen source are monitored at single cell or molecular level which promotes the identification of a degrader in a complex microbial community. These methods include stable isotope probing (SIP) and single cell probing.

7.2.2.1 Stable isotope probing (SIP)

Radiolabeled chemicals are used to assess biodegradability as well as to elucidate biotransformation pathways of hazardous pollutants by microbial strains or communities. Major elements have one common isotope found in nature and others in much less abundance. For example, 98.93% of carbon in nature is ^{12}C, whereas only 1.07% of carbon occurs as ^{13}C. Similarly, 99.9885% of hydrogen is found as ^{1}H, but the second stable isotope deuterium (^{2}H, or D) is present in 0.0115% (de Laeter *et al.* 2003).

SIP employs pollutants containing heavier stable atoms such as ^{13}C, ^{15}N or even ^{18}O to label DNA of organisms that can utilize that compound in a microbial community. First, the pollutant enriched with stable isotopes is prepared. Then, the sample taken from the environment is fed with the isotope-enriched pollutant. When the isotope-enriched pollutant is consumed by the degraders, it is assimilated into the cellular components including lipids, nucleic acids, carbohydrates and proteins. In SIP, generally nucleic acids (DNA or RNA) are used since they give phylogenetic information about the microorganisms. As isotope-enriched pollutant is consumed by the degraders, DNA of the degraders is also enriched with heavy elements. As a result, the degraders' DNA becomes heavier than the DNA of other microorganisms present in the microbial community. Separation of the heavy

DNA (^{13}C-DNA) which belongs to the degraders from the light DNA (^{12}C-DNA) is achieved using ultracentrifugation. Amplification of the 16S rRNA gene in the heavy DNA by PCR following cloning, PhyloChip technology or metagenomics is used to reveal the identity of the degraders in the community (Figure 7.6).

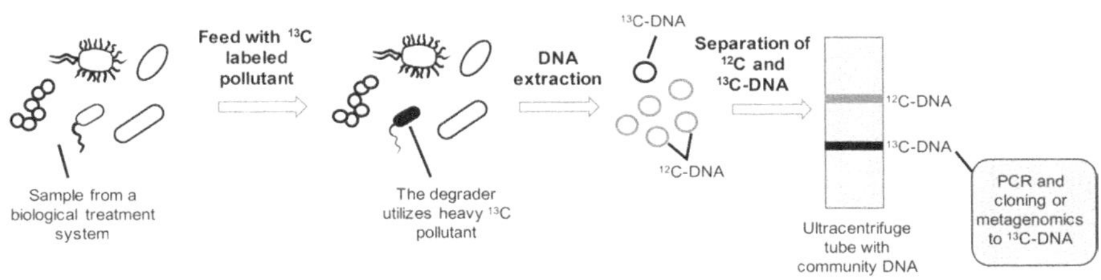

Figure 7.6 Experimental pipeline used to identify pollutant degraders in a microbial community using SIP.

Relatively recently, Lee and co-workers identified the microorganisms that degrade triclosan in activated sludge. ^{13}C-labeled triclosan was fed into a triclosan enrichment culture derived from activated sludge. Resulting heavy DNA was isolated by equilibrium ultracentrifugation in CsCl-EtBr density gradients and used for 16S rRNA PCR following clone library preparation. They identified 11 clones enriched in heavy DNA which contributed to triclosan degradation in activated sludge. None of these clones were similar to known triclosan-degraders previously identified by culture-dependent methods. As a result, application of SIP led to successful identification of diverse uncultivable triclosan-degrading microorganisms in an activated sludge enrichment culture (Lee *et al.* 2014).

SIP is a powerful tool to identify pollutant degraders in the environment without the enrichment bias. Identification of many uncultivable degraders has been achieved by SIP to date (Cupples, 2016). Detailed protocols for DNA and RNA based SIP are given elsewhere (Whiteley *et al.* 2007; Dunford & Neufeld, 2010).

7.2.2.2 Isotope microarray

The isotope array, which is a very similar method to SIP, aims to study the function and activity of microorganisms in their natural environment by employing radiolabeled pollutants and rRNA-targeted oligonucleotide probes. In this method, the sample taken from a wastewater treatment system is incubated with a ^{14}C-labeled pollutant. Next, the community RNA is extracted, covalently linked with a fluorescent dye, fragmented and hybridized with a 16S rRNA targeted oligonucleotide microarray such as PhyloChip (Andersen *et al.* 2010). Subsequently, fluorescence and radioactivity probe signals are quantified with a fluorescence scanner and a beta-imager, respectively. Isotope array is a high-throughput methodology which can simultaneously determine the structure of a complex microbial community and identify species that can degrade a pollutant. On the other hand, the effectiveness of the isotope array approach strongly depends on the availability of suitable microarrays and their analytical performance (i.e. their specificity and sensitivity) (Dumont & Murrell, 2005; Wagner *et al.* 2006).

7.2.2.3 Fluorescent in situ hybridization – microautoradiography (FISH-MAR)

Another method that uses isotope-labeled pollutant and fluorescent oligonucleotide-probes to link biodegradation to microbial identity is fluorescent in situ hybridization-microautoradiography (FISH-MAR). The method involves short-term incubations of environmental samples with a specific radiolabeled pollutant, fixation of thin sections of these samples to glass slides, hybridization of 16S rRNA targeted probes and subsequent analysis of the slide by fluorescence and inverse confocal laser scanning microscopy. The microorganism that uptakes the labeled-pollutant is identified by the silver aggregates in the cells by MAR and its identity is determined by the specific 16S rRNA-targeted FISH probes (Dumont & Murrell, 2005; Wagner *et al.* 2006).

Although it is a powerful method which is more sensitive than SIP, FISH-MAR requires the use of oligonucleotide probes with defined specificity. Therefore, a priori knowledge about phylogenetic affiliation of organisms present in the sample is necessary. This information can be obtained via genomic methods aiming to determine the microbial community structure in the sample such as metagenomics or SIP. If SIP is combined with the FISH-MAR, SIP is used to obtain the rRNA gene sequences of the organisms that are potential degraders. Based on the rRNA sequences of potential degraders, specific oligonucleotide probes can be designed and used for FISH-MAR after incubation of the environmental sample with the same pollutant, which must now be radioactively labeled to confirm the biodegradation capability of the degrader. In a recent study, DNA-SIP was combined with FISH-MAR to identify triclosan degraders in samples taken from a wastewater treatment plant. Results showed that novel *Methylobacillus* spp. were active triclosan degraders in activated sludge (Lolas *et al.* 2012).

FISH-MAR protocols and troubleshooting tips were discussed in detail elsewhere (Nielsen *et al.* 2002; Nielsen & Nielsen, 2005; Alonso, 2012).

7.2.2.4 Single cell probing

New technologies to isolate microorganisms at single cell resolution from complex microbial communities and environments without culturing enable us to identify uncultivable degraders in the environment and elucidate genes involved in the biotransformation of pollutants by those microorganisms. Single cell isolation methods (SCIMs) that may be used for identification of degraders in the microbial communities involve fluorescence activated cell sorting (FACS), Raman activated cell sorting (RACS) and microfluidics (Hu *et al.* 2016c).

In all SCIMs, degrader cells must be labeled based on their biotransformation activity, so that degraders can be separated from non-degraders with the help of the label they are carrying using optics, magnetics or microfluidics. Labeling can be done either on the gene that codes the enzyme degrading the pollutant in the case where it is already known, or alternatively on the substrate, the assimilation of which may give a specific property to the degrader cell (Figure 7.7).

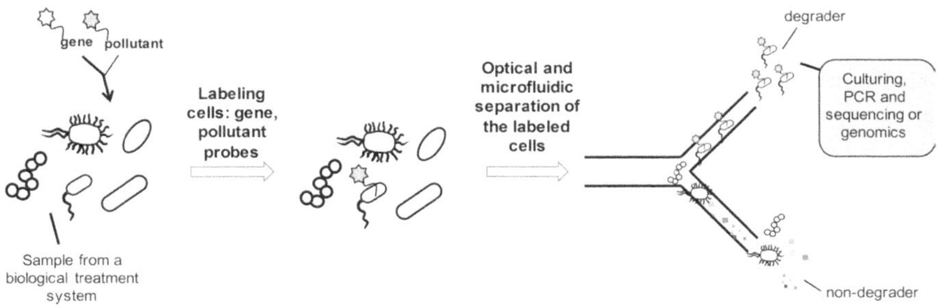

Figure 7.7 Pipeline of single cell probing to identify degraders in a microbial community.

FACS is a specialized type of flow cytometry with sorting capacity. FACS is a high throughput system which allows simultaneous quantitative and qualitative multi-parametric analyses of single cells. Before separation, a cell suspension is prepared and the target cells are labeled with fluorescent probes. The probes are fluorescent oligonucleotide probes that hybridize into the gene that codes the enzyme that catalyzes biotransformation of the target pollutant. In order to label a degrader, the degrading gene has to be known. The nucleotide sequence of many catalytic genes that are involved in biotransformation of pollutants can be accessed through the Swiss Federal Institute of Aquatic Science and Technology's Biocatalysis/Biodegradation Database (EAWAG-BBD) and the links supplied therein. As the cell suspension which contains the fluorescently labeled cells passes the cytometry, each cell is exposed to a laser, which allows the fluorescence detectors to identify fluorescent cells. The instrument applies a charge (positive or negative) to the droplet containing the fluorescent cell of interest and an electrostatic deflection system facilitates the collection of the charged droplets into appropriate collection tubes for later analysis such as culturing, PCR or genomics.

Similarly, an emerging approach to link biodegradation and microorganisms to identify degraders in a microbial community based on single cell probing is RACS. RACS overcomes the external labeling of a microorganism. Since single cell Raman spectra (SCRS) measure the vibration of molecular bonds, this is sensitive to stable isotope compounds and SCRS undergoes Raman shift when cells incorporate stable isotope compounds. Therefore, cells enriched with stable isotopes, in a sample incubated with stable isotope-labeled pollutant, can be separated from others using RACS. Isolated cells, which are degraders, can be further processed using PCR and genomics for phylogenetic classification (Huang *et al.* 2015).

Microfluidics coupled with micromanipulation results in manual isolation of single cells in a microbial community. Labeling of cells as a result of biodegradation activity helps to identify the degraders in the cells sorted by microfluidics. Recently, Nakamura and co-workers developed a water-in-oil microdroplet platform where individual cells of microorganisms were encapsulated in droplets containing a fluorescent substrate and sorted with a microfluidic device. Cells which degraded the fluorescent substrate were isolated manually by micromanipulation and phylogenetically classified (Nakamura *et al.* 2016). This methodology resulted in identification of uncultivable microorganisms and genes involved in the biotransformation of the pollutant.

When the phylogenetic identity of a degrader is determined or the gene encoding the enzyme that degrades a pollutant is known, it is easy to determine the abundance of the degrader in a microbial community. Fluorescent oligonucleotide probes specific to the degrader or the degradation gene are designed based on the nucleotide sequence of the 16S rRNA gene of the degrader or the catalytic gene. A sample where the abundance of the degrader is measured is treated with the probe and the fluorescent particles are counted either using fluorescent microscopy or flow cytometry.

7.3 ASSESSMENT OF POLLUTANT IMPACT

7.3.1 Microbial community structure

Understanding the microbial community structure and its dynamics upon perturbations caused by a pollutant in both natural and engineered systems is crucial to determine the extent of the impact and the response of the environment to the pollutant. In this section, we cover major techniques that can be applied in situ to examine community structure and function when an environmental perturbation occurs due to pollutant exposure.

7.3.1.1 16 S rRNA gene cloning

Clone library construction and sequencing of the cloned amplicons has long been a standard method for the analysis of phylogenetic diversity in microbial communities. Earlier molecular cloning studies to determine bacterial diversity were based on the construction of 16S rDNA libraries cloned and introduced to model hosts such as *E. coli* (Sambrook *et al.* 1989; Ward *et al.* 1990; Dunbar *et al.* 1999). These libraries were screened and sequenced, though this route was expensive, labor intensive and prone to errors leading to under-represented community analyses. With the advent of next generation sequencing (NGS) systems, the cloning step is no longer necessary as PCR products can be directly sequenced. This feature has increased the number of sequences tremendously, allowing detection and identification of organisms with low abundance in mixed communities with greater biodegradation potential, and has become indispensable to metagenomics studies. Therefore, the details of library construction protocols, analysis of sequence data, construction of rarefaction curves and examples studies are provided in Section 7.3.1.5 on metagenomics.

7.3.1.2 DGGE and TGGE

One of the most traditional methods to identify phylotypes (observed similarities among organisms used to classify them based on features obtained by evolutionary mechanisms or genetic material exchange) in a mixed microbial community is denaturing gradient gel electrophoresis (DGGE). DGGE can also be used for rapid comparison of different microbial communities. For example, Mertoglu and co-workers used DGGE (combined with real-time PCR, cloning and slot-blot hybridization techniques) to monitor and evaluate the changes in nitrifying bacterial population under cadmium (Cd) loadings in a laboratory-scale continuous-flow enriched nitrification system. In this study, population shifts from *Nitrosomonas* and *Nitrosococcus sp.* to *Nitrosospira sp.* were observed to tolerate higher Cd loadings (Mertoglu *et al.* 2008). The technique is based on the separation of genes of the same size differing in their melting temperature profiles because of differences in their base sequences (please remember that adenine-thymine (A-T) base pairs have two hydrogen bonds whereas guanine-cytosine (G-C) have three). Abundance of A-T and G-C base pairs in a gene sequence causes different melting profiles. Once applied to a polyacrylamide gel containing gradients of denaturant (mixture of urea and formamide) via electrophoresis, DNA fragments are obtained allowing differentiation among the samples. Temperature gradient gel electrophoresis (TGGE) is a modification of DGGE, where there is a temperature gradient across the gel and the method relies on temperature related changes in structure to separate nucleic acids (Thatcher & Hodson, 1981).

At the end of DGGE, patterns of DNA fragments can be compared to a control (isolated from a known system) ran along with unknown samples to match the profiles. Alternatively, individual bands can be excised and sequenced to determine the phylotypes or identify actual species present in the community if the 16S rRNA gene is used as target. The same method applies to seek for specific gene of interest responsible for the degradation of a pollutant to determine whether it exists in the gene pool of an unknown community/environment.

DGGE Protocol

Materials required:

(1) Deionized formamide
    (a) Prepare by dissolving 10 g of mixed bed resin in 100 mL formamide for 30–60 min.

    (b) Decant or filter the formamide to separate the resin.
    (c) Store 35 mL aliquots of formamide at –20°C.

(2) Acrylamide/bis-acrylamide stock solution.
    (a) Prepare a 37.5:1; 40% (w/v) solution from acrylamide powder.
    (b) Filter and store at 4°C in an amber bottle.
(3) DGEE acrylamide/bis-acrylamide solutions (please note that you need to filter and store these solutions at 4°C, use 0.45 μm filter).
    (a) Prepare a 10% acrylamide – 0% denaturant solution by mixing the following: 2.5 mL 50X Tris-acetate-ethylenediaminetetraacetic acid (TAE); 62.5 mL acrylamide/bis-acrylamide 40% stock solution; and 185 mL dH_2O.
    (b) Prepare an 8% acrylamide – 0% denaturant solution by mixing the following: 2.5 mL 50X TAE; 50 mL acrylamide/bis-acrylamide 40% stock solution; and 197.5 mL dH_2O.
    (c) Prepare an 8% acrylamide – 100% denaturant solution by mixing the following: 2.5 mL 50X TAE; 50 mL acrylamide/bis-acrylamide 40% stock solution; 100 mL formamide; 105 g urea; and 250 mL dH_2O.
    (d) Prepare a 10% acrylamide – 100% denaturant solution by mixing the following: 2.5 mL 50X TAE; 62.5 mL acrylamide/bis-acrylamide 40% stock solution; 100 mL formamide; 105 g urea; and 250 mL dH_2O.
(4) Mixed DGGE gel solutions (please note that you need to degas these solutions for 15 min under vacuum and store at 4°C in amber bottles. TEMED (tetramethylethylenediamine); and APS (ammonium persulfate) should be added before loading the gel).
    (a) Prepare a 30% denaturing solution by mixing the following: 7.2 mL of solution 3c (8% acrylamide – 100% denaturant solution); 16.8 mL of solution 3b; 32 μL TEMED; and 75 μL 10% APS.
    (b) Prepare a 65% denaturing solution by mixing the following: 15.6 mL of solution 3d; 8.4 mL of solution 3a; 32 μL TEMED; and 75 μL 10% APS.
    (c) Prepare a stacking gel solution by mixing the following: 10 mL of solution 3c; 32 μL TEMED; and 75 μL 10% APS.
(5) Gel loading solution (10X)
    (a) Mix bromophenol blue (0.025 g; 0.25% w/v); xylene cyanole (0.025 g; 0.25% w/v); glycerol (5 mL; 100% v/v); and 5 mL dH_2O.

Methodology:
DNA extraction and PCR steps

(1) Extract genomic DNA from environmental samples.
(2) Perform PCR amplification of 16S rRNA by using DGGE specific primers. Group specific examples can be found in notable studies (Boon *et al.* 2002; O'Callaghan *et al.* 2006; Mühling *et al.* 2008).

(3) Prepare 50 μL PCR rxn mix including 1.0 U of *Taq* polymerase; 0.5 μL from each forward and reverse primer; 50 ng template DNA.
(4) Perform following PCR program:
   (a) 94°C for 12 min
   (b) 94°C for 1 min
   (c) 50°C for 1 min
   (d) 72°C for 2 min
   (e) Repeat b-d for 24 cycles
   (f) 72°C for 7.5 min
   (g) Keep at 15°C
(5) Check your PCR amplicons on 1% agarose gel for quality control before DGGE.

DGGE steps

(1) Perform DGGE by loading purified PCR products onto 8% (w/v) polyacrylamide gel with a gradient of 30–65% denaturant (100% denaturant contained 7 M urea and 40% formamide).
(2) Run the gel at 30 V for 15 min and then 130 V for 4.5 h at 60°C. Use 1X TAE buffer.
(3) Stain the gels with EtBr solution (0.5 mg/L).
(4) Destain twice in 0.5X TAE for 15 min.
(5) Capture image using a Gel-doc system.

Extraction of DNA from Acrylamide Gels

(1) Cut and excise central portions (appx. 1 mm^2) of the strongest DGGE bands and soak in 50 μL ultrapure water kept at 4°C overnight.
(2) Prepare the following mix and use 1 μL of the supernatant as template in PCR-DGGE rxn.
   (a) 1 U Advantage 2 polymerase mix
   (b) 2 μL 10X buffer
   (c) 2 μL 2 mM $MgCl_2$
   (d) 1.5 μL dNTP (1.5 μM each)
   (e) 10 picomoles of primers 341F and 534R (each)
   (f) 0.5 μL formamide
   (g) 1 μL template DNA
   (h) 25 μL ultrapure water
(3) Perform following PCR program:
   (a) 96°C for 20 s
   (b) 60°C for 45 s
   (c) 72°C for 45 s
   (d) Repeat a-c for 34 cycles
   (e) 72°C for 5 min
   (f) Keep at 15°C

(4) Check the PCR products on 1% agarose/1X Tris-Borate-ethylenediaminetetraacetic acid (TBE) gel.
(5) Purify by PCR clean-up kits or centrifuge.

Sequencing of purified PCR products and cloning steps

(1) Sequence PCR products obtained from the strongest DGGE bands and free of minor bands directly using the purified PCR products.
(2) Clone amplification products into plasmid vectors (pCR II or pCR4- TOPO).
(3) Transform ligated products into *E. coli.*
(4) Extract the DNA from the recombinant colonies and use the aqueous DNA solution as template in the PCR rxn.
(5) Analyze products by using gel electrophoresis and digest with EcoRI for the presence of different bands.
(6) Choose high frequency sequences in the cloned libraries for sequence analysis.

Sequence analysis and phylogenetics

(1) Perform BLAST analysis of the DNA sequences.
(2) Use Clustal X software to align sequences.
(3) Construct phylogenetic tree and analyze the results with respect to species association.

Please note that TGGE requires minor modifications to DGGE protocol and requires a temperature gradient to be applied to the gel. Comparative details of the protocol can be found in other resources (Muyzer & Smalla, 1998; Muyzer, 1999).

7.3.1.3 T-RFLP and ARISA

Other methods for rapid microbial community analysis are terminal restriction fragment length polymorphism (T-RFLP) and automated ribosomal intergenic spacer analysis (ARISA). T-RFLP is similar to DGGE in theory, however, in T-RFLP the target gene is amplified with primers that are labeled with fluorescent dyes. At the end of PCR, end products are cut by DNA restriction enzymes at specific locations and the fragments of DNA varying in size and number are separated on gel electrophoresis. Fluorescently labeled fragments are further separated in a DNA sequencer allowing the sequencer to select for only terminal ends. Observed patterns allow determination of sequence variations present in the sampled community.

Originally developed by Avaniss-Aghajani and co-workers (Avaniss-Aghajani *et al.* 1996) to identify mycobacterial species, T-RFLP applied to 16S rRNA genes from different microbes to obtain information on microbial community analyses was successfully achieved by many others for glycogen and polyphosphate accumulating organisms in activated sludge (Liu *et al.* 1998), and for mercury resistance genes in soils and sediments (Bruce, 1997).

T-RFLP Protocol

Materials required:

(1) Extraction buffer (100 mM Tris, 100 mM ethylenediaminetetraacetic acid (EDTA), 100 mM sodium phosphate buffer). Adjust pH to 8.0
(2) Sodium Dodecyl Sulfate (10%)
(3) Isopropanol
(4) Ethanol (70%)
(5) 5M NaCl
(6) Lysozyme (100 mg/L)
(7) 5% CTAB (hexadecylmethylammonium bromide)
(8) Proteinase K (10 mg/mL)
(9) 10M ammonium acetate (adjust pH to 7.5)
(10) Phenol:chloroform:isoamylalcohol (25:24:1, v/v)
(11) Deionized formamide (prepare as described in previous section)
(12) PCR reagents
(13) Loading buffer
(14) 5% polyacrylamide gel
(15) ROX-labeled GS500 internal size standard

Methodology:

DNA extraction

(1) Suspend the environmental sample (approx. 5g) in 12 mL extraction buffer.
(2) Add 100 mL proteinase K and 180 mL of lysozyme.
(3) Shake the sample at 37°C for 30 min.
(4) Add 3 mL of sodium dodecyl sulfate (SDS), 4.5 mL of NaCl, and 1.5 mL of 5% CTAB/1.5 M NaCl.
(5) Incubate at 65°C for 15 min.
(6) Freeze the sample using liquid nitrogen.
(7) Thaw the sample at 65°C for 15 min, repeat twice.
(8) Centrifuge at 4,200 rpm for 10 min at 48°C.
(9) Transfer supernatant to a sterile tube.
(10) Resuspend the pellet by vortexing for 10 s in 4 mL of extraction buffer, 1 mL of SDS.
(11) Incubate at 65°C for 10 min.
(12) Repeat step 8.
(13) Combine the supernatant with the other supernatant obtained in step 9.
(14) Add equal volume of phenol:chloroform:isoamylalcohol (25:24:1) mixture to the combined supernatants. Mix by inversion.
(15) Centrifuge the sample at 4,200 rpm for 10 min at 48°C and transfer the upper layer to a sterile tube.
(16) Add an equal volume of chloroform:isoamylalcohol (24:1 mixture).
(17) Mix by inversion and centrifuge at 4,200 rpm for 10 min.

(18) Transfer the aqueous phase and add isopropanol (0.7 by vol.) and 0.3 vol. of 10M ammonium acetate at room temperature to the supernatant.
(19) Centrifuge at 12,000 rpm for 30 min.
(20) Discard the supernatant and resuspend the pellet in 1.5 ml of 70% ethanol in an Eppendorf tube.
(21) Centrifuge at 14,000 rpm for 15 min.
(22) Discard the supernatant and the air-dry the pellet.
(23) Resuspend the pellet in 250 mL of sterile dH_2O.

PCR Steps

(1) Perform PCR using 2.5 U of *Taq* polymerase with primers of your choice.
(2) Perform following PCR program:
    (a) 96°C for 5 min
    (b) 95°C for 15 s
    (c) 49°C for 30 s
    (d) 72°C for 1 min
    (e) Repeat b-d for 24 cycles
    (f) 72°C for 5 min
    (g) Keep at 15°C
(3) Check your PCR products on 0.8% agarose TAE gel.
(4) Purify your PCR products using kits or centrifugal devices.
(5) Elute final DNA (50 μl).

T-RFLP Analysis steps

(1) Digest purified PCR products (10 μl) with 20 U of either *Alu*I or *Hha*I in a total volume of 15 mL at 37°C for 3 h.
(2) Mix this restriction digestion (2 mL) with 2 mL of deionized formamide, 0.5 ml of ROX-labeled GS500 internal size standard, and 0.5 mL of loading buffer.
(3) Denature each sample by heating at 95°C for 5 min and immediately transfer to ice.
(4) Perform electrophoresis on each digested product (1.5 mL) in 5% polyacrylamide gel containing 7M urea for 6 h at 3,000 V.
(5) Visualize the RFLPs after staining with SYBER green I dye using a gel-doc system.
(6) Determine the lengths of fluorescently labeled fragments by comparison with internal standards using GeneScan software.

Data analysis

(1) Estimate the similarity of communities by visual comparison of the electropherograms or by numerically analyzing T-RFLP patterns in gel images using GelCompare software.

(2) Generate a stacked band pattern of T-RFLPs for each community from the same terminus following the normalization (no background subtraction) and image analysis steps of the software.
(3) Produce the similarity matrix for the pattern fragments in the samples by calculating Jaccard coefficient (Liu *et al.* 1997). The Jaccard coefficient considers the presence or absence of bands and the number of terminal fragments in common in communities and the total number of observed T-RFLPs (Morisita, 1959).
(4) Obtain a similarity dendrogram by the unweighted pair group method using average linkages.
(5) Analyze the results, as value of 0 means that community fingerprints are completely different whereas value of 1 indicates that they are identical.

Another variation of T-RFLP capable of providing more detailed information on community analyses is referred to as (automated) ribosomal intergenic spacer analysis (ARISA). As can be inferred, this technique targets PCR based amplification of a specific region between the 16S and 23S rRNA genes commonly referred to as the intergenic spacer region (ISR) or internal transcribed spacer (ITS). These spacer regions differ in sequence length among different species of a mixed community. This technique was originally used to analyze diversity in soils (Borneman & Triplett, 1997) and automated refers to utilization of a DNA sequencer assigning sizes to identified fragments automatically (Fisher & Triplett, 1999).

ARISA has been successfully applied to determine short-term effects of several heavy metals (copper, cadmium, and mercury) on soil microbial communities and demonstrated cumulative effects of metal toxicity (Ranjard *et al.* 2006). In another example, a non-automated version of ribosomal intergenic spacer analysis (RISA) was applied in a laboratory-scale membrane bioreactor (MBR) system used to degrade bromoamine acid (BAA), and showed the increased dominance of BAA-degrading populations (*Sphingomonas xenophaga*) compared to the non-augmented control MBR (Qu *et al.* 2009). Applicability of RISA has also been verified for determination of archaeal community structure changes during anaerobic digestion (Ciesielski *et al.* 2013). A deeper analysis of the benefits and limitations of ARISA can be found elsewhere (Popa *et al.* 2009) along with tested primer sets (Jones *et al.* 2007; Purahong *et al.* 2015).

7.3.1.4 Microarrays: GeoChip and Phylochip

Despite the loss of popularity under the shadow of the rising NGS trend, microarrays are still very useful tools for rapid analyses of biodiversity and the functional potential of microbial communities. In simplest terms, microarrays are spotted microscope slides containing arrayed samples of DNA, RNA, or protein. These samples are carefully designed and printed probes chemically (usually amines or aldehydes) or physically (using electrostatic interactions) stabilized onto the slides.

Laboratories can design and print microarrays by using specialized printers or obtain commercial microarrays commonly referred to as "chips".

Microarrays designed for biodiversity studies are called *Phylochips* and used for searching for specific microbes and population changes in mixed communities by using rRNA probes affixed on the chips. There are also gene microarrays such as *GeoChip*, used to detect genes encoding functions of biogeochemical significance, such as production and consumption of methane, alternative respiratory systems (e.g. dissimilative metal reduction, halorespiration), heavy metal resistance, degradation of recalcitrant chlorinated pollutants, and common oxidative and reductive steps in the nitrogen, carbon, and sulfur cycles. Microarrays are very robust and provide a significant amount of data as several hundreds to thousands of genes can be surveyed with a single microarray slide or chip. For example, a Phylochip can survey a mixed community to seek for all known sulfate-reducing bacteria (more than 100 species) to be detected. Similarly, a GeoChip can contain around 50,000 genes representing key metabolic functions including degradative pathways against pollutants of interest.

Nemir and co-workers conducted a comparative phylogenetic microarray analysis of microbial communities in TCE-contaminated soil environments and determined threshold concentrations where microbial diversity started to change significantly (Nemir *et al.* 2010). Kang and co-workers utilized GeoChips targeting multiple functional genes in heavy metal polluted sediments where sulfate reduction genes (dsr) and a suite of metal-resistant genes were abundant and exhibited strong correlations with heavy metal concentrations (Kang *et al.* 2013). In another study, comprehensive DNA microarrays containing 148 genes for xenobiotic metabolizing enzymes, DNA repair enzymes, heat shock proteins, cytokines, and housekeeping genes were used to examine gene expression patterns in mice models as a response to cadmium chloride, benzo(*a*)pyrene, and TCE (Bartosiewicz *et al.* 2001). Handley and co-workers utilized a PhyloChip that identified hundreds to thousands of operational taxonomic units (OTUs) present during acetate amendment, and revealed close similarity among highly enriched taxa during uranium bioremediation experiments. They also found that OTUs within the Fe(III)- and sulfate-reducing lineages, *Desulfuromonadales* and *Desulfobacterales*, were also stimulated (Handley *et al.* 2012). These studies show that microarrays can be applied to evaluate the potential hazards of hazardous chemicals and/or chemical mixtures with different concentrations to provide effective assessments of the potential toxicity of these pollutants as well changes in microbial diversity as a response to the pollutant exposure.

7.3.1.5 Metagenomics: targeted and non-targeted

As mentioned frequently in earlier sections, understanding microbial communities in natural or engineered systems requires tools beyond culturing. Metagenomics refers to direct sequencing of genes and/or genomes present in complex microbial communities. Metagenomics was first introduced in the 1990s to extract genomic materials from complex environments and analyze to answer questions such as "who is there?", "what do they (potentially) do?" and "what are the interactions within themselves and the ecosystem?" Once NGS systems were introduced, metagenomics studies became much more prominent and in depth compared to traditional Sanger-based sequencing approaches. Modern metagenomics combines systems biology tools powered by genomics and bioinformatics.

The first and one of the most important steps in metagenomics studies is the extraction of high quality and purity DNA from environmental samples. Extracted DNA must well-represent the microbial community and be free of contaminants that may interfere with downstream sequencing steps. Therefore, it is highly recommended to optimize DNA extraction based on sample characteristics and environmental conditions (Bouhajja *et al.* 2016).

Once high-quality and well-represented DNA is extracted from environmental samples, there are several pathways forward depending on the final target outcomes. One can choose generating 16S rRNA clone libraries and choose a targeted community sampling approach (Dafale *et al.* 2010; Yang *et al.* 2014). You can also apply digestion to total DNA with restriction enzymes and then choose a shotgun sequencing approach, or conduct direct sequencing without cloning for untargeted analyses (Ju *et al.* 2014). Often, a combination of approaches is preferred to minimize disadvantages due to the limitations of sampling and sequencing coverage (Kakirde *et al.* 2010; Bouhajja *et al.* 2016).

There are excellent protocols for 16S rRNA gene library construction (Vieites *et al.* 2010) and metagenomics libraries construction for large- (cosmid, fosmid, bacterial artificial chromosomes [BACs]) and short-insert clone libraries (Simon & Daniel, 2017). The last important component of metagenomics studies is the bioinformatics where a massive amount of sequence data needs to be sorted, analyzed, and evaluated. Several tools and resources are available for accurate computational analysis of metagenomics data (Wooley *et al.* 2010; Teeling & Glöckner, 2012; Thomas *et al.* 2012).

Notable metagenomics studies on the functional analysis of species in polluted environments include that of Tyson and co-workers, where *Leptospirillum* group II and *Ferroplasma* type II genomes were recovered from biofilm in acid mine drainage (Tyson *et al.* 2004) and methanogenic hydrocarbon-degrading enrichment cultures from an oil sands tailings pond and a hydrocarbon-contaminated aquifer (Tan *et al.* 2015). Using a targeted metagenomics approach, a novel dibenzofuran-degrading consortium consisting of *Agromyces sp.*, *Bacillales sp.*, and *Comamonadaceae sp.* were isolated which completely degraded dibenzofuran (Johnson, 2013). This consortium was also found to cometabolize chlorinated dioxins including 2,3-dichlorodibenzo-p-dioxin (Johnson, 2013). Using high-throughput metagenomics analysis Wang and co-workers successfully isolated and characterized three *Dehalococcoides mccartyi* strains that respire on commercial persistent polychlorinated biphenyls (Wang *et al.* 2014). Lykidis *et al.* utilized metagenomics tools to characterize metabolic interactions among a methanogenic consortium and found that *Pelotomaculum sp.* are responsible for terephthalate (TA) degradation (Lykidis *et al.* 2011).

With respect to impacts of hazardous pollutants on environmental communities, Kao and co-workers studied the changes in the composition of the bacterial community in PCE/TCE-contaminated groundwater before and after stimulation with slow polycolloid-releasing substrate (SPRS) containing vegetable oil, cane molasses, and surfactants (Kao *et al.* 2016). The found increased amounts of *Dehalogenimonas, Pseudomonas*, and *Sulfuricurvum* where most bacterial community was decreased following the biostimulation (Kao *et al.* 2016). On a much larger scale, the response of ocean floor sediment (Mason *et al.* 2014) and beach sands (Rodriguez-R *et al.* 2015) to crude oil was investigated during the Deepwater Horizon Well accident in 2010. *Gammaproteobacterium* and *Colwellia* species were dominant in the most heavily oil-impacted sediments (Mason *et al.* 2014), whereas *Gamma-* and *Alphaproteobacteria* were enriched in beach sand (Rodriguez-R *et al.* 2015). Using metagenomic and 16S rRNA gene amplicon techniques, researchers have found that the beach community was re-established showing low evidence of petroleum hydrocarbon degradation potential, rather enriched in archaeal taxa known to be sensitive to xenobiotics. However, the community differed significantly before the oil spill and a clear succession pattern was observed (Rodriguez-R *et al.* 2015). In another large-scale study, Quero and co-workers collected sediments from two of the most chemically polluted sites along the coast of Italy (Po River Prodelta and Mar Piccolo of Taranto), and the metagenomics analyses showed that species richness decreased with higher contamination levels of 11 metals, polychlorinated biphenyls (PCBs) and polycyclic aromatic hydrocarbons (PAHs), where the communities evolved toward the selection of species that are able to use the organic pollutants as carbon and energy sources (Quero *et al.* 2015).

7.3.2 Mutations, gene transfer and antibiotic resistance

Increasing presence of antibiotics in the environment promotes the acquisition and/or evolution of resistance elements in the absence of intrinsic antibiotic production (D'Costa *et al.* 2006). Exposure to hazardous contaminants might trigger co-evolution of resistance to both antimicrobials and pollutants or result in mutations selecting for more resistant ecotypes against contaminants. In addition, selective pressures for resistance to hazardous compounds in the presence of high organic content and high bacterial diversity, such as wastewater settings, promote gene exchange mechanisms via conjugative plasmids with bacterial chromosomes, integrons and transposons. These eventually result in the acquisition and accumulation of various antibiotic resistance genes in plasmids (Rahube & Yost, 2010).

Muller has shown that *Pseudomonas aeruginosa* exposed to pentachlorophenol (PCP), a common environmental pollutant arising from agricultural activities, can cope with the PCP by up-regulating its multidrug efflux pumps (a known mechanism of antibiotic resistance). Meanwhile, PCP-mediated induction of efflux pumps resulted in simultaneous increase in antibiotic resistance. It was shown that the resistance nodulation-cell division (RND) efflux pump, MexAB-OprM, was responsible for the PCP-induced increase in antibiotic resistance (Muller, 2006).

There has been strong correlation between antibiotic resistance genes (ARGs) and heavy metals (Cu, Zn and Hg) in manures and manure-amended agricultural soils associated with heavy metal concentration and independent of antibiotics (Ji *et al.* 2012). Hu and co-workers also concluded similar studies where long-term nickel (Hu *et al.* 2016a) and copper contamination (Hu *et al.* 2016b) induced occurrence of antibiotic resistance genes in agricultural soils. Su and co-workers utilized quantitative PCR and 16S rRNA gene based sequencing to investigate the dynamics of ARGs in composting sewage sludge, and reported increases in abundance and diversity, distinct patterns, and enriched ARGs during composting (Su *et al.* 2015). Growth promoting antimicrobials (GPAs) in livestock feed have important impacts on selection and dissemination of antimicrobial resistance (You & Silbergeld, 2014).

7.4 GENETIC INSIGHTS INTO BIODEGRADATION

7.4.1 Metatranscriptomics

Integration of molecular methods to determine key biogeochemical mechanisms is vital to the bioremediation of hazardous pollutants. As microbial cells contain a very small amount of mRNA which is degraded within a few minutes, significant challenges still exist to capture the biodegradative potential of a microbial community in situ. Metatranscriptomics can provide small, yet comprehensive clues to discover novel genes and enzymes relevant for biodegradation. In addition, the effects of a pollutant on the regulation of the transcripts encoding relevant functions can also be determined by metatranscriptomics.

For example, De Menezes and co-workers evaluated soil microbial community responses during biodegradation of phenanthrene and observed up-regulation in transcripts encoding aromatic compound metabolism, respiration and stress responses, and concurrent decreases in virulence, carbohydrate, DNA metabolism and phosphorus metabolism transcripts (de Menezes *et al.* 2012). In another study, Zakrzewski and co-workers used metatranscriptomics tools to determine the transcriptionally active part of a microbial community in a production-scale biogas plant and successfully identified transcripts encoding enzymes involved in substrate hydrolysis, acidogenesis, acetate formation and methanogenesis (Zakrzewski *et al.* 2012). Wu and co-workers investigated short-term effects of temperature changes (i.e. 35°C, 21°C and 13°C) on a partial nitrification-Anammox (PN/A) system by using a metatranscriptomics approach, and identified aerobic and anaerobic catabolic activities of ammonia oxidation and the expression of related functional gene markers to develop specific insights on the effects of temperature at microbiological scales (Wu *et al.* 2016). The effectiveness of this approach is exemplified in the study of Embree and co-workers. Metatranscriptomics data from less than 500 pg of mRNA were generated and used to highlight metabolically active genes present in a methanogenic consortium capable of oxidizing saturated hydrocarbons under anoxic conditions (Embree *et al.* 2014).

Readers can refer to detailed procedures on sample preparation (Stewart *et al.* 2010; Shi, 2011; Quaiser *et al.* 2014; Peimbert & Alcaraz, 2016) and the bioinformatics pipeline (Martinez *et al.* 2016; Westreich *et al.* 2016) for data analyses in metatranscriptomics studies.

7.4.2 Metaproteomics

Metaproteomics aims to measure the entire diversity and abundance of proteins in particular environments using advanced mass spectroscopy (MS). Due to certain post-transcriptional mechanisms, metaproteomics accounts for certain limitations, however its high sensitivity allows phenotypic modifications to be detected at the cell level, in turn better capturing the proteome at meta-level. One of the key challenges in metaproteomics is the mapping of extracted proteins from the environmental communities inhabiting different matrices and identifying novel metabolic pathways, genes and enzymes. This shortcoming can be overcome by combining metatranscriptomics, metaproteomics, and metagenomics techniques together (DeLong, 2013).

In metaproteomics studies, environmental samples are analyzed by 2-D gel electrophoresis (2DE) usually combined with high-throughput MS, followed by bioinformatics analyses for identification. The procedure starts with the extraction of proteins and an untargeted protein map obtained by 2DE as it provides protein separation information and facilitates downstream identification. Following this, the proteins of interest can be extracted from the gels and analyzed in MS for further identification and quantification. The details of these procedures in environmental metaproteomics can be found elsewhere (Wilmes & Bond, 2006a; Alain *et al.* 2007; Chauhan & Jain, 2010).

Wilmes and Bond applied metaproteomics tools on samples from two activated sludges in laboratory-scale sequencing batch reactors (SBRs) with dissimilar phosphorus removal performances and compared protein expression patterns of good enhanced biological phosphorus removal (EBPR) performance to the non-EBPR. The comparison of protein expressions showed that the metaproteomes of the two sludges were substantially different and this was reflected in their microbial community structures and metabolic transformations (Wilmes & Bond, 2006b). In another study, Ram *et al.* identified key activities and examined metabolic functions in a natural acid mine drainage (AMD) microbial biofilm community where they predicted 2033 proteins from the five most abundant species in the biofilm, where half of the predicted proteins were from the dominant *Leptospirillum* group II (Ram *et al.* 2005). Benndorf and co-workers developed a protein separation method from contaminated soil and groundwater systems, and applied this method to a soil matrix enriched in chlorophenoxy acid-degrading bacteria by incubation with 2,4-dichlorophenoxyacetic acid (2,4-D), and to groundwater of a chlorobenzene-contaminated site. Chlorocatechol dioxygenases were consistently detected in addition to the expected bacterial metabolic pathways being highly expressed (Benndorf *et al.* 2007). Guazzaroni and co-workers combined metaproteomic analyses of the microbial community of a highly PAH-contaminated soil with metagenomics tools. Naphthalene was metabolized through two different routes via salicylate in biostimulated soil systems, whereas a gentisate route was found in unstimulated samples (Guazzaroni *et al.* 2013).

In a combined approach of metagenomics and metaproteomics, Kantor and co-workers studied metabolic potential and temporal dynamics of a community degrading thiocyanate (SCN^-) (Kantor *et al.* 2017). They sampled a continuous-flow bioreactor fed with increasing loadings of SCN^- and a second reactor fed with ammonium sulfate to mimic breakdown products of SCN^-. Metagenomics analyses yielded the set of genomes for the abundant communities in both reactors. Metaproteomics revealed the metabolic potential and temporal dynamics of identified organisms where *Thiobacillus* strains capable of SCN^- degradation were highly abundant, and expressed proteins involved in SCN^- degradation, sulfur oxidation, carbon fixation, and nitrogen removal were high in the SCN^- reactor. However, nitrifiers and heterotrophs capable of nitrate reduction were found to be dominating the ammonium sulfate reactor.

All these studies provided means to study the diversity of environmental microbial communities while addressing functional aspects in these samples relevant to biodegradation.

7.4.3 Comparative genomics

Comparison of the genomic features of microorganisms capable of degrading hazardous compounds can provide critical information on the cellular and molecular mechanisms of biodegradation. Being exposed to pollutants in the environment, the intra- and interspecies relationship among microorganisms promote their biodegradative potential. As most degrader strains gain these capabilities by mobile elements via horizontal gene transfer or conjugation, their comparison to other non-degrading strains provides useful information for the identification of genes and enzymes responsible for biodegradation and related mechanisms. Niche-specific metabolic functions encoded by certain genomic islands could also be identified by comparative genomics tools.

There are certain examples that prove the usefulness of comparative genomics. In one of the earlier studies, West and co-workers compared the genome of *Dehalococcoides ethenogenes* strain 195 to a microbial consortium containing *Dehalococcoides* (referred as ANAS), and detected genes responsible for biodegrading tetrachloroethene (PCE) and trichloroethene (TCE) by reductive dehalogenation in strain 195 while missing in the ANAS consortium (West *et al.* 2008). Nelson and co-workers compared the genome of *Pseudomonas putida* KT2440, a plasmid-free toluene degrading strain to the *Pseudomonas aeruginosa* PAO1 genome as well as other microbial genomes and determined key open reading frames (ORFs) along with metabolic pathways for the transformation of a variety of aromatic compounds (Nelson *et al.* 2002). They also identified oxygenases and oxidoreductases playing key roles in the chemical transformation of recalcitrant compounds (Nelson *et al.* 2002). In a similar study, Wu and co-workers compared the genomes of four *P. putida* strains: KT2440, W619, aromatic hydrocarbon-degrading strain F1 and the manganese-oxidizing strain GB-1, in an effort to gain insights into the adaptation of these strains to specific metabolisms and environmental niches including but not limited to heavy metal resistance and breakdown of aromatic compounds (Wu *et al.* 2011).

Makarova and co-workers conducted comparative genomics on extremely radiation-resistant bacterium *Deinococcus radiodurans* towards ongoing development of its use for bioremediation of radioactive wastes. Analysis of paralogs in *Deinococcus* revealed specific expansions of phosphatases, proteases, acyltransferases, Nudix family pyrophosphohydrolases, and genes that affect DNA repair and recombination, and stress responses were also detected (Makarova *et al.* 2001).

7.5 CONCLUDING REMARKS

In the last 30 years or so, microbial and molecular tools have drastically changed and improved at the dispense of environmental scientists and engineers. Now we can utilize culture-dependent and independent techniques either separately or in combination, often merged with the power of next generation sequencing systems, in order to achieve high resolution to isolate and identify microorganisms, and understand their functions and interactions within the environmental matrices. More and more user friendly biocomputational tools are also available to assist in analyzing the massive amounts of data generated during performing some of these tasks. We hope techniques provided in this section assist in your projects to study microbes in situ and ex situ, and assess their biodegradation potential along with the impacts on the environment.

7.6 REFERENCES

Accashian J. V., Vinopal R. T., Kim B.-J. and Smets B. F. (1998). Aerobic growth on nitroglycerin as the sole carbon, nitrogen, and energy source by a mixed bacterial culture. *Applied and Environmental Microbiology*, **64**(9), 3300–3304.

Aitken M. D., Stringfellow W. T., Nagel R. D., Kazunga C. and Chen S.-H. (1998). Characteristics of phenanthrene-degrading bacteria isolated from soils contaminated with polycyclic aromatic hydrocarbons. *Canadian Journal of Microbiology*, **44**(8), 743–752.

Alonso C. (2012). Tips and tricks for high quality MAR-FISH preparations: Focus on bacterioplankton analysis. *Systematic and Applied Microbiology*, **35**(8), 503–512.

Andersen G. L., He Z. L., DeSantis T. Z., Brodie E. L. and Zhou J. Z. (2010). The use of microarrays in microbial ecology. In: Environmental Molecular Microbiology, Liu W. and Jansson J. K. (eds), Caister Academic Press, Norfolk, UK, pp. 87–110.

Avaniss-Aghajani E., Jones K., Holtzman A., Aronson T., Glover N., Boian M., Froman S. and Brunk C. F. (1996). Molecular technique for rapid identification of mycobacteria. *Journal of Clinical Microbiology*, **34**(1), 98–102.

Aziz R. K., Albertsen M., Karst S. M., Ziegler A. S., Kirkegaard R. H. and Nielsen P. H. (2015). Back to basics – the influence of DNA extraction and primer choice on phylogenetic analysis of activated sludge communities. *PLoS One*, **10**(7), e0132783.

Bartosiewicz M., Penn S. and Buckpitt A. (2001). Applications of gene arrays in environmental toxicology: fingerprints of gene regulation associated with cadmium chloride, benzo (a) pyrene, and trichloroethylene. *Environmental Health Perspectives*, **109**(1), 71.

Benndorf D., Balcke G. U., Harms H. and von Bergen M. (2007). Functional metaproteome analysis of protein extracts from contaminated soil and groundwater. *The ISME Journal*, **1**(3), 224–234.

Benson D. A., Cavanaugh M., Clark K., Karsch-Mizrachi I., Lipman D. J., Ostell J. and Sayers E. W. (2017). GenBank. *Nucleic Acids Research*, **45**(D1), D37–D42.

Bielefeldt A. R. and Stensel H. D. (1999). Biodegradation of aromatic compounds and TCE by a filamentous bacteria-dominated consortium. *Biodegradation*, **10**(1), 1–13.

Blasco R., Moore E., Wray V., Pieper D., Timmis K. and Castillo F. (1999). 3-Nitroadipate, a metabolic intermediate for mineralization of 2, 4-dinitrophenol by a new strain of a *Rhodococcus* species. *Journal of Bacteriology*, **181**(1), 149–152.

Boon N., Windt W., Verstraete W. and Top E. M. (2002). Evaluation of nested PCR–DGGE (denaturing gradient gel electrophoresis) with group-specific 16S rRNA primers for the analysis of bacterial communities from different wastewater treatment plants. *FEMS Microbiology Ecology*, **39**(2), 101–112.

Borneman J. and Triplett E. W. (1997). Molecular microbial diversity in soils from eastern Amazonia: evidence for unusual microorganisms and microbial population shifts associated with deforestation. *Applied and Environmental Microbiology*, **63**(7), 2647–2653.

Bouhajja E., Agathos S. N. and George I. F. (2016). Metagenomics: probing pollutant fate in natural and engineered ecosystems. *Biotechnology Advances*, **34**(8), 1413–1426.

Bruce K. D. (1997). Analysis of mer Gene subclasses within bacterial communities in soils and sediments resolved by fluorescent-PCR-restriction fragment length polymorphism profiling. *Applied and Environmental Microbiology*, **63**(12), 4914–4919.

Burland S. M. and Edwards E. A. (1999). Anaerobic benzene biodegradation linked to nitrate reduction. *Applied and Environmental Microbiology*, **65**(2), 529–533.

Chang J. H., Rhee S. K., Chang Y. K. and Chang H. N. (1998). Desulfurization of diesel oils by a newly isolated Dibenzothiophene-degrading Nocardia sp. strain CYKS2. *Biotechnology Progress*, **14**(6), 851–855.

Chauhan A. and Jain R. K. (2010). Biodegradation: gaining insight through proteomics. *Biodegradation*, **21**(6), 861–879.

Chess B. (2015). Laboratory Applications in Microbiology. McGraw-Hill Education, Columbus, OH, USA.

Ciesielski S., Bułkowska K., Dabrowska D., Kaczmarczyk D., Kowal P. and Możejko J. (2013). Ribosomal intergenic spacer analysis as a tool for monitoring methanogenic archaea changes in an anaerobic digester. *Current Microbiology*, **67**(2), 240–248.

Clara M., Kreuzinger N., Strenn B., Gans O. and Kroiss H. (2005). The solids retention time-a suitable design parameter to evaluate the capacity of wastewater treatment plants to remove micropollutants. *Water Research*, **39**(1), 97–106.

Cupples A. M. (2016). Contaminant-degrading microorganisms identified using stable isotope probing. *Chemecal Engineering Technology*, **39**(9), 1593–1603.

Cutter L., Sowers K. R. and May H. D. (1998). Microbial dechlorination of 2, 3, 5, 6-tetrachlorobiphenyl under anaerobic conditions in the absence of soil or sediment. *Applied and Environmental Microbiology*, **64**(8), 2966–2969.

D'Costa V. M., McGrann K. M., Hughes D. W. and Wright G. D. (2006). Sampling the antibiotic resistome. *Science*, **311**(5759), 374–377.

Dafale N., Agrawal L., Kapley A., Meshram S., Purohit H. and Wate S. (2010). Selection of indicator bacteria based on screening of 16S rDNA metagenomic library from a two-stage anoxic–oxic bioreactor system degrading azo dyes. *Bioresource Technology*, **101**(2), 476–484.

de Laeter J. R., Böhlke J. K., De Bièvre P., Hidaka H., Peiser H., Rosman K. and Taylor P. (2003). Atomic weights of the elements. Review 2000 (IUPAC Technical Report). *Pure and Applied Chemistry*, **75**(6), 683–800.

DeLong, E. (2013). Microbial Metagenomics, Metatranscriptomics, and Metaproteomics. Academic Press, San Diego, CA, USA.

de Menezes A., Clipson N. and Doyle E. (2012). Comparative metatranscriptomics reveals widespread community responses during phenanthrene degradation in soil. *Environmental Microbiology*, **14**(9), 2577–2588.

Dhillon J. and Shivaraman N. (1999). Biodegradation of cyanide compounds by a *Pseudomonas* species (S1). *Canadian Journal of Microbiology*, **45**(3), 201–208.

Dumont M. G. and Murrell J. C. (2005). Stable isotope probing – linking microbial identity to function. *Nature Reviews Microbiology*, **3**(6), 499–504.

Dunbar J., Takala S., Barns S. M., Davis J. A. and Kuske C. R. (1999). Levels of bacterial community diversity in four arid soils compared by cultivation and 16S rRNA gene cloning. *Applied and Environmental Microbiology*, **65**(4), 1662–1669.

Dunford E. A. and Neufeld J. D. (2010). DNA stable-isotope probing (DNA-SIP). *Journal of Visualized Experiments*, (42), 2027, doi: 10.3791/2027.

Embree M., Nagarajan H., Movahedi N., Chitsaz H. and Zengler K. (2014). Single-cell genome and metatranscriptome sequencing reveal metabolic interactions of an alkane-degrading methanogenic community. *The ISME Journal*, **8**(4), 757–767.

Fisher M. M. and Triplett E. W. (1999). Automated approach for ribosomal intergenic spacer analysis of microbial diversity and its application to freshwater bacterial communities. *Applied and Environmental Microbiology*, **65**(10), 4630–4636.

Gantner S., Andersson A. F., Alonso-Saez L. and Bertilsson S. (2011). Novel primers for 16S rRNA-based archaeal community analyses in environmental samples. *Journal of Microbiology Methods*, **84**(1), 12–18.

Garnier P., Auria R., Augur C. and Revah S. (1999). Cometabolic biodegradation of methyl t-butyl ether by *Pseudomonas aeruginosa* grown on pentane. *Applied Microbiology and Biotechnology*, **51**(4), 498–503.

Gieg L. M., Coy D. L. and Fedorak P. M. (1999). Microbial mineralization of diisopropanolamine. *Canadian Journal of Microbiology*, **45**(5), 377–388.

Griffin J. S. and Wells G. F. (2017). Regional synchrony in full-scale activated sludge bioreactors due to deterministic microbial community assembly. *The ISME Journal*, **11**(2), 500–511.

Guazzaroni M.-E., Herbst F.-A., Lores I., Tamames J., Pelaez A. I., Lopez-Cortes N., Alcaide M., Del Pozo M. V., Vieites J. M., von Bergen M., Gallego J. L. R., Bargiela R., Lopez-Lopez A., Pieper D. H., Rossello-Mora R., Sanchez J., Seifert J. and Ferrer M. (2013). Metaproteogenomic insights beyond bacterial response to naphthalene exposure and bio-stimulation. *The ISME Journal*, **7**(1), 122–136.

Guo F. and Zhang T. (2013). Biases during DNA extraction of activated sludge samples revealed by high throughput sequencing. *Applied Microbiology and Biotechnology*, **97**(10), 4607–4616.

Häggblom M. M. and Young L. (1999). Anaerobic degradation of 3-halobenzoates by a denitrifying bacterium. *Archives of Microbiology*, **171**(4), 230–236.

Handley K. M., Wrighton K. C., Piceno Y. M., Andersen G. L., DeSantis T. Z., Williams K. H., Wilkins M. J., N'Guessan A., Peacock A. and Bargar J. (2012). High-density PhyloChip profiling of stimulated aquifer microbial communities reveals a complex response to acetate amendment. *FEMS Microbiology Ecology*, **81**(1), 188–204.

Harms G., Zengler K., Rabus R., Aeckersberg F., Minz D., Rosselló-Mora R. and Widdel F. (1999). Anaerobic oxidation of o-xylene m-xylene and homologous alkylbenzenes by new types of sulfate-reducing bacteria. *Applied and Environmental Microbiology*, **65**(3), 999–1004.

He J., Holmes V. F., Lee P. K. and Alvarez-Cohen L. (2007). Influence of vitamin B12 and cocultures on the growth of Dehalococcoides isolates in defined medium. *Applied and Environmental Microbiology*, **73**(9), 2847–2853.

Holoman T. R. P., Elberson M. A., Cutter L. A., May H. D. and Sowers K. R. (1998). Characterization of a defined 2,3,5,6-tetrachlorobiphenyl-ortho-dechlorinating microbial community by comparative sequence analysis of genes coding for 16S rRNA. *Applied and Environmental Microbiology*, **64**(9), 3359–3367.

Hu H., Wang J.-T., Li, J., Shi X., Ma, Y., Chen D. and He J.-Z. (2016a). Long-term nickel contamination increases the occurrence of antibiotic resistance genes in agricultural soils. *Environmental Science and Technology*, **51**(2), 790–800.

Hu H. W., Wang J. T., Li J., Li J. J., Ma Y. B., Chen D. and He J. Z. (2016b). Field-based evidence for copper contamination induced changes of antibiotic resistance in agricultural soils. *Environmental Microbiology*, **18**(11), 3896–3909.

Hu P., Zhang W., Xin H. and Deng G. (2016c). Single cell isolation and analysis. *Front Cell. Dev. Biol.*, **4**, 116.

Huang W. E., Song Y. Z. and Xu J. (2015). Single cell biotechnology to shed a light on biological 'dark matter' in nature. *Microbiology Biotechnology*, **8**(1), 15–16.

Hubert C., Shen Y. and Voordouw G. (1999). Composition of toluene-degrading microbial communities from soil at different concentrations of toluene. *Applied and Environmental Microbiology*, **65**(7), 3064–3070.

Iwaki H., Shimizu M., Tokuyama T. and Hasegawa Y. (1999). Biodegradation of cyclohexylamine by Brevibacterium oxydans IH-35A. *Applied and Environmental Microbiology*, **65**(5), 2232–2234.

Ji X., Shen Q., Liu F., Ma J., Xu G., Wang Y. and Wu M. (2012). Antibiotic resistance gene abundances associated with antibiotics and heavy metals in animal manures and agricultural soils adjacent to feedlots in Shanghai; China. *Journal of Hazardous Materials*, **235236**, 178–185.

Johnson A. (2013). Microbial Community Response to Anthropogenic Pollution: Antibiotic Resistance Genes and Dioxin Biodegradation. Michigan State University.

Jones S. E., Shade A. L., McMahon K. D. and Kent A. D. (2007). Comparison of primer sets for use in automated ribosomal intergenic spacer analysis of aquatic bacterial communities: an ecological perspective. *Applied and Environmental Microbiology*, **73**(2), 659–662.

Ju F., Guo F., Ye L., Xia Y. and Zhang T. (2014). Metagenomic analysis on seasonal microbial variations of activated sludge from a full-scale wastewater treatment plant over 4 years. *Environmental Microbiology Reports*, **6**(1), 80–89.

Kakirde K. S., Parsley L. C. and Liles M. R. (2010). Size does matter: application-driven approaches for soil metagenomics. *Soil Biology & Biochemistry*, **42**(11), 1911–1923.

Kang S., Van Nostrand J. D., Gough H. L., He, Z., Hazen T. C., Stahl D. A. and Zhou J. (2013). Functional gene array–based analysis of microbial communities in heavy metals-contaminated lake sediments. *FEMS Microbiology Ecology*, **86**(2), 200–214.

Kantor R. S., Huddy R. J., Iyer R., Thomas B. C., Brown C. T., Anantharaman K., Tringe S., Hettich R. L., Harrison S. T. and Banfield J. F. (2017). Genome-resolved meta-omics ties microbial dynamics to process performance in biotechnology for thiocyanate degradation. *Environmental Science and Technology*, **51**(5), 2944–2953.

Kao C.-M., Liao H.-Y., Chien C.-C., Tseng Y.-K., Tang P., Lin C.-E. and Chen S.-C. (2016). The change of microbial community from chlorinated solvent-contaminated groundwater after biostimulation using the metagenome analysis. *Journal of Hazardous Materials*, **302**, 144–150.

Kengen S. W., Breidenbach C. G., Felske A., Stams A. J., Schraa G. and de Vos W. M. (1999). Reductive Dechlorination of Tetrachloroethene tocis-1, 2-Dichloroethene by a Thermophilic anaerobic enrichment culture. *Applied and Environmental Microbiology*, **65**(6), 2312–2316.

Kleerebezem R., Pol L. W. H. and Lettinga G. (1999). Anaerobic degradation of phthalate isomers by methanogenic consortia. *Applied and Environmental Microbiology*, **65**(3), 1152–1160.

Kleinheinz G., Bagley S., St. John W., Rughani J. and McGinnis G. (1999). Characterization of alpha-pinene-degrading microorganisms and application to a bench-scale biofiltration system for VOC degradation. *Arch. Environ. Con. Tox.*, **37**(2), 151–157.

Kniemeyer O., Probian C., Rosselló-Mora R. and Harder J. (1999). Anaerobic mineralization of quaternary carbon atoms: isolation of denitrifying bacteria on dimethylmalonate. *Applied and Environmental Microbiology*, **65**(8), 3319–3324.

Kontchou C. Y. and Gschwind N. (1999). Biodegradation of s-Triazine compounds by a stable mixed bacterial community. *Ecotox. Environ. Safe.*, **43**(1), 47–56.

Le Campion L., Vandais A. and Ouazzani J. (1999). Microbial remediation of NTO in aqueous industrial wastes. *FEMS Microbiology Letters*, **176**(1), 197–203.

Lee D. G., Cho K. C. and Chu K. H. (2014). Identification of triclosan-degrading bacteria in a triclosan enrichment culture using stable isotope probing. *Biodegradation*, **25**(1), 55–65.

Lie T. J., Clawson M. L., Godchaux W. and Leadbetter E. R. (1999). Sulfidogenesis from 2-aminoethanesulfonate (taurine) fermentation by a morphologically unusual sulfate-reducing bacterium Desulforhopalus singaporensis sp. nov. *Applied and Environmental Microbiology*, **65**(8), 3328–3334.

Ling L. L., Schneider T., Peoples A. J., Spoering A. L., Engels I., Conlon B. P., Mueller A., Schaberle T. F., Hughes D. E., Epstein S., Jones M., Lazarides L., Steadman V. A., Cohen D. R., Felix C. R., Fetterman K. A., Millett W. P., Nitti A. G., Zullo A. M., Chen C. and Lewis K. (2015). A new antibiotic kills pathogens without detectable resistance. *Nature*, **517**(7535), 455–459.

Liu W.-T., Marsh T. L., Cheng H. and Forney L. J. (1997). Characterization of microbial diversity by determining terminal restriction fragment length polymorphisms of genes encoding 16S rRNA. *Applied and Environmental Microbiology*, **63**(11), 4516–4522.

Liu W.-T., Marsh T. L. and Forney L. J. (1998). Determination of the microbial diversity of anaerobic-aerobic activated sludge by a novel molecular biological technique. *Water Science and Technology*, **37**(4–5), 417–422.

Lolas I. B., Chen X. J., Bester K. and Nielsen J. L. (2012). Identification of triclosan-degrading bacteria using stable isotope probing, fluorescence in situ hybridization and microautoradiography. *Microbiol-Sgm*, **158**, 2796–2804.

Lomans B. P., Maas R., Luderer R., den Camp H. J. O., Pol A., van der Drift C. and Vogels G. D. (1999). Isolation and characterization of Methanomethylovorans hollandica gen. nov., sp. nov., isolated from freshwater sediment, a methylotrophic methanogen able to grow on dimethyl sulfide and methanethiol. *Applied and Environmental Microbiology*, **65**(8), 3641–3650.

Lurescia S., Marconi A. M., Tofani D., Gambacorta A., Housernò A., Devirgiliis C., van der Werf M. J. and Zennaro E. (1999). Identification and sequencing of β-myrcene catabolism genes from *Pseudomonas* sp. strain M1. *Applied and Environmental Microbiology*, **65**(7), 2871–2876.

Lykidis A., Chen C.-L., Tringe S. G., McHardy A. C., Copeland A., Kyrpides N. C., Hugenholtz P., Macarie H., Olmos A., Monroy O. and Liu W.-T. (2011). Multiple syntrophic interactions in a terephthalate-degrading methanogenic consortium. *The ISME Journal*, **5**(1), 122–130.

Maidak B. L., Olsen G. J., Larsen N., Overbeek R., McCaughey M. J. and Woese C. R. (1997). The RDP (Ribosomal Database Project). *Nucleic Acids Research*, **25**(1), 109–111.

Makarova K. S., Aravind L., Wolf Y. I., Tatusov R. L., Minton K. W., Koonin E. V. and Daly M. J. (2001). Genome of the extremely radiation-resistant bacterium Deinococcus radiodurans viewed from the perspective of comparative genomics. *Microbiology and Molecular Biology Reviews*, **65**(1), 44–79.

Manti A., Boi P., Falcioni T., Canonico B., Ventura A., Sisti D., Pianetti A., Balsamo M. and Papa S. (2008). Bacterial cell monitoring in wastewater treatment plants by flow cytometry. *Water Environment Research*, **80**(4), 346–354.

Mao D.-P., Zhou Q., Chen C.-Y. and Quan Z.-X. (2012). Coverage evaluation of universal bacterial primers using the metagenomic datasets. *BMC Microbiology*, **12**(1), 66.

Martinez X., Pozuelo M., Pascal V., Campos D., Gut I., Gut M., Azpiroz F., Guarner F. and Manichanh C. (2016). MetaTrans: an open-source pipeline for metatranscriptomics. *Scientific Report*, **6**, 26447.

Mason O. U., Scott N. M., Gonzalez A., Robbins-Pianka A., Balum J., Kimbrel J., Bouskill N. J., Prestat E., Borglin S., Joyner D. C., Fortney J. L., Jurelevicius D., Stringfellow W. T., Alvarez-Cohen L., Hazen T. C., Knight R., Gilbert J. A. and Jansson J. K. (2014). Metagenomics reveals sediment microbial community response to Deepwater Horizon oil spill. *The ISME Journal*, **8**(7), 1464–1475.

Mertoglu B., Semerci N., Guler N., Calli B., Çeçen, F. and Saatci A. M. (2008). Monitoring of population shifts in an enriched nitrifying system under gradually increased cadmium loading. *Journal of Hazardous Materials*, **160**(2), 495–501.

Mikosch C. A., Denger K., Schäfer E.-M. and Cook A. M. (1999). Anaerobic oxidations of cysteate: degradation via L-cysteate: 2-oxoglutarate aminotransferase in *Paracoccus pantotrophus*. *Microbiology*, **145**(5), 1153–1160.

Molina M., Araujo R. and Hodson R. E. (1999). Cross-induction of pyrene and phenanthrene in a *Mycobacterium* sp. isolated from polycyclic aromatic hydrocarbon contaminated river sediments. *Canadian Journal of Microbiology*, **45**(6), 520–529.

Morisita M. (1959). Measuring of interspecific association and similarity between communities. *Memoirs of the Faculty of Science, Kyushu University Series E*, **3**, 65–80.

Mühling M., Woolven-Allen J., Murrell J. C. and Joint I. (2008). Improved group-specific PCR primers for denaturing gradient gel electrophoresis analysis of the genetic diversity of complex microbial communities. *The ISME Journal*, **2**(4), 379–392.

Muller J. F. (2006). The Role of Multidrug Efflux Pumps in the Stress Response of *Pseudomonas aeruginosa* to Organic Contamination. Virginia Polytechnic Institute and State University, Virginia Tech at Blacksburg, VA.

Muyzer G. (1999). DGGE/TGGE a method for identifying genes from natural ecosystems. *Current Opinion in Microbiology*, **2**(3), 317–322.

Muyzer G. and Smalla K. (1998). Application of denaturing gradient gel electrophoresis (DGGE) and temperature gradient gel electrophoresis (TGGE) in microbial ecology. *Antonie van Leeuwenhoek*, **73**(1), 127–141.

Nakamura K., Iizuka R., Nishi S., Yoshida T., Hatada Y., Takaki Y., Iguchi A., Yoon D. H., Sekiguchi T., Shoji S. and Funatsu T. (2016). Culture-independent method for identification of microbial enzyme-encoding genes by activity-based single-cell sequencing using a water-in-oil microdroplet platform. *Bioscience Reports*, **6**, 22259.

Nelson K., Weinel C., Paulsen I., Dodson R., Hilbert H., Martins dos Santos V., Fouts D., Gill S., Pop M. and Holmes M. (2002). Complete genome sequence and comparative analysis of the metabolically versatile *Pseudomonas putida* KT2440. *Environmental Microbiology*, **4**(12), 799–808.

Nemir A., David M. M., Perrussel R., Sapkota A., Simonet P., Monier J.-M. and Vogel T. M. (2010). Comparative phylogenetic microarray analysis of microbial communities in TCE-contaminated soils. *Chemosphere*, **80**(5), 600–607.

Nichols D., Cahoon N., Trakhtenberg E. M., Pham L., Mehta A., Belanger A., Kanigan T., Lewis K. and Epstein S. S. (2010). Use of Ichip for high-throughput in situ cultivation of "uncultivable" microbial species. *Applied and Environmental Microbiology*, **76**(8), 2445–2450.

Nielsen J. L., Juretschko S., Wagner M. and Nielsen P. H. (2002). Abundance and phylogenetic affiliation of iron reducers in activated sludge as assessed by fluorescence in situ hybridization and microautoradiography. *Applied and Environmental Microbiology*, **68**(9), 4629–4636.

Nielsen P. H. and Nielsen J. L. (2005). Microautoradiography: recent advances within the studies of the ecophysiology of bacteria in biofilms. *Water Science and Technology*, **52**(7), 187–194.

O'Callaghan M., Lorenz N. and Gerard E. M. (2006). Characterization of phylloplane and rhizosphere microbial populations using PCR and denaturing gradient gel electrophoresis (DGGE). In: Molecular Approaches to Soil, Rhizosphere and Plant Microorganism Analysis, Cooper J. E., Rao J. R. (eds), CABI, Oxfordshire, UK, pp. 99–115.

O'loughlin E. J., Sims G. K. and Traina S. J. (1999). Biodegradation of 2-methyl, 2-ethyl, and 2-hydroxypyridine by an *Arthrobacter* sp. isolated from subsurface sediment. *Biodegradation*, **10**(2), 93–104.

Park H.-S., Lim S.-J., Chang Y. K., Livingston A. G. and Kim H.-S. (1999). Degradation of chloronitrobenzenes by a coculture of *Pseudomonas putida* and a *Rhodococcus* sp. *Applied and Environmental Microbiology*, **65**(3), 1083–1091.

Peimbert M. and Alcaraz L. D. (2016). Field Guidelines for Genetic Experimental Designs in High-Throughput Sequencing. Springer, Switzerland, pp. 313–342.

Phelps C. D. and Young L. (1999). Anaerobic biodegradation of BTEX and gasoline in various aquatic sediments. *Biodegradation*, **10**(1), 15–25.

Pierre-Alain M., Christophe M., Séverine S., Houria A., Philippe L. and Lionel R. (2007). Protein extraction and fingerprinting optimization of Bacterial Communities in Natural Environment. *Microbial Ecology*, **53**(3), 426–434.

Popa R., Popa R., Mashall M. J., Nguyen H., Tebo B. M. and Brauer S. (2009). Limitations and benefits of ARISA intra-genomic diversity fingerprinting. *Journal of Microbiological Methods*, **78**(2), 111–118.

Purahong W., Stempfhuber B., Lentendu G., Francioli D., Reitz T., Buscot F., Schloter M. and Krüger D. (2015). Influence of commonly used primer systems on automated ribosomal intergenic spacer analysis of bacterial communities in environmental samples. *PLoS One*, **10**(3), e0118967.

Qu Y.-Y., Zhou J.-T., Wang J., Xing L.-L., Jiang N., Gou M. and Uddin M. S. (2009). Population dynamics in bioaugmented membrane bioreactor for treatment of bromoamine acid wastewater. *Bioresource Technology*, **100**(1), 244–248.

Quaiser A., Bodi X., Dufresne A., Naquin D., Francez A.-J., Dheilly A., Coudouel S., Pedrot M. and Vandenkoornhuyse P. (2014). Unraveling the stratification of an iron-oxidizing microbial mat by metatranscriptomics. *PLoS One*, **9**(7), e102561.

Quast C., Pruesse E., Yilmaz P., Gerken J., Schweer T., Yarza P., Peplies J. and Glockner F. O. (2013). The SILVA ribosomal RNA gene database project: improved data processing and web-based tools. *Nucleic Acids Research.*, **41**(Database issue), D590–596.

Quero G. M., Cassin D., Botter M., Perini L. and Luna G. M. (2015). Patterns of benthic bacterial diversity in coastal areas contaminated by heavy metals, polycyclic aromatic hydrocarbons (PAHs) and polychlorinated biphenyls (PCBs). *Frontiers in Microbiology*, **6**, 1053.

Rahube T. O. and Yost C. K. (2010). Antibiotic resistance plasmids in wastewater treatment plants and their possible dissemination into the environment. *African Journal of Biotechnology*, **9**(54), 9183–9190.

Ram R. J., VerBerkmoes N. C., Thelen M. P., Tyson G. W., Baker B. J., Blake R. C., Shah M., Hettich R. L. and Banfield J. F. (2005). Community Proteomics of a Natural Microbial Biofilm. *Acta Science*, **308**(5730), 1915–1920.

Ramanathan M. P. and Lalithakumari D. (1999). Complete mineralization of methylparathion by *Pseudomonas* sp. A3. *Applied Biochemistry and Biotechnology*, **80**(1), 1–12.

Ranjard L., Lignier L. and Chaussod R. (2006). Cumulative effects of short-term polymetal contamination on soil bacterial community structure. *Applied and Environmental Microbiology*, **72**(2), 1684–1687.

Roberts S., Walker A., Cox L. and Welch S. (1998). Isolation of isoproturon-degrading bacteria from treated soil via three different routes. *Journal of Applied Microbiology*, **85**(2), 309–316.

Rodriguez-R L. M., Overholt W. A., Hagan C., Huettel M., Kostka J. E. and Konstantinidis K. T. (2015). Microbial community successional patterns in beach sands impacted by the Deepwater Horizon oil spill. *The ISME Journal*, **9**(9), 1928–1940.

Sahoo A., Sethunathan N. and Sahoo P. (1998). Microbial degradation of carbosulfan by carbosulfan-and carbofuran-retreated rice soil suspension. *Journal of Environmental Science and Health. Part B*, **33**(4), 369–379.

Sambrook J., Fritsch E. F. and Maniatis T. (1989). Molecular Cloning: a Laboratory Manual. Cold Spring Harbor Laboratory Press, Cold Spring Harbor, NY, USA.

Saunders A. M., Albertsen M., Vollertsen J. and Nielsen P. H. (2016). The activated sludge ecosystem contains a core community of abundant organisms. *The ISME Journal*, **10**(1), 11–20.

Selifonov S. A., Chapman P. J., Akkerman S. B., Gurst J. E., Bortiatynski J. M., Nanny M. A. and Hatcher P. G. (1998). Use of 13C nuclear magnetic resonance to assess fossil fuel biodegradation: fate of [1-13C] acenaphthene in creosote polycyclic aromatic compound mixtures degraded by bacteria. *Applied and Environmental Microbiology*, **64**(4), 1447–1453.

Shi Y. (2011). Microbial Metatranscriptomics: towards Understanding Microbial Gene Expression and Regulation in Natural Habitats. Massachusetts Institute of Technology, Cambridge, MA.

Simon C. and Daniel R. (2017). Construction of small-insert and large-insert metagenomic libraries. In: Metagenomics: Methods and Protocols, Streit W. R. and Daniel R. (eds), Springer, NY, USA, pp. 1–12.

Singh S. N. and Jauhari N. (2017). Degradation of Atrazine by plants and microbes. In: Microbe-Induced Degradation of Pesticides, Singh S. N. (ed.), Springer International Publishing, Cham, pp. 213–225.

Stewart F. J., Ottesen E. A. and DeLong E. F. (2010). Development and quantitative analyses of a universal rRNA-subtraction protocol for microbial metatranscriptomics. *The ISME Journal*, **4**(7), 896–907.

Stoesser G., Sterk P., Tuli M. A., Stoehr P. J. and Cameron G. N. (1997). The EMBL Nucleotide Sequence Database. *Nucleic Acids Research*, **25**(1), 7–14.

Su J.-Q., Wei B., Ou-Yang W.-Y., Huang F.-Y., Zhao Y., Xu H.-J. and Zhu Y.-G. (2015). Antibiotic resistome and its association with bacterial communities during sewage sludge composting. *Environmental Science and Technology*, **49**(12), 7356–7363.

Sukesan S. and Watwood M. (1998). Effects of hydrocarbon enrichment on trichloroethylene biodegradation and microbial populations in finished compost. *Journal of Applied Microbiology*, **85**(4), 635–642.

Suyama T., Hosoya H. and Tokiwa Y. (1998). Bacterial isolates degrading aliphatic polycarbonates. *FEMS Microbiology Letters*, **161**(2), 255–261.

Tan B., Fowler S. J., Laban N. A., Dong X., Sensen C. W., Foght J. and Gieg L. M. (2015). Comparative analysis of metagenomes from three methanogenic hydrocarbon-degrading enrichment cultures with 41 environmental samples. *The ISME Journal*, **9**(9), 2028–2045.

Tanghe T., Dhooge W. and Verstraete W. (1999). Isolation of a bacterial strain able to degrade branched nonylphenol. *Applied and Environmental Microbiology*, **65**(2), 746–751.

Teeling H. and Glöckner F. O. (2012). Current opportunities and challenges in microbial metagenome analysis – a bioinformatic perspective. *Briefings in Bioinformatics*, bbs039.

Thatcher D. R. and Hodson B. (1981). Denaturation of proteins and nucleic acids by thermal-gradient electrophoresis. *Biochemical Journal*, **197**(1), 105–109.

Thomas T., Gilbert J. and Meyer F. (2012). Metagenomics-a guide from sampling to data analysis. *Microb. Inform. Exp.*, **2**(1), 3.

Tyson G. W., Chapman J., Hugenholtz P., Allen E. E., Ram R. J., Richardson P. M., Solovyev V. V., Rubin E. M., Rokhsar D. S. and Banfield J. F. (2004). Community structure and metabolism through reconstruction of microbial genomes from the environment. *Nature*, **428**(6978), 37–43.

van Ginkel C. G., Gancet C., Hirschen M., Galobardes M., Lernaire P. and Rosenblom J. (2008). Improving ready biodegradability testing of fatty amine derivatives. *Chemosphere*, **73**(4), 506–510.

van Hylckama Vlieg J. E., Kingma J., van den Wijngaard A. J. and Janssen D. B. (1998). A glutathione S-transferase with activity towards cis-1, 2-dichloroepoxyethane is involved in isoprene utilization by *Rhodococcus* sp. strain AD45. *Applied and Environmental Microbiology*, **64**(8), 2800–2805.

van Schie P. M. and Young L. (1998). Isolation and characterization of phenol-degrading denitrifying bacteria. *Applied and Environmental Microbiology*, **64**(7), 2432–2438.

Veiga M. C., Fraga M., Amor L. and Kennes C. (1999). Biofilter performance and characterization of a biocatalyst degrading alkylbenzene gases. *Biodegradation*, **10**(3), 169–176.

Vieites J. M., Guazzaroni M.-E., Beloqui A., Golyshin P. N. and Ferrer M. (2010). Molecular methods to study complex microbial communities. *Metagenomics: Methods and Protocols*, 1–37.

Wackett L. P. and Hershberger C. D. (2001a). Biocatalysis and Biodegradation. ASM Press, Washington, DC, USA.

Wackett L. P. and Hershberger C. D. (2001b). Biocatalysis and Biodegradation: Microbial Transformation of Organic Compounds. ASM Press, Washington, DC, USA.

Wagner M., Nielsen P. H., Loy A., Nielsen J. L. and Daims H. (2006). Linking microbial community structure with function: fluorescence in situ hybridization-microautoradiography and isotope arrays. *Current Opinion in Biotechnology*, **17**(1), 83–91.

Wagner-Döbler I., Bennasar A., Vancanneyt M., Strömpl C., Brümmer I., Eichner C., Grammel I. and Moore E. (1998). Microcosm enrichment of biphenyl-degrading microbial communities from soils and sediments. *Applied and Environmental Microbiology*, **64**(8), 3014–3022.

Wang S., Chng K. R., Wilm A., Zhao S., Yang K.-L., Nagarajan N. and He J. (2014). Genomic characterization of three unique Dehalococcoides that respire on persistent polychlorinated biphenyls. *PNAS*, **111**(33), 12103–12108.

Ward D. M., Weller R. and Bateson M. M. (1990). 16S rRNA sequences reveal numerous uncultured microorganisms in a natural community. *Nature*, **345**(6270), 63.

Watanabe K., Teramoto M., Futamata H. and Harayama S. (1998). Molecular detection, isolation, and physiological characterization of functionally dominant phenol-degrading bacteria in activated sludge. *Applied and Environmental Microbiology*, **64**(11), 4396–4402.

West K. A., Johnson D. R., Hu P., DeSantis T. Z., Brodie E. L., Lee P. K., Feil H., Andersen G. L., Zinder S. H. and Alvarez-Cohen L. (2008). Comparative genomics of "Dehalococcoides ethenogenes" 195 and an enrichment culture containing unsequenced "Dehalococcoides" strains. *Applied and Environmental Microbiology*, **74**(11), 3533–3540.

Westreich S. T., Korf I., Mills D. A. and Lemay D. G. (2016). SAMSA: a comprehensive metatranscriptome analysis pipeline. *BMC Bioinformatics*, **17**(1), 399.

Whiteley A. S., Thomson B., Lueders T. and Manefield M. (2007). RNA stable-isotope probing. *Nature Protocols*, **2**(4), 838–844.

Wilmes P. and Bond P. L. (2006a). Metaproteomics: studying functional gene expression in microbial ecosystems. *Trends in Microbiology*, **14**(2), 92–97.

Wilmes P. and Bond P. L. (2006b). Towards exposure of elusive metabolic mixed-culture processes: the application of metaproteomic analyses to activated sludge. *Water Science and Technology*, **54**(1), 217–226.

Wooley J. C., Godzik A. and Friedberg I. (2010). A primer on metagenomics. *PLoS Computational Biology*, **6**(2), e1000667.

Wu S., Bhattacharjee A. S., Dulith B. E. and Goel R. (2016). Whole Community Metatranscriptomics To Study Dynamics of Functional Genes in Partial Nitritation-Anammox Systems Under Low Temperatures. *Proceedings of the Water Environment Federation*, **2016**(11), 1992–2007.

Wu X., Monchy S., Taghavi S., Zhu W., Ramos J. and van der Lelie D. (2011). Comparative genomics and functional analysis of niche-specific adaptation in *Pseudomonas putida*. *FEMS Microbiology Reviews*, **35**(2), 299–323.

Xia S. Q., Duan L. A., Song Y. H., Li J. X., Piceno Y. M., Andersen G. L., Alvarez-Cohen L., Moreno-Andrade I., Huang C. L. and Hermanowicz S. W. (2010). Bacterial Community Structure in Geographically Distributed Biological Wastewater Treatment Reactors. *Environmental Science and Technology*, **44**(19), 7391–7396.

Yan J., Im J., Yang Y. and Loffler F. E. (2013). Guided cobalamin biosynthesis supports Dehalococcoides mccartyi reductive dechlorination activity. *Philosophical Transactions of the Royal Society B*, **368**(1616), DOI: 10.1098/rstb.2012.0320.

Yang Y., Yu K., Xia Y., Lau F. T., Tang D. T., Fung W. C., Fang H. H. and Zhang T. (2014). Metagenomic analysis of sludge from full-scale anaerobic digesters operated in municipal wastewater treatment plants. *Applied Microbiology and Biotechnology*, **98**(12), 5709–5718.

You Y. and Silbergeld E. K. (2014). Learning from agriculture: understanding low-dose antimicrobials as drivers of resistome expansion. *Frontiers in Microbiology*, (5), 1–10.

Zakrzewski M., Goesmann A., Jaenicke S., Jünemann S., Eikmeyer F., Szczepanowski R., Al-Soud W. A., Sørensen S., Pühler A. and Schlüter A. (2012). Profiling of the metabolically active community from a production-scale biogas plant by means of high-throughput metatranscriptome sequencing. *EJB Electronic Journal of Biotechnology*, **158**(4), 248–258.

Zhao J.-S. and Ward O. P. (1999). Microbial degradation of nitrobenzene and mono-nitrophenol by bacteria enriched from municipal activated sludge. *Canadian Journal of Microbiology*, **45**(5), 427–432.

Chapter 8

Biological removal of hazardous pollutants in drinking water treatment

Jess Brown[1], R. Scott Summers[2], Eric Dickenson[3] and Giridhar Upadhyaya[1]

[1]Carollo Engineers, Inc., 3150 Bristol Street, Suite 500, Costa Mesa, CA 92626; E-mail: jbrown@carollo.com; gupadhyaya@carollo.com.

[2]University of Colorado Boulder, College of Engineering & Applied Science, 1111 Engineering Drive, Boulder, CO 80309–0428; Email: R.Summers@colorado.edu.

[3]Southern Nevada Water Authority, PO Box 99954, Las Vegas, NV, 89193-9954; Email: eric.dickenson@snwa.com

8.1 BIOLOGICAL DRINKING WATER TREATMENT BACKGROUND

Intentional biological drinking water treatment (biotreatment) has been documented since the early 1800s, and considerable science, process, and engineering work has expanded its application since then. Biological drinking water treatment processes can be classified in several ways: (1) the degree of engineering involved, (2) the design hydraulic loading rate, (3) the redox potential during treatment, (4) the type of natural filtration used, (5) pollutant applicability, and (6) whether the process is typically applied to surface water or groundwater. Table 8.1 (modified from Evans (2010) and reprinted with permission from Journal of American Water Works Association (JAWWA)) combines all but the groundwater/surface water approach to biotreatment classification. The table also highlights one of the main drivers for the increased interest in and application of biotreatment

Ferhan Çeçen, Ulaş Tezel
doi: 10.2166/9781780407708_313

Table 8.1 Pollutant removal potential for various biological drinking water treatment classifications.

System Classification	**Generalized Potential Effectiveness**				
			Specific Organic Pollutants		
	Biological Stability[a] (e.g. Low Assimilable Organic Carbon)	**Natural Organic Matter, Disinfection By-Product Precursors**	**Naturally Occurring Taste & Odor Causing Compounds (e.g. MIB & Geosmin) and Microtoxins**	**Synthetic Compounds (e.g. Ozonation By-Products, Pharmaceutical and Personal Care Products)**	**Inorganic Compounds (e.g. Iron, Manganese, Ammonia, Nitrate, Perchlorate)**
Engineered Systems					
Slow sand filtration (1°)	High	Moderate	Moderate	High to none	Moderate
Rapid rate filtration					
a) Without pre-ozonation	Moderate (1°)	Low to moderate (2°)	Moderate to high (1°–2°)	Moderate to none (1°–2°)	High to none (1°–2°)
b) With pre-ozonation	Moderate (1°)	Moderate (2°)	High (1°–2°)	High to none (1°–2°)	High to none (1°–2°)
Anoxic biological treatment	Low	Low	Low	Low	High (1°)
Natural Systems					
Riverbank filtration (1°–2°)	High	Moderate	Moderate	High to none	Low
Aquifer filtration (1°–2°)	High	Moderate	Moderate	High to none	Low

1° – primary removal process – biotreatment process is solely relied upon to control the problem of concern.

2° – secondary removal process – requires other processes to resolve the problem of concern.

[a]Biological stability indicates the potential for constituents in a given water matrix to serve as a nutrient and/or substrate source for microbial metabolism. Biologically unstable water has a high potential to promote biological growth and vice versa (LeChevallier *et al.* 2015; Prest *et al.* 2016).

Note: The potentials for treatment effectiveness are intended to be used solely as a general guide and not for design. Actual treatment effectiveness is affected by many factors, including the specific pollutant, source water quality and conditions, treatment process design and operation, and presence of specific strains of bacteria.

in recent years: the ability to treat a wide variety of regulated and unregulated pollutants, including macro-organic compounds, naturally occurring trace organic compounds, synthetic organic compounds, and inorganic compounds. Regulated constituents include pollutants that are hazardous, such as haloacetic acids, nitrate, perchlorate, chromium, selenium, perchlorate, volatile organic compounds (VOCs), benzopyrene, 2,4-dichlorophenoxyacetic acid, or nuisance pollutants. Nuisance pollutants can cause taste, odor, color, corrosion, or scale issues (i.e. iron, manganese, 2-methylisoborneol [MIB], geosmin) or biological instability (i.e. assimilable organic carbon, ozone by-products, ammonia). Unregulated pollutants include pharmaceuticals and potential endocrine disrupting chemicals. So far there has been no evidence of human health risk from the consumption of water containing these chemicals on an individual compound basis. However, there is concern regarding the synergistic effects of a mixture of these chemicals causing adverse health effects. Other drivers toward biotreatment include the increasing cost and complexity of handling concentrated water treatment plant residuals (i.e. biotreatment often converts pollutants to harmless end-products) and the push toward greener processes (e.g. low energy, multi-pollutant removal, minimized concentrated waste streams).

This chapter focuses on engineered biotreatment and groups processes according to whether they are typically applied to surface water or groundwater. The groundwater section is further divided between aerobic and anoxic/anaerobic applications. While these biotreatment categories overlap, their fundamental process, design, and operational characteristics are unique.

8.2 SURFACE WATER

The primary objective of surface water treatment is to control pathogenic microorganisms. As such, filtration is mandated in the United States and other countries. Filters that remove pathogenic microorganisms and turbidity can also be used to control biodegradable compounds found in the source water or created by oxidation processes like ozonation. Typically, the source water contains enough dissolved oxygen (DO) to serve as the electron acceptor such that biodegradable organic or inorganic compounds can be oxidized under aerobic conditions.

There are low levels of biological activity in all surface water treatment systems, even in the presence of disinfectant residuals. However, to effectively harness the capacity of biological filtration, the approach should be to recognize, design, and operate a filter so it is a biologically active process that occurs in the absence of disinfectant residuals. Biological acclimation, which is a critical component of all biological treatment processes, is the accumulation of a critical mass of microbial communities capable of degrading pollutants. Indigenous microorganisms, present in the biofilter influent, are all that is needed to populate the biofilter. To ensure pathogen-safe water, downstream disinfection is always provided, since inactivating pathogenic microorganisms always takes precedence. This section focuses on biodegradable organic compounds. Although inorganic compounds like iron and manganese can be present in surface waters, they are discussed in the groundwater section (Section 8.3.1).

8.2.1 Biodegradable organic compounds

A range of biodegradable organic compounds is present in surface water, some identifiable and others not. Table 8.1 shows an overview of biodegradable compounds. Biodegradable organic matter (BOM) makes up a fraction of the overall dissolved organic matter (DOM) and includes identifiable specific biodegradable organic compounds (SBOC) as well as non-identifiable BOM. Biodegradable dissolved organic carbon (BDOC) and assimilable organic carbon (AOC) are the most commonly used parameters to measure BOM. Many specific compounds are not biodegradable under the conditions of drinking water biofilters, while others can be completely removed. For these compounds, the range of biodegradability is not entirely related to their origins. Biodegradable compounds can be of natural origins or from anthropogenic sources. Natural organic matter (NOM), fractions of which can serve as disinfection by-product (DBP) precursors, is partially biodegradable, as are some naturally occurring specific organic pollutants such as MIB, geosmin (Metz *et al.* 2006), and microcystins. Compounds of anthropogenic origins, such as pharmaceuticals and personal care products (PPCPs), can also contribute to the BOM. In addition, in-plant ozonation creates ozonation by-products (OBPs), many of which are biodegradable (Servais *et al.* 2005). The BDOC fraction of DOM ranges from less than 10% to 30%, with an average of 22%. If ozonation is used during the treatment process, that range can increase to 25 to 40%, with an average of 32%. Thus, an effective biofilter can remove up to 10 to 25% of the DOM (Terry & Summers, 2017).

8.2.2 Process configurations

Configurations used for the biotreatment of surface waters can be divided into three groups: rapid rate filters (RRFs), slow sand filters (SSFs), and river bank filters (RBFs). The RRF and SSF are dominated by aerobic conditions in which oxygen serves as the electron acceptor, while RBF can include both aerobic and anoxic conditions. Other differences in these filters are the hydraulic loading rates, such as filter velocity, filter lengths, and the amount of biomass within the system. The process residence time, often termed empty bed contact time (EBCT), is the ratio of the filter's length to the hydraulic loading rate. EBCT is a primary design criterion for biofiltration (Carlson & Amy, 1998). The EBCTs of the RRF, SSF, and RBF are typically 5 to 20 min, 5 to 10 h, and days to months, respectively. The amount of biomass within the system depends on the amount of BOM removed. Typically, a RBF achieves full BOM removal, SSF achieves near complete removal, and RRF achieves only partial removal.

RRFs are the most common biofilters in drinking water treatment. Most have been converted from conventional surface water treatment plant filters by moving the point of chlorination downstream of the filters or by using granular activated carbon (GAC) as the filter media. RRFs are biologically active if there is no disinfectant in the filter effluent and can be made more effective by optimizing their operating conditions. Anthracite and/or sand, called "inert media," are the most common media.

After start-up of an RRF with new media, an acclimation time occurs as the biomass accumulates on the media, influenced by temperature and the type of BOM. For most of the BOM measured by BDOC, the acclimation time is relatively short, a few weeks. Figure 8.1 shows this concept for an SBOC with inert media in which C_0 and C are the influent and effluent concentrations, respectively. For this case, the SBOC has a steady-state bioremoval of 45%, which depends on the compound and filter operating condition. Some SBOC may take several months to acclimate (Zearley & Summers, 2012). If the filter media is GAC and the target compound is adsorbable, removal is initially complete, or near, complete. As the sorption sites are exhausted, the effluent concentration increases with time (i.e. unsteady-state operation). For nonbiodegradable compounds, complete breakthrough eventually occurs, as illustrated in Figure 8.1, with the length of the run determined by the compound's adsorbability. For compounds that are both adsorbable and biodegradable, the initial removal is dominated by adsorption. When the adsorption capacity is exhausted, biodegradation is the dominant removal mechanism, as illustrated in Figure 8.1. At this steady-state stage, the RRFs with GAC are often termed "biological activated carbon (BAC) filters." If the compound is not adsorbable, the GAC is effectively inert media and the effluent history is that of sand or anthracite.

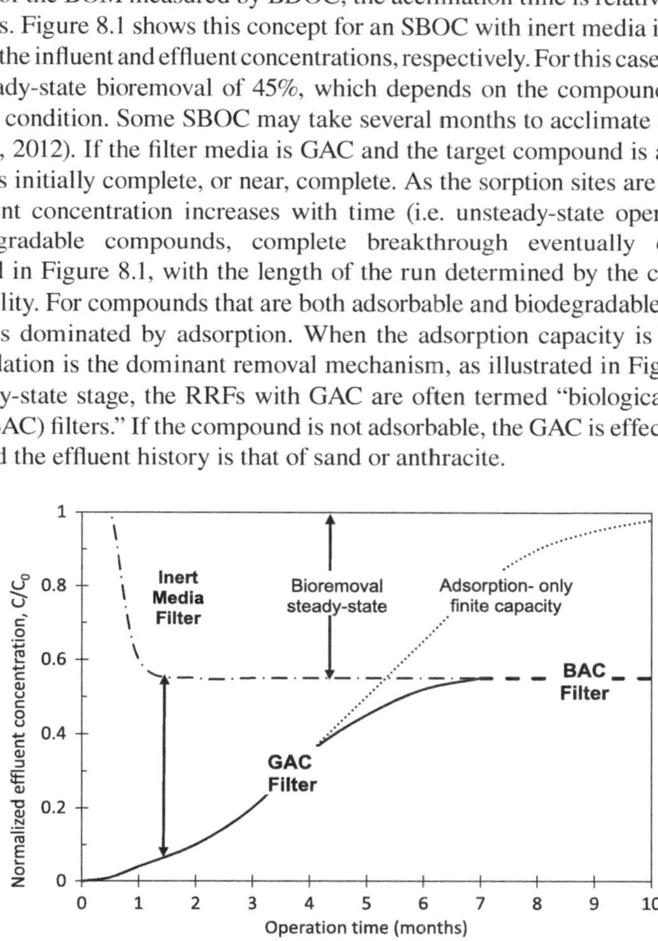

Figure 8.1 RRF performance with inert media and GAC.

Table 8.2 summarizes the performance of inert media (anthracite or sand) RRFs for the bioremoval of 83 organic compounds in 10 studies. Summers *et al.* (2014) reviewed 7 studies; Melin and Ødegaard (2000) evaluated OBP bioremoval; Xie and Zhou (2002) evaluated the disinfectant by-product haloacetic acids (HAAs) bioremoval; Richter *et al.* (2008) evaluated sulfonamides bioremoval. A complete range (0 to 100%) of removal was found. Some trends were observed. For example, OBPs and HAAs were very well removed, and many herbicides were not.

Table 8.2 Range of bioremoval and associated pseudo first-order rate constants (based on a 7.5 min EBCT) by inert media biofilters for 83 organic compounds in 10 studies.

>85% Bioremoval	**85%–50% Bioremoval**	**50%–15% Bioremoval**	**<15% Bioremoval**
k′ > 0.2 min^{-1}	**k′ = 0.2–0.1 min^{-1}**	**k′ = 0.1–0.01 min^{-1}**	**k′ < 0.01 min^{-1}**
Benzo[a]pyrene MIB Molinate Hydrocodone Fluoxetine Gemfibrozil Geosmin p-toluenesulfonamide Aldehydes (formaldehyde, acetaldehyde, glyoxal, methyl glyoxal) Ketoacids (glyoxylic, pyruvic and ketomalonic acids) Haloacetic acids (monochloro-, monobromo-, dichloro-, dibromo-, and trichloro-acetic acids)	2,4-Dichloro-phenoxyacetic acid (2,4-D) Idicarb DDT Dimethoate Acetaminophen Triclosan Bisphenol A Ibuprofen Naproxen Caffeine	Diazinon Malaoxon Erythromycin Trimethoprim Dilantin Galaxolide Pentoxifylline Androstenedione Progesterone Testosterone Clofibric acid Fluorene Tributyl phosphate Diclofenac Cotinine Anilines (several halogenated, methyl- and nitro- anilines)	Acetochlor Atrazine Carbaryl DEET Diuron Methomyl Metolachlor Prometon Simazine Oxybenzone Sulfamethoxazole Carbamazepine TCEP Musk Ketone Estradiol Estriol Estrone 17 α-ethinylestradiol (EE2) Benzenesulfonamide o-toluenesulfonamide Diazepam Meprobamate Iopromide

Source: Melin and Ødegaard (2000); Xie and Zhou (2002); Richter *et al.* (2008); Summers *et al.* (2014)

Although no standardized design values currently exist for RRFs, typical values are no different from those for conventional surface water treatment plants filters, with respect to filter configuration (underdrains, backwash configuration), media diameter and depth, and hydraulic loading rate. The main difference is that EBCT is not a design parameter for conventional surface water treatment plants filters, while it is important for biological RRF, and higher values in the range of 5 to 20 min EBCT may improve performance.

For SSFs, the media is sand with about the same filter depth as an RRF. However, the main difference is that raw untreated surface water is fed directly to the filter at a much lower load hydraulic loading rate, such that the EBCT is about 50 times longer. For RBFs, the media is alluvial till dominated by sand and gravel, and the flow path is from a few metres to hundreds of metres. These long flow paths and a relatively low travel velocity (hydraulic loading rate) lead to long residence times of days to months.

8.2.3 Factors affecting performance

A number of design, operating, and environmental conditions are known to impact the effectiveness of biofilters under steady-state operating conditions, including the compound type, temperature, EBCT of the filters, and available biomass. While not applicable in all cases, a pseudo or observed first-order rate model can illustrate the impact of these design, operating, and environmental conditions.

The first-order model can be derived from the Monod model, as the concentrations of the SBOC and the collective BOM are low, and secondary utilization mechanisms (secondary substrate or cometabolism) control (Alexander, 1999; Rittmann & McCarty, 2001). For biofilter influent water with a relatively high DOC concentration of 3 mg/L and a biodegradable fraction of 25%, the BDOC concentration would be 750 μg/L. The typical concentration of SBOCs would be in the low microgram per litre range: <25 μg/L.

Assuming secondary substrate utilization conditions for the SBOCs in which the collective BOM serves as the primary substrate, the utilization rate describing pollutant degradation at low concentrations in biological filters is described by the following reaction equation:

$$r = dC/dt = -k \cdot X \cdot C \tag{8.1}$$

where r is the utilization rate, mass per volume per time, C is the pollutant mass concentration, t is the time, X is the biomass concentration, and k is a second-order utilization rate constant per biomass concentration per time. Integrating this rate equation over filter depth or total EBCT results in the following integrated rate equation, which describes the resulting pollutant concentration for a filter with a set EBCT:

$$C_{eff}/C_{inf} = \exp(-k \cdot X \cdot EBCT) \tag{8.2}$$

where C_{inf} and C_{eff} are the biofilter influent and effluent concentrations, respectively. Often the biomass concentration is not known and the equation can be expressed as:

$$C_{eff}/C_{inf} = \exp\left(-k' * EBCT\right) \tag{8.3}$$

where k′, the product of the biomass concentration and k, and is termed the pseudo or observed first-order rate constant, which has units that are the inverse of the EBCT, typically min^{-1}.

These equations illustrate the impact of EBCT, biomass concentration, compound concentration, and temperature. Table 8.2 shows k′ values calculated for the compounds based on literature removal values and assuming a 7.5 min EBCT. The higher end of the k′ value range, >0.2 min^{-1}, is typically associated with easily biodegradable compounds like OBPs, HAAs, MIB and geosmin.

Figure 8.2 shows the compound bioremoval in an RRF as a function of EBCT for a range of observed rate constants. The EBCT range, 3 to 20 min, corresponds to typical design and operating hydraulic loading rate values. As the EBCT grows from 3 to 20 min, bioremoval of SBOCs is expected to increase. For 19 pollutants, including pharmaceuticals and personal care products, Zearley and Summers (2012) observed increased removal at a longer EBCT of 15.8 min than at 7.9 min at a temperature of 20°C, though the increases were marginal. For three pollutants, *N,N*-Diethyl-meta-toluamide (DEET), naproxen and ibuprofen, Hallé *et al.* (2015) observed increased removal at an EBCT of 14 min compared to 5 min EBCT, at <5°C. In addition, Wu and Xie (2005) observed increased removal at an EBCT of 20 min compared to 5 min for HAAs at <10°C. Interestingly, at a temperature of 20°C or higher, increasing the EBCT beyond 5 min yielded little benefit for HAAs, DEET, naproxen and ibuprofen removals (Wu & Xie, 2005; Zearley & Summers, 2012; Hallé *et al.* 2015).

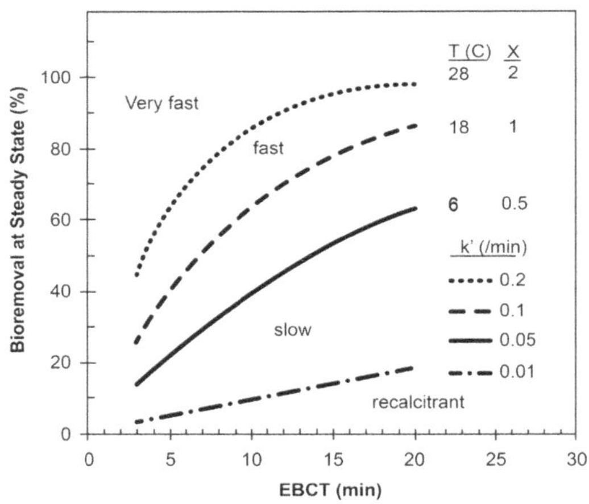

Figure 8.2 Bioremoval in an RRF as impacted by empty bed contact time (EBCT), observed rate constant (k′), temperature (T) and relative biomass (X).

Figure 8.2 shows that for a given compound, temperature can impact the relative performance of a biofilter. For an RFF with an EBCT of 10 min, a compound with an observed rate constant of 0.1 min^{-1} at 18°C yields bioremoval of 63%. If the temperature increases to 28°C, the observed rate constant is 0.2 min^{-1}, which yields bioremoval of 89%. If the temperature decreases to 6°C, the observed rate constant is 0.05 min^{-1}, yielding bioremoval of only 39%.

Wu & Xie (2005) and Hallé *et al.* (2015) have demonstrated temperature's impact. Wu & Xie (2005) showed that increasing temperature from 4 to 30°C greatly increased HAA removal during BAC filtration. Hallé *et al.* (2015) determined that pollutants, DEET, naproxen and ibuprofen, had higher observed pseudo-first-order rate constants at ~20°C than <12°C for both 5 and 14 min EBCT RRFs. Interestingly, these rate constants at ~20°C were a factor of three to four times higher than those shown in Table 8.2. This difference suggests that the rate of biodegradation depends on site-specific factors, meaning that caution is needed if applying rate constants reported in the literature. Considering both EBCT and temperature, a longer EBCT could be used to compensate for the effect of low water temperature.

Figure 8.2 shows a similar sensitivity analysis for the impact of biomass, X. For a compound with an observed rate constant of 0.1 min^{-1} at 18°C yields a bioremoval of 63%, and if the biomass is increased by a factor of two the observed rate constant is 0.2 min^{-1}, which yields a bioremoval of 89%. If the relative biomass decreases by 50%, the observed rate constant is 0.05 min, which yields a bioremoval of only 39%. The amount of biomass can be linked to the primary substrate concentration and utilization.

The recent Dowdell (2012) study exemplified the impact of biomass concentration. Filter media with higher biomass levels (measured as phospholipids: 50 nmol PO_4/dry gram of media) in a 15 min EBCT filter removed the emerging pollutants acetaminophen, gemfibrozil, ibuprofen, and trimethoprim more effectively than a biofilter with less biomass (17 nmol PO_4/dry gram of media) (Dowdell, 2012).

Figure 8.3 extends the EBCT and observed rate constant impacts to the other systems, SSF and RBF. EBCT is plotted on a logarithmic scale. From the SSF and RBF systems, compounds with low observed rate constants and little removal in the RRF can be effectively removed. However, some organic compounds are not aerobically biodegradable.

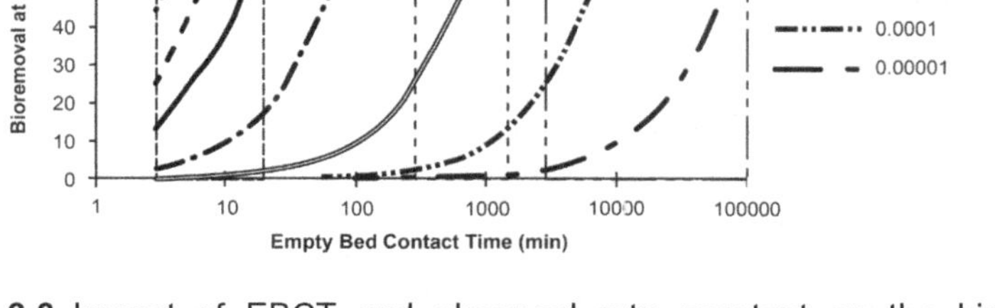

Figure 8.3 Impact of EBCT and observed rate constant on the bioremoval performance of RRF, SSF and RBF systems.

8.3 GROUNDWATER – AEROBIC TREATMENT

Objectives for groundwater treatment significantly differ from those of surface water treatment. Particle removal is a prime objective in surface water treatment, whereas removing dissolved pollutants, such as ferrous iron (Fe[II]), manganese (Mn[II]), and ammonia (NH_3), is typically the goal in groundwater treatment. In general, surface water has significant levels of DO, and aerobic treatment processes may need limited to no aeration. Conversely, groundwater generally contains low levels of DO. Therefore, treating groundwater requires aeration, especially when the objective is to remove dissolved reduced pollutant species.

Biological groundwater treatment processes can be categorized as (i) aerobic and (ii) anoxic/anaerobic, depending on the redox conditions used to remove the target pollutants. In aerobic groundwater treatment, reduced pollutant species such as Fe(II), Mn(II), and NH_3 are removed through biologically mediated oxidation, using oxygen as the electron acceptor. Typically, autotrophic microorganisms that use CO_2 as the carbon source mediate the electron transfer reaction. However, heterotrophic bacteria that grow on carbon sources other than CO_2 can coexist in the bioreactors, using the organic by-products secreted by the autotrophic bacteria and oxidizing the reduced pollutants (Viswanathan & Boettcher, 1991; Mouchet, 1992; Vandenabeele *et al.* 1995). The most common reduced pollutant species removed in biological processes include Fe(II), Mn(II), and NH_3. Aerobic biotreatment processes also can remove some VOCs, such as tetrachloroethylene (PCE) and trichloroethylene (TCE), and other trace organic pollutants including methyl tertiary butyl ether (MTBE).

To create the optimal redox conditions for degrading the pollutant of interest, aeration may be applied upstream of or in the biological reactor. In this case, the presence of other reduced compounds that may consume oxygen, such as sulfides and methane, significantly affects treatment efficiency. Co-pollutants, if any, must be simultaneously removed or their negative effects on the treatment process minimized.

The following sections describe biologically mediated aerobic removal of Fe(II), Mn(II), and NH_3 from groundwater. Factors that may affect biological pollutant removal, including hydraulic flow rate, DO, temperature, pH, redox potential, pollutant concentration, and the presence of other co-pollutants, are also discussed.

8.3.1 Process configurations

Typically, biological systems used to remove reduced pollutants fall into three categories: (1) rapid sand filters, (2) trickling filters, and (3) biological aerated filters (BAFs). Both pressure vessel and open basin flow reactor configurations have been implemented for aerobic groundwater treatment.

8.3.1.1 *Pressure vessel reactors*

Pressure vessels (Figure 8.4) made of steel and fitted with internal distribution headers, collection laterals, and automated control valves to form a reactor configuration are typically used to remove dissolved reduced pollutants from groundwater sources. The vessel is packed with a granular media, such as sand, anthracite, or GAC, on which microorganisms grow in the biofilm. Water is then pumped into the pressure vessel, typically in down-flow mode. Depending on the pollutant of interest, aeration is applied upstream of the pressure vessel to enhance aerobic biological activity in the pressure vessel.

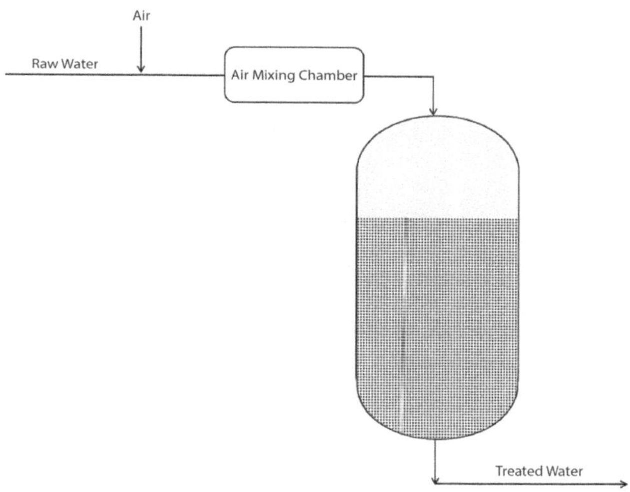

Figure 8.4 Pressure vessel configuration for biological removal of reduced pollutants.

In addition to providing an attachment surface for microorganism growth, the media bed filters those solids generated during the reactor operation. Over time, the filter bed gets clogged, resulting in headloss across the bed and consequent decline in filter production. To maintain filter production, the media bed is routinely cleaned by backwashing, which removes excess biomass and the deposited solids from the filter bed.

8.3.1.2 *Open basin reactors*

As the name suggests, the typically rectangular open basin reactor chambers are exposed to the atmosphere and rely on gravity flow. Similar to pressure vessels, the basin is packed with granular media to encourage microorganism growth and retain generated solids. Open basin reactors commonly are larger than pressure vessels. As shown in Figure 8.5, raw water can be aerated with cascades or by injecting air into the raw water line.

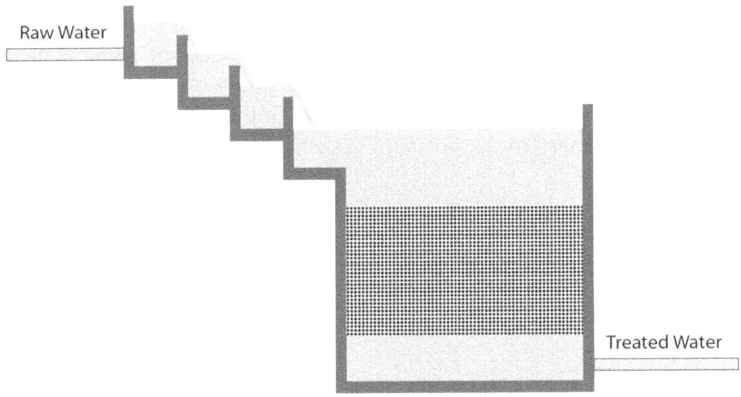

Figure 8.5 Simplified scheme of gravity flow configuration for biological removal of reduced pollutants.

8.3.1.3 Single-stage versus multi-stage reactors

A single-stage bioreactor consists of a single filter bed, whereas multi-stage reactors implement a series of two or more filters. Choosing to implement a two-stage reactor configuration depends on the co-pollutants. When iron, ammonia, and manganese coexist, pollutants are removed in the following order: iron > ammonia > manganese. This is because complete iron and ammonia oxidation generally occur before manganese oxidation is initiated (Mouchet, 1992; Vandenabeele *et al.* 1995; Kohl & Dixon, 2012). As a result, a multi-stage biofilter system may be required to remove all three pollutants. Recent studies, however, have shown that a single-stage bioreactor can effectively remove these pollutants simultaneously (Tekerlekopoulou & Vayenas, 2008; Yulan *et al.* 2010; Brown *et al.* 2013; Hasan *et al.* 2013; Tekerlekopoulou *et al.* 2013).

8.3.2 Factors affecting aerobic treatment of groundwater

Biological treatment processes rely on the growth and activity of microorganisms. Therefore, aerobic biotreatment of groundwater can be affected by anything that influences the growth of microorganisms, and creating the right environmental conditions is critical to ensure optimal aerobic degradation of pollutants.

The prime environmental factors for biological iron oxidation include an upper limit of pH (pH < 7) (Mouchet, 1992), redox, and a lower limit for temperature

(Tekerlekopoulou *et al.* 2013). Similarly, there are lower limits of pH (>7.5) (Sparrow & Uren, 1987), DO (>5 mg/L) (Mouchet, 1992), redox (>300 mV) (Mouchet, 1992; Vandenabeele *et al.* 1995), and temperature for manganese removal, below which manganese-removing biological activity may be limited. Lower limits of pH, alkalinity, DO, and temperature for biological ammonia removal are approximately 6, 11 mg/L as HCO_3^- per mg/L of NH_4, 3.55 mg/L O_2 per mg/L NH_4, and 5°C, respectively. However, studies have shown pollutant removal beyond these limits (Sparrow & Uren, 1987), indicating environment-specific acclimation and diversity of pollutant-oxidizing bacteria. The following sections describe the environmental and operating factors that may affect iron, manganese, and ammonia in a biofilter.

8.3.2.1 DO concentration

DO is directly related to redox conditions, with a higher DO resulting in a higher redox potential. For biological water treatment (Dong *et al.* 2009) and for minimizing the growth of filamentous bacteria such as *Sphaerotilus natans* (Liu & Liu, 2006), the minimum reported DO requirement is 2 mg/L. However, DO and redox requirements for a specific application depend on the pollutant of interest. Stoichiometric relationships suggest 0.14, 0.29, and 4.57 mg/L DO is required for converting 1 mg/L of Fe, Mn, and NH_4, respectively.

Recent studies have demonstrated that biological iron oxidation can occur under a fairly broad range of environmental conditions (Hanert *et al.* 2006; de Vet *et al.* 2011). Hanert *et al.* (1992) observed optimal biological iron oxidation with redox potential between +200 and +320 mV and 0.1 to 1 mg/L DO. Viswanathan and Boettcher (1991) demonstrated successful removal of 2.6 to 3 mg/L iron from groundwater when the raw water DO was 0.1 to 0.3 mg/L with redox potential and pH of 50 to 70 mV and 5.2 to 5.5, respectively. The system was inoculated with iron-oxidizing bacteria, presumably *Gallionella*. While Mn oxidation requires DO levels >1 mg/L, increasing the DO level does not necessarily lead to better Mn removal (Basu *et al.* 2016).

When multiple pollutants coexist, consider DO requirements for the biological degradation of all target pollutants. In addition, DO requirements for any other reduced species present, such as methane (CH_4) and hydrogen sulfide (H_2S), may need to be considered to ensure complete removal of the target pollutants. Stoichiometric DO requirements for 1 mg/L of CH_4 and H_2S oxidation are 4 and 1.9 mg/L O_2, respectively. In general, average DO levels in the treated effluent should not be lower than 3 mg/L to avoid developing anaerobic conditions in any part of the reactor (Tekerlekopoulou *et al.* 2013).

Literature does not define a level of DO required for simultaneously removing ammonium and manganese. For example, Fu *et al.* (2011) removed ammonium and manganese to below maximum contaminant level (MCL) with a continuous feed of 4.5 mg/L DO, whereas Štembal *et al.* (2005) reported very high levels of DO required for simultaneous removal of these compounds. In evaluating the possibility of using intermittent aeration to maintain a steady state DO level of 3 mg/L, Hasan *et al.* (2012) reported that ammonium and manganese were removed to less than 1.5 mg/L NH_4-N and 0.1 mg/L Mn from initial levels of ~9 mg/L NH_4-N and ~0.4 mg/L Mn, respectively.

8.3.2.2 Empty bed contact time (EBCT)

In general, increasing the EBCT may improve pollutant removal up to a threshold beyond which treatment performance is independent of EBCT. For example, Wert *et al.* (2008) did not observe any additional nitrification with EBCT past 3.2 min. When a biofilter packed with ceramsite was evaluated to simultaneously remove iron, manganese, and ammonia, removal within the upper 80 cm of the bed (total bed depth was 160 cm) declined when the flow rate rose from 20 L/h (equivalent to 64.5 min EBCT) to 80 L/h (EBCT 16 min) (Yulan *et al.* 2010). However, Mn removal throughout the bed depth (with 160 cm bed height; EBCT 129 min) improved with the increased flow rate, suggesting that while a 16 min EBCT was not enough, 129 min EBCT was too long to minimize the effect of increased flow rate.

Regardless of the flow rate, nearly all of the iron and ammonia were removed in the upper 80 cm of the bed, suggesting that an EBCT of 16 min sufficiently removed iron and ammonia. This further indicates that manganese limits performance when iron, manganese, and ammonia coexist in the influent water. In four pilot-scale studies using different raw water sources and quartz sand-packed bioreactors operated with a difference in only the aeration applied, Štembal *et al.* (2005) simultaneously removed iron, manganese, and ammonium with an EBCT as low as 4.8 min. However, other water quality and operating parameters, such as temperature, pollutant concentration, coexisting pollutants, and flow rate may affect the EBCT required for pollutant removal. This concept is discussed below.

8.3.2.3 Pollutant concentration

In general, higher influent concentrations require longer EBCT. In addition, pollutant concentration above a certain threshold may affect biological activity. For example, Vandenbeele *et al.* (1995) reported that bacterial activity was inhibited at 3 mM Mn(II). Thus, when iron, manganese, and ammonia coexist in the groundwater, using two or more stages of biofilters may be necessary, depending on the influent concentrations and flow rate (Tekerlekopoulou *et al.* 2013).

8.3.2.4 Presence of co-pollutants

Competitive inhibition of pollutant removal may occur when multiple pollutants coexist. While the presence of iron may affect manganese and ammonia removal, iron removal is independent of the presence of ammonia or manganese (Tekerlekopoulou

& Vayenas, 2008). Thermodynamically, ammonia oxidation is more favorable (Gibb's free energy generated under standard conditions is –350.7 kJ/mol) compared to the oxidation of manganese (–78.45 kJ/mol) (Vandenabeele *et al.* 1995). Therefore, considerable ammonia removal occurs before manganese is removed, and an ammonia-removing zone precedes a manganese-removing zone in the biofilter.

The inhibition of Mn removal in the presence of ammonia is likely related to the poor redox potential when ammonia is present. Gouzinis *et al.* (1998) realized that ammonia levels higher than 2 mg/L inhibited manganese removal. When iron, manganese, and ammonia coexist, typically manganese is the rate-limiting pollutant, which drives the design criteria for the bioreactor (Tekerlekopoulou & Vayenas, 2008; Yulan *et al.* 2010). If the raw water contains hydrogen sulfide, maintaining a high redox potential can be a challenge (Mouchet, 1995).

The effects of nitrite and nitrate on manganese (Mn) removal are not clearly understood. While nitrite may reduce MnO_2 abiotically, especially under acidic conditions, Vandenabeele *et al.* (1995) suggested that although nitrate may stabilize MnO_2, it may prevent the reduction of MnO_2. They also showed that Mn was biologically removed only after complete oxidation of ammonium and nitrite. The accumulation of nitrite in the system reduced MnO_2. Therefore, if nitrite oxidizers are well established with efficient nitrite oxidation, the chemical reduction of MnO_2 by nitrite can be minimized or eliminated.

Under low DO and oligotrophic conditions, they suggested that biologically mediated manganese oxidation may occur using nitrate as the electron donor. However, recent studies demonstrate that biological oxidation of Mn(II) coupled with nitrate- or nitrite-reduction is not possible (Clement *et al.* 2009).

8.3.2.5 Temperature

Temperature affects the EBCT required for pollutant removal, and longer EBCTs may be required at lower temperatures (Hozalski *et al.* 1999). Furthermore, temperature is directly related to the growth and activity of microorganisms. Big declines in temperature may affect microbial enzymes, which may result in slower pollutant degradation kinetics.

Optimal pollutant removal occurs within a specific temperature range. For example, the optimum temperature range for iron oxidation by the members of *Gallionella* genus is 10 to 15°C (Hanert, 1981), whereas the members related to *Sphaerotilus* and *Leptothrix* genera require a temperature range of 20 to 25°C (Mouchet, 1992). Similarly, temperatures lower than 15°C may affect biological nitrification (Lytle *et al.* 2007). Andersson *et al.* (2001) saw the strong effect of temperature on ammonia removal: with temperature >10°C, 40 to 90% of ammonia was removed in a pilot-scale biologically active carbon (BAC) bioreactor, whereas only 10 to 40% was removed when the temperature was 4 to 10°C. Temperature lower than 14°C also may result in lower Mn oxidation due to slower microbial activity and chemical oxidation (Berbenni *et al.* 2000).

Cai *et al.* (2016) evaluated simultaneous removal of Fe, Mn, and NH_3 at 3 to 4°C in a pilot-scale biofilter packed with fresh Mn-coated sand. The filter was inoculated with Mn-coated sand collected from a full-scale biofilter operated at 10°C. Fe removal was observed immediately after starting the system, suggesting it was likely due to chemical oxidation. Iron oxyhydroxide (FeOOH) was also likely generated through chemical oxidation acting as the catalyst for further Fe(II) oxidation and subsequent removal. Nitrite accumulation was observed after 45 days, suggesting that ammonia oxidizing bacteria (AOB) could rapidly acclimate. Consequently, complete ammonia removal (>90%) was observed only after 155 days, suggesting that the activity of nitrite oxidizing bacteria (NOB) is the rate-limiting factor for complete ammonia removal in cold conditions. The longer acclimation period likely was required for the adaptation of the microbial community that originated in the system operated at 10°C. The negative effects of low temperatures may be minimized with a bacterial consortium previously acclimated to lower temperatures.

8.3.2.6 Nutrient availability

Nitrogen and phosphorus are essential nutrients for microbial growth and activity. Phosphorus typically exists as dissolved organic complexes or particulate inorganic or organic complexes (de Vet *et al.* 2012). If the raw water is devoid of nutrients, a supplemental minimal dose of nutrients may boost pollutant removal. Even when the raw water contains considerable levels of phosphorus as orthophosphate, the influent of a bioreactor may become phosphorus-limited due to adsorption and co-precipitation of orthophosphate with iron hydroxides in upstream flocculation and sedimentation processes (de Vet *et al.* 2012; Lauderdale *et al.* 2012). Furthermore, when iron, manganese, and ammonia coexist in the influent water, competitive biological growth of iron- and ammonia-oxidizing microbial populations may limit nutrients for the manganese-oxidizing bacteria.

Interestingly, pre-ozonation may increase concentration of biologically available phosphorus, which in turn may remove additional pollutants. In a study with full-scale nitrifying biofilters, de Vet *et al.* (2012) observed removal of most orthophosphate along with iron upstream of the filters. Out of the remaining orthophosphate in the biofilter influent, only a fraction was bioavailable since the majority of the orthophosphate likely existed as calcium and magnesium complexes. A good correlation was reported between the influent phosphorus concentration and nitrification: nitrification was negatively affected when the influent phosphorus concentration was less than 100 μg/L; less than 10 μg/L completely stopped nitrification. Nitrification linearly increased with the increase in phosphorus in the range of 10 to 50 μg/L. However, when a bench-scale GAC/sand biofilter was supplemented with 20 to 200 μg/L phosphorus, Granger *et al.* (2014) did not see any improvement in Mn removal, even when DOC removal improved by approximately 11%. This suggests that the effects of phosphorus-supplementation can be case-specific.

8.3.2.7 Media selection

Media type and size are two components to carefully consider during media selection to ensure that pollutant removal goals can be met. Primary characteristics to look for when selecting filter media include low density, high specific surface area and resistance to attrition.

A wide variety of granular media including sand, anthracite, GAC, and expanded clay has been used for removing iron, manganese, or ammonia from groundwater. While both anthracite and GAC have been equally effective for removing pollutants, GAC may be advantageous under operating conditions that affect the growth and activity of microorganisms, such as higher pollutant loading, colder temperatures, or the presence of oxidants.

GAC's advantage over inert media can be attributed to several factors, including high specific surface area, surface roughness, and adsorption/desorption capacity. Higher biomass density can be achieved in GAC media compared to anthracite or sand, which can lead to more efficient pollutant removal. Furthermore, the adsorption/desorption capacity of GAC may lower the potential toxicity of the pollutants of interest when the influent concentration is high.

Desorption of pollutants, driven by concentration gradient, results in bioregeneration of the GAC, which may extend its life (Seredyńska-Sobecka *et al.* 2006). Comparing the performance of an anthracite biofilter with a GAC biofilter, Kohl and Dixon (2012) reported 40% manganese removal in the GAC biofilter and less than 10% removal in the anthracite biofilter.

However, pollutant removal performance can vary depending on the GAC's characteristics. For example, Andersson *et al.* (2001) observed more than 90% ammonia removal in a biofilter with opened superstructure GAC operated at 4 to 10°C, but only 45% removal in a biofilter with closed superstructure GAC.

Media size determines the efficiency of pollutant removal, primarily because of the surface area available for adsorption (in the case of GAC), the attachment of microorganisms, and the exclusion of suspended particles. The smaller the media size, the larger the surface area per unit media volume for biofilm attachment and the greater the particle retention. Consequently, headloss across the bed will be higher with smaller media size, requiring more frequent backwashing to remove excess biomass and deposited solids. Media size should be determined based on the water quality and hydraulic performance goals.

8.3.2.8 Biomass maintenance

Any fixed-bed biological reactor requires periodic backwashing to remove excess biomass and accumulated solids sludge. Backwashing also redistributes media and associated biomass throughout the filter bed. Implementing air scour or supplementing oxidants to the backwash water may improve its effectiveness. However, to minimize negative effects on the growth and activity of microorganisms, the air scour rate and oxidant dose should be carefully selected considering the media. GAC biofilters have been less sensitive to the presence of oxidants (LeChevallier *et al.* 1992; Weinberg *et al.* 1993).

8.3.3 Pollutant removal performance

Table 8.3 summarizes pollutant removal for selected aerobic biological groundwater treatment systems implemented at bench-, pilot-, and full-scales. While iron, manganese, and ammonia are not generally considered directly hazardous in a drinking water setting, their presence can lead to hazardous water quality in drinking water distribution systems. For example, nitrification can lead to the accumulation of nitrite; all three inorganic compounds can cause biologically-induced corrosion, which can release toxic and/or carcinogenic metals from distribution pipe scales (Beech & Sunner, 2004; Gu *et al.* 2011; Hu *et al.* 2016).

8.4 GROUNDWATER – ANOXIC/ANAEROBIC TREATMENT

Anoxic/anaerobic biological groundwater treatment is similar to aerobic treatment since it focuses solely on the degradation of dissolved constituents and not on particle removal. However, anoxic/anaerobic biotreatment targets oxidized pollutants, such as nitrate, perchlorate, chromate, and/or selenate, which typically requires the addition of an electron donor/carbon source (substrate). Nitrate is an endocrine disrupting compound (EDC) (Jannat *et al.* 2014; Guillette & Edwards, 2015), a probable human carcinogen, and can be reduced to nitrite, which decreases the oxygen carrying capacity of blood. Perchlorate can inhibit the production of thyroid hormones and is a probable human carcinogen (Leung *et al.* 2015; Steinmaus, 2016). Chromate is a known human carcinogen (Cardenas-Gonzalez *et al.* 2016). Selenate can be toxic at high levels, causing gastrointestinal disorders, hair loss, sloughing of nails, fatigue, irritability, and neurological damage (Vinceti *et al.* 2001; Vinceti *et al.* 2010). Some volatile organic compounds (VOCs) also have been degraded under reducing conditions (Haston & McCarty, 1999; Scheutz *et al.* 2011; Koenig *et al.* 2012; Mitra & Gupta, 2014). VOCs can harm the central nervous system, kidneys, or liver, can cause irritation when they contact the skin, and several are probable or known human carcinogens (Chen & Wu, 2017). Heterotrophic anoxic bioreactors use an external organic carbon source, such as acetic acid or ethanol, which also serves as the electron donor. Autotrophic anoxic bioreactors use carbon dioxide as the carbon source, and an inorganic electron donor, such as hydrogen or sulfur, is typically added.

Though full-scale anoxic biological drinking water treatment plants have operated in Europe for nitrate removal for more than 30 years, such applications in the United States have been limited until recently. However, recent trends are starting to change that, such as: (1) the rising costs and increasing complexities of handling water treatment residuals (for example, ion exchange brine or reverse osmosis

Table 8.3 Summary of selected bench-, pilot-, and full-scale hazardous pollutant removal associated with aerobic biological groundwater treatment.

Bioreactor Configuration	Scale	Media	Environmental/Operating Conditions			Influent Concentration (mg/L)			Percent Removal			
			EBCT (min)	Temp (°C)	pH	Fe	Mn	NH_3	Fe	Mn	NH_3	References
Two-stage fixed-bed biofilter	full	Mn-coated sand		8	6.9	15	1.2	1.3	98	95	80	Li *et al.* (2013)
Single-stage BAF	pilot	ceramic bead		9.5–12	6.8–7.3	0.65–2.82	3.33–5.41	2.45–4.15	90	99	87	Fu *et al.* (2011)
Biofilter	bench	silica sand	20	18–22	7.5–7.8	0.5–2.0	0.6–2.2	0.8–1.7	100	100	100	Cai *et al.* (2015)
Roughing filter	pilot	gravel and sand			7.0–7.3	0.19–0.44	0.18–0.37	–	95	94	–	Pacini *et al.* (2005)
Roughing filter	pilot	gravel			6.8–7.4	0.09–0.37	1.08–1.83	–	98	98	–	Pacini *et al.* (2005)
Two-stage filter	pilot	polystyrene beads	7.5		7.2	0.98–2.6	0.34–1.57	–	100	>90	–	Katsoyiannis and Zouboulis (2004)
Biofilter	pilot	ceramsite	64.5			0.82–3.97	0.95–3.92	1.72–4.22	>98	>99	>93	Yulan *et al.* (2010)
Single stage Biofilter			6.75	9	6.8	0.01–0.5	0.58–3.05	–	>99	>99	–	Li *et al.* (2005)
Trickling filter	pilot	crushed foamed slag				2	0.7	0.6	100	100	100	Frischherz *et al.* (1985)
Filter	pilot	anthracite/ sand				1.3	0.7	0.5	>96	86	92	Frischherz *et al.* (1985)
Three-stage filter	pilot	sand		11	6.7–6.8	2.5–4	0.5–0.6	0.6	100	>95	93	Bourgine *et al.* (1994)
Biofilter	pilot	Mn-coated sand		13.4	7.48	2.45	0.224	1.01	>95	>95	>95	Štembal *et al.* (2005)

(*Continued*)

Table 8.3 Summary of selected bench-, pilot-, and full-scale hazardous pollutant removal associated with aerobic biological groundwater treatment (*Continued*).

Bioreactor Configuration	Scale	Media	Environmental/Operating Conditions			Influent Concentration (mg/L)			Percent Removal			
			EBCT (min)	Temp (°C)	pH	Fe	Mn	NH_3	Fe	Mn	NH_3	References
Biofilter	pilot	Mn-coated sand		11.2	7.29	–	1.06	0.02	100	>95	100	Štembal *et al.* (2005)
Biofilter	pilot	Mn-coated sand		13.8	7.00	1.05	1.06	0.3	>81	>95	100	Štembal *et al.* (2005)
Biofilter	pilot	Mn-coated sand		14.6	8.09	0.98	0.1	2.62	>80	>50	>81	Štembal *et al.* (2005)
Biofilter	pilot	quartz sand		16.5	7.3	1.04	0.147	0.84	>81	>66	>41	Štembal *et al.* (2004)
Biofilter	pilot	quartz sand		12.9	6.27	0.04	0.425	0.018	>50	>80	>100	Štembal *et al.* (2004)
Trickling filter	pilot	silicic gravel	9.1	22	7.5–8	1.9	–	1.1	>97	–	100	Tekerlekopoulou and Vayenas (2007)
Trickling filter	pilot	silicic gravel/gravel	9.1	22	6.8–7.8	0.9	0.45	0.6	>95	>94	100	Tekerlekopoulou and Vayenas (2007)
Trickling filter	pilot	silicic gravel/gravel	9.1	25	6.8–7.8	0.5–4.0	0.5–1.3	0.5–3.0	>94	>89	>83	Tekerlekopoulou and Vayenas (2008)
Trickling filter	full	silicic gravel	11.25	20–21	6.8–7.5	0.2	0.02–0.07	0.8	88 ± 12	69 ± 14	85 ± 11	Tekerlekopoulou *et al.* (2010)
Trickling filter	full	anthracite/sand	54.5	20 ± 2	7.24–7.41	5.4	0.54	4.5	>98	>96	>90	de Vet *et al.* (2009)

Trickling filter	pilot	sand	8.2	13	7.25 ± 0.15	3.3 ± 0.7	–	–	>95	–	–	de Vet *et al.* (2011)
Trickling filter	full	sand		11.6 ± 0.3	7.33 ± 0.04	5.5 ± 0.6	–	–	>95	–	–	de Vet *et al.* (2011)
Trickling filter	full	sand		11.5 ± 0.2	7.1 ± 0.05	8.5 ± 0.8	–	–	>95	–	–	de Vet *et al.* (2011)
Biofilter	full	quartz sand	25	–	7.3–7.5	4.5–6.9	0.18–2.5	0.29–0.78	–	100	–	Bruins *et al.* (2015a)
Biofilter	full	quartz sand	13.2	–	7.5–7.6	0.03–0.14	0.12–0.18	< 0.23	–	100	–	Bruins *et al.* (2015a)
Biofilter	pilot	quartz sand	3.5	10.5–12.5	7.5–7.9	0.03–0.1	0.1–0.15	0–0.2	–	98	–	Bruins *et al.* (2015b)
Biofilter	pilot	quartz sand	10–12	–	7.7 ± 0.1	–	0.07 ± 0.01	0.38 ± 0.06	–	>75	100	Vandenbeele *et al.* (1995)
BAF	bench	plastic beads	30	–	5.94–6.84	0.3	1.45	5	–	95.9	94.1	Hasan *et al.* (2011)
Biofilter	full	sand	21.2	13	7.2	1.49 ± 0.14	0.43 ± 0.06	1.56 ± 0.25	>98	>98	100	Madoni *et al.* (2001)
BAF	bench	Mn coated sand		8	6.9	15	12	13	>98	>95	>81	Li *et al.* (2013)
Biofilter	bench	Anthracite/Mn coated sand	24	7–9	6.7–7	10–17	0.8–1.4	1.7–2.3	>97	75–95	88–97	Du *et al.* (2016)

concentration), (2) water scarcity and the need for agencies to better leverage their own water sources, (3) the push for green technologies (biotreatment processes typically use low energy and efficiently destroy pollutants instead of concentrating them), and (4) increasing regulatory attention and support for anoxic biological processes.

8.4.1 Process configurations

8.4.1.1 Fixed-bed bioreactor

Fixed-bed bioreactors use a stationary bed of media on which biofilms develop, similar to conventional aerobic biofiltration. The granular media can be contained in pressure vessels or open basins. Raw water is amended with an organic or inorganic substrate and then flows by gravity or is pumped across the media bed. As constituents are degraded, growing biofilms accumulate and cause a pressure drop across the bed. Thus, fixed-bed bioreactors are routinely taken off-line and backwashed to remove excess biomass.

8.4.1.2 Fluidized-bed bioreactor

In fluidized-bed anoxic bioreactor configurations, water is pumped up-flow through the reactor to fluidize the biofilm-coated granular media, an approximately 25 to 30% expansion relative to the resting bed height. A portion of the bioreactor effluent flow is recycled and blended with raw water to reach the necessary fluidization velocity. An organic or inorganic substrate is dosed to the combined feed flow. Excess biomass is removed from the media bed through fluid shearing forces and/or by in-line mechanical biomass/media separation devices. Thus, though fluidized-bed bioreactors require higher feed-flow capacity, they do not require off-line backwashing.

8.4.1.3 Membrane biofilm reactors

Membranes also can be used to support biofilm growth. In this case, hydrogen (electron donor) is delivered across a gas-transfer membrane and diffuses directly into the biofilm attached to the other side of the membrane. Carbon dioxide also can be delivered through the membrane as a carbon source, but it may not be necessary. The membranes are submerged in a reactor vessel through which raw water passes and constituents diffuse from the bulk water into the biofilms where they are degraded. Occasionally, the membranes are backwashed and/or chemically cleaned in place to remove excess biomass.

8.4.1.4 Ion exchange membrane bioreactors

An ion exchange membrane bioreactor (IXMB) integrates the advantages of a membrane system and a biological treatment process. Charged pollutants are transported through a membrane from a water-containing compartment (Donnan

Dialysis) to a biocompartment, where pollutant-specific microorganisms degrade the ionic pollutants.

8.4.1.5 CSTR-based bioreactors

A CSTR is a suspended-growth reactor which continuously maintains the microorganisms in suspension. CSTRs are generally used for high-concentration water sources, such as industrial and domestic wastewater. Biological perchlorate removal for drinking water has been evaluated at bench- and pilot-scale CSTR (Coppala, 2000; Interstate Technology & Regulatory Council [ITRC], 2007).

8.4.2 Process considerations

8.4.2.1 Media selection

For anoxic biotreatment applications, various media types are used to support the growth of biofilms, including granular activated carbon, sand, plastic, expanded clay, membranes, or specialized media. The selection of a specific reactor configuration also may dictate media selection.

8.4.2.2 Substrate selection and dose

Anoxic biotreatment processes require dosing an electron donor and sometimes a separate carbon source. For heterotrophic processes (processes that use an organic carbon source), the electron donor and carbon source are the same chemical, commonly acetic acid. Some full-scale anoxic biotreatment plants in Europe use ethanol as the electron donor and carbon source. For autotrophic processes (those that use an inorganic carbon source), carbon dioxide may be dosed as the carbon source, and hydrogen is often dosed as the electron donor, although other electron donors such as sulfur and iron can be considered. The electron donor dose is typically calculated based on the concentration of dissolved oxygen and the raw water concentration of the pollutant of interest, such as nitrate or perchlorate. To avoid over- or under-dosing an electron donor, feed-forward and/or feedback chemical dosing control should be considered (i.e. tying chemical dosing to real-time raw water or treated water quality monitoring).

8.4.2.3 Nutrient availability

As discussed in Section 8.3.2.6, nitrogen and phosphorus are essential nutrients for microbial growth and activity. When nutrients are limiting, pollutant removal may diminish and extracellular polymeric substance (EPS) production may increase (Liu *et al.* 2006), which can disrupt process hydraulics (Mauclaire *et al.* 2004). Figures 8.6 and 8.7 illustrate the potential impacts of nutrient limitation on treatment and hydraulic performance in two separate fixed-bed, anoxic biotreatment systems.

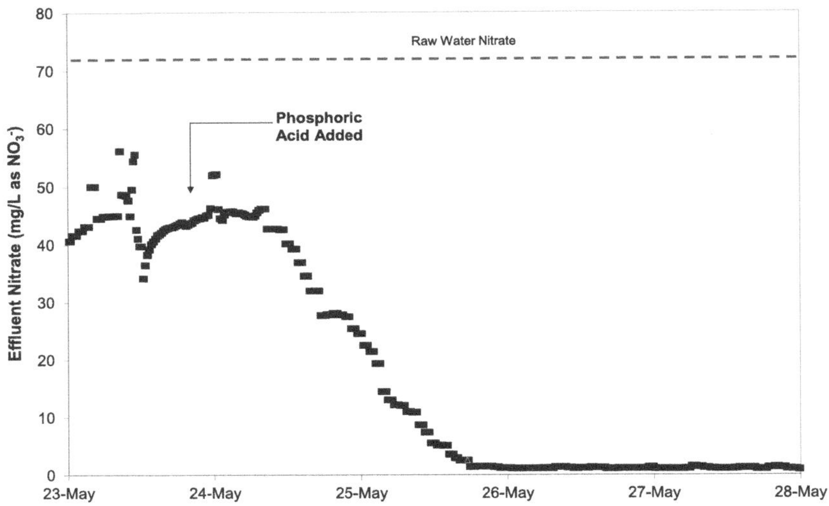

Figure 8.6 Impact of phosphorus limitation on nitrate removal across a fixed-bed bioreactor (Brown *et al.*, 2009).

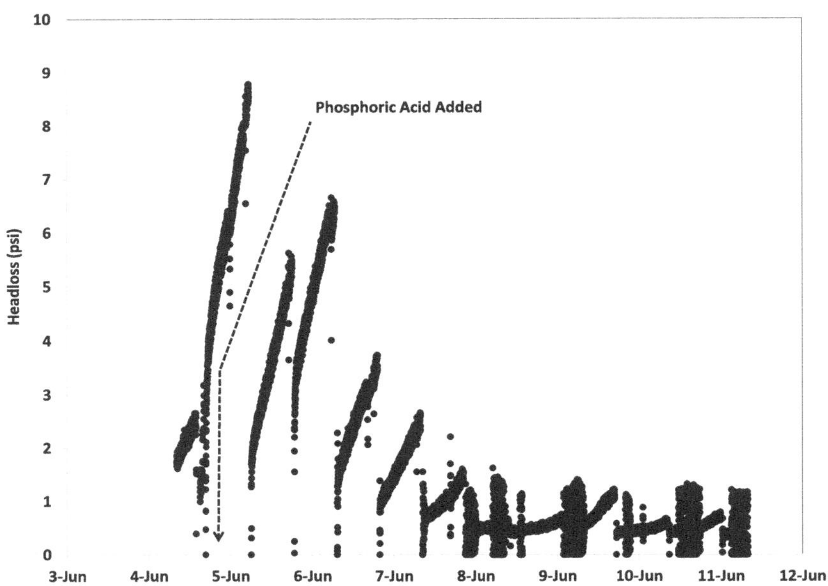

Figure 8.7 Impact of phosphorus limitation on terminal headloss across a perchlorate-reducing fixed-bed bioreactor (Brown, 2012).

8.4.2.4 Contact time

For most biotreatment processes, constituent removal is strongly affected by the length of time the water is in contact with the microbial community in a bioreactor. The design reactor volume/contact time must consider any seasonal changes in constituent loading, temperature, and water demand. It also may be affected by the bioreactor configuration.

8.4.2.5 Post-treatment

Another important consideration for anoxic applications is implementing downstream treatment. In general, downstream treatment is designed to reoxygenate the water, remove any residual electron donor (particularly applicable to heterotrophic processes), remove turbidity such as biomass that sloughs from the anoxic bioreactor process, and disinfect the water. The local regulatory authority may require a second stage filtration step for biomass control.

8.4.2.6 Biomass control

Key elements of any anoxic biotreatment process are the method and tools used to control biomass within the bioreactors. Since an external electron donor is used to enhance the growth of microorganisms, anoxic biotreatment generally results in higher biomass concentrations in the bioreactor compared to aerobic biotreatment. Therefore, anoxic biotreatment processes require special biomass control considerations. The control strategy can include mechanical tools, such as air scour, fluidization, and mechanical media cleaning, as well as chemical tools such as hydrogen peroxide and chlorine. The strategy should maintain uniform hydraulic conditions without affecting the overall biological activity/constituent degradation capacity of the process.

If the filter effluent discharges to a clearwell a multi-compartment clearwell could be considered to facilitate periodic removal of accumulated biomass. A well-designed and operated system, however, should not accumulate significant biomass in the clearwell.

8.4.2.7 Biomass waste handling

Biomass control strategies generate a waste stream that contains microbial cells and other biofilm components. Unless the application involves hexavalent chromium or selenate removal, this waste stream is not likely to contain any chemical constituents of concern. Without implementing a solid/biomass recovery step with the waste stream, such as, biomass settling or flotation, both of which may involve adding a polymer or coagulant, wastage rates can be 4 to 10% of the treated water flow. With a biomass recovery step, wastage rates can be less than 1% of the treated water flow. If additional water recovery is critical, biomass concentrated through settling or floatation can be further concentrated using a mechanical dewatering process.

Waste biomass storage time is a critical issue because storage for more than a few hours can generate offensive odors. Along with shortening the waste biomass storage time, aeration also minimizes odors before the wastewater is disposed of or treated. Treated or untreated wastewater disposal options include discharge to a local sewer or an on-site lagoon or percolation basin. A portion of treated wastewater may be recirculated into the treatment process.

When constituents such as hexavalent chromium or selenate are reduced across an anoxic bioreactor process, they tend to precipitate and get filtered out. To identify appropriate handling and discharge options, the biomass waste handling process must characterize and quantify these inorganic solids and determine if they are resolubilized during wastewater handling.

8.4.2.8 Monitoring and control

Multiple parameters can be monitored in-line for real-time feedback on anoxic bioreactor health and performance. For example, nitrate/nitrite and perchlorate analyzers can directly quantify constituent removal and determine the appropriate substrate dose. Raw water DO can be monitored as part of the substrate dose calculation and treated water DO monitored to track reoxygenation efficacy. Turbidity is typically measured in the system effluent to ensure that biomass stays within the treatment process, thereby limiting demand in the final disinfection step. Hydraulic parameters such as headloss across a fixed-bed bioreactor should also be monitored to track whether the biomass control strategy is working or needs to be modified. Microbiological parameters such as adenosine triphosphate (ATP) can also be monitored as a direct measure of biological activity. However, these analyses tend to cost more and require a concerted effort to correlate data to observed treatment performance.

Anoxic bioreactor processes rely on adding an external substrate (and sometimes nutrients such as phosphorus) to meet constituent removal goals. Therefore, a robust chemical feed system must be integrated to ensure consistent substrate delivery, monitor raw or treated water quality, and adjust the substrate dose as necessary.

Biomass control strategy is also an important consideration since it impacts bioreactor hydraulics, which in turn drives the efficiency and efficacy of treatment. Depending on the specific bioreactor configuration, these methods may include backwashing, in-line biomass/media separators, or in-place chemical cleans.

8.4.2.9 Operator training

Anoxic biological treatment processes can be fully automated to minimize on-site operator demands. However, specialized training is recommended to familiarize operators with hydraulic and water quality performance trends, so adjustments can be made on a timely basis to maintain system performance and reliability. Specific trends will vary based on the anoxic biotreatment process/system in operation, for example, clean-bed headloss for a fixed-bed process.

8.4.2.10 *Pilot testing*

Site-specific pilot testing is recommended ahead of full-scale anoxic biotreatment implementation. In fact, the administrative authority may require it. Pilot testing should address (1) the range of temperatures and water quality conditions anticipated for the full-scale system, and (2) the range of operating conditions anticipated for the full-scale system, including, continuous or intermittent operation, extended shut-down periods and so forth.

To better understand the effects of seasonal and operating variations, pilot tests should run for a year, if possible. Key design parameters to identify or confirm include contact time, chemical doses and dosing strategy, biomass control strategy, backwash wastewater solids settling/removal strategy (if this is planned for the full-scale application), post-treatment (reoxygenation, substrate removal, and filtration), and disinfection strategy. In addition to monitoring the target constituent(s), potential by-products such as nitrite, hydrogen sulfide, and disinfection by-products should be monitored.

8.4.3 Pollutant removal performance

Table 8.4 summarizes nitrate, perchlorate, and hexavalent chromium removal for selected anoxic biological groundwater treatment systems implemented at bench-, pilot-, and full-scale. Table 8.5 summarizes VOC removal for selected anoxic groundwater treatment systems implemented at bench-, pilot-, and full-scales.

8.5 TESTING BIOTREATMENT PROCESSES

Before implementing any full-scale biotreatment process, testing should be performed to (1) prove process efficacy, (2) develop or confirm full-scale design criteria, and (3) identify any process limitations, failure scenarios, and/or unintended consequences.

8.5.1 Scale

Deciding on the testing scale is an important step of the evaluation phase. The scale will be determined by the testing goals and available budget. Bench-scale testing is typically used for proof-of-concept evaluations and may indicate whether pilot- or full-scale testing should be performed. Since the characteristics of a bench-scale system do not allow direct translation of the results to a full-scale implementation, interpret results with caution. Results generated through bench-scale testing should be verified in a pilot-scale or full-scale demonstration system before implementing a full-scale treatment system. Bench-scale testing can be used to answer questions such as these:

- Are there benefits to selecting one process configuration over others?
- What type of biogrowth support media should be selected?

Table 8.4 Summary of selected bench-, pilot-, and full-scale hazardous pollutant removal associated with anoxic/anaerobic biological groundwater treatment.

Bioreactor Configuration	Scale	Environmental/Operating Conditions				Influent Concentration			Percent Removal			References
		Substrate	Contact Time (min)	Temp (°C)	pH	NO_3^--N (mg/L)	ClO_4^- (μg/L)	Cr(VI) (μg/L)	NO_3^-–N	ClO_4^-	Cr(VI)	
Packed-bed	bench	H_2	1.5	23	7	–	79 ± 3	–	–	29 ± 3	–	Logan and LaPoint (2002)
Packed-bed	bench	H_2	1.5	23	7	22 ± 2	78 ± 3	–	10 ± 5	17 ± 3	–	Logan and LaPoint (2002)
Packed-bed	bench	H_2	1.5	23	7	21 ± 2	73 ± 2	–	10	25 ± 5	–	Logan and LaPoint (2002)
Two-stage MBR	bench	ethanol	150	20 ± 2	7–7.5	33.8	–	–	>99	–	–	Ravnjak *et al.* (2013)
Hollow fiber MBR	bench	methanol	13	–	7	200	–	–	>99	–	–	Ergas and Rheinheimer (2004)
Fixed-bed	pilot	acetic acid	30	–	–	4–5	70–80	–	>99	>99	–	Logan *et al.* (2004)
Fixed-bed	bench	acetic acid	20	18	7.5–7.9	5.6	75	–	>99	>99	–	Li *et al.* (2010)
Fixed-bed	pilot	acetic acid	18	18–22	7.6–7.8	5.6	50	–	>99	>99	–	Li *et al.* (2010)
Fixed-bed	bench	acetate or ethanol	9	22	6.5–9	0–2	50	–	>99	>99	–	Brown *et al.* (2003)
Fluidized-bed	full	acetate or ethanol	61.5	21–24	–	1.9	14,700	–	–	>99		Polk and Murray (2001)
Fluidized-bed	pilot	ethanol	126	19–21	–	20	400,000		100	>99.9	>99.9	Hatzinger *et al.* (2000)
Fluidized-bed	full	ethanol	–	–	–	1.5	6000–8000	–	100	>99.9	–	Hatzinger *et al.* (2000)
Fixed-bed	bench	H_2	22.2	–	7	–		–	–	38 ± 9	–	Miller and Logan (2000)
Fluidized-bed	full	acetic acid	12.2	18.5	7.7	6.1–6.3	50–53	–	>99.5	>99.5	–	Webster and Togna (2009)
Fluidized-bed	bench	acetate	16	–	7.2– 7.4	–	3000–5000	–	–	>99.5	–	Nozawa-inoue *et al.* (2010)
Fixed-bed	bench	acetate	18	28	7.0–7.6	20	800	–	100	100	–	Losi *et al.* (2002)
Fixed-bed	pilot	acetate	71.5	10–30	7.6 ± 0.26	0.9 ± 0.27	50 ± 0.39	–	>97	>96	–	Dugan *et al.* (2009)
Fixed-bed	pilot	sulfur	714–1068	9.7	6.7 ± 0.2	–	84.5 ± 113	–	–	73–96	–	Boles *et al.* (2001)
Fixed-bed	pilot	acetate	5.8	19.9 ± 0.8	6.74 ± 0.1	0.9 ± 0.09	77 ± 13	–	>99	>95	–	Min *et al.* (2004)
Fixed-bed	full	ethanol	22–38.3	10	7.2	19.7	–	–	65–79	–	–	Richard (1988)
Fluidized-bed	full	ethanol	10	19	7.2	0.34	2600	–	>99.4	>99.9	–	McCarty and Meyer (2005)
Fixed-bed	pilot	acetate	10	17.5–25.4	7.2–7.7	7.1–9.0	–	20–29	>87	–	>80	Najm *et al.* (2014)
Fixed-bed (two-stage)	full	ethanol	12 (1st stage) 30 (2nd stage)	12–13	7.2–7.4	9–15	–	–	57–77			Rogalla *et al.* (1990)

Table 8.5 Summary of selected bench-, pilot-, and full-scale hazardous pollutant removal associated with anoxic/anaerobic biological groundwater treatment.

Bioreactor Configuration	Scale	Environmental/Operating Conditions				Target VOC	Influent Conc. (μg/L)	Percent Removal	References
		Substrate	Contact Time (min)	Temp (°C)	pH				
Fixed-bed	bench	lactate	1440	20		PCE	1490	95–98	de Bruin *et al.* (1992)
Fixed-bed	full	DCA		8–12	7.1–7.6	DCA	2000–15000	100	Stucki and Thuer (1995)
Fluidized	bench	methanol and sodium acetate	1440	25–27		1,2-DCP	3000	>90	Hauck and Hegemann (2000)
Fixed-bed two-stage – anaerobic column	bench	acetate	7360	22					Fathepure and Vogel (1991)
						PCE	500–1000	80	
						CF	500–1000	32	
Fixed-bed	bench	acetate	3600			PCE	18 ± 3	86 ± 7	Bouwer and Wright (1988)
						CF	17 ± 4	95 ± 3	
						1,1,1-TCA	10 ± 2	>99	
						CT	7 ± 2	>99	
						DBCP	17 ± 2	>99	
						BDCM	13 ± 1	89 ± 8	
						BF	18 ± 3	>99	
						EDB	20 ± 4	>99	
						HCA	4 ± 1	>99	
Fixed-bed	bench	acetate	3600			CF	17 ± 3	−39 ± 37	Bouwer and Wright (1988)
						PCE	8 ± 4	13 ± 45	
						1,2-DBA	19 ± 3	63 ± 9	

(Continued)

Table 8.5 Summary of selected bench-, pilot-, and full-scale hazardous pollutant removal associated with anoxic/anaerobic biological groundwater treatment (*Continued*).

Bioreactor Configuration	Scale	Environmental/Operating Conditions				Target VOC	Influent Conc. (μg/L)	Percent Removal	References
		Substrate	Contact Time (min)	Temp (°C)	pH				
						TCA	9 ± 2	72 ± 8	
						BDCM	16 ± 2	96 ± 5	
						DBCP	12 ± 4	98 ± 7	
						CT	7 ± 1	>99	
						BF	19 ± 3	>99	
						HCA	7 ± 2	>99	
Fixed-bed	bench	acetate	3600			CF	28 ± 2	−13 ± 16	Bouwer and Wright (1988)
						PCE	7 ± 2	0 ± 40	
						DBCP	37 ± 3	14 ± 11	
						1,2-DBA	30 ± 2	23 ± 7	
						TCA	27 ± 3	30 ± 7	
						HCA	4 ± 2	80 ± 17	
						BDCM	24 ± 2	90 ± 8	
						BF	36 ± 2	>99	
						CT	17 ± 2	>99	
Attached-film extended-bed	bench	sucrose	108–240	15	~7	PCE	8000–12000	>98	Carter and Jewell (1993)
Fixed-bed	bench	lactate	1080	14	7	TCE	300	100	Schollhorn *et al.* (1997)
		formate						74	
		acetate						70–90	
		methanol						>80	
		H_2/CO_2						>75	

UASB			1440		7.1–7.4	PCE	3000	38 ± 7	Hwu and Lu (2008)
			5760		7.1–7.4			76 ± 4	
Fixed-bed two-stage	bench	acetate	15	22	6.9–7.3	PCE	7.9 ± 3.7	99 ± 1	Upadhyaya and Brown (2016)
						TCE	11.8 ± 2.9	99 ± 4	
						1,1-DCE	3.9 ± 0.9	100	
						cis 1,2-DCE	0.48 ± 0.11	98 ± 14	
						DBCP	0.22 ± 0.07	98 ± 4	
						CF	0.32 ± 0.06	100	
						CT	0.17 ± 0.03	100	
Fixed-bed two-stage	bench	acetate	25	20–24	6.9–7.5	TCE	2.4 ± 3.4	70 ± 44	Upadhyaya *et al.* (2013)
						PCE	87.6 ± 180.8	94 ± 16	
						TCP	82.86 ± 171.9	76 ± 63	
Fixed-bed two-stage	pilot	acetate	7–12			TCE	189 ± 25	93 ± 4	Brown *et al.* (2015)
						1,1-DCE	4.4 ± 0.5	67 ± 16	
						TCP	6.4 ± 0.7	93 ± 7	

BDCM – bromodichloromethane; BF – bromoform; CF – chloroform; DBA – dibromoethane; DBCP – dibromochloropropane; DCE – dichloroethylene; EDB – ethylene dibromide; HCA – hexachloroethane; PCE – tetrachloroethylene; TCE – trichloroethylene; TCP – trichloropropane; TCF – trichlorofluoromethane; UASB – up-flow anaerobic sludge blanket.

- What substrate and nutrients should be used and at what doses?
- Can target pollutants be removed at contact times and temperatures anticipated for the full-scale system?
- How should "complete" bioacclimation be defined, how long does it take, and what methods could accelerate acclimation?
- Can pollutant removal be enhanced, for example with media sizing or nutrient addition?
- What is the treated water quality, and thus what post-treatment processes might be required?
- What unintended consequences of implementing biotreatment might be expected? Examples could be manganese release, production of unwanted degradation by-products, odors, or increased disinfectant demand.

Pilot testing is typically used to develop design criteria and characterize operational requirements for a full-scale installation. Pilot testing is also an excellent opportunity to engage regulators that will be involved in permitting the full-scale treatment process/plant. Since their questions, concerns, and requirements will be identified at this beginning stage, they can be incorporated into the pilot testing scope of work. Questions to consider during a pilot study include:

- How do different process configurations compare with respect to pollutant removal performance and hydraulic characteristics?
- How does contact time and temperature impact water quality and hydraulic performance?
- What bioacclimation time can be expected?
- Can pollutant removal be enhanced, for example, through media selection, media sizing, or and nutrient addition?
- How do treatment and/or hydraulics change with fluctuations in water quality characteristics or operating conditions? Examples are intermittent operating scenarios or temporary substrate feed disruptions.
- What's the strength and composition of any waste streams from the process?
- What's an appropriate biomass control strategy?
- Is a reacclimation period required immediately following a biomass control step, which necessitates treatment-to-waste (i.e. water is treated but not sent to distribution while the biological community regrows)?
- Can wasted biomass be further concentrated for disposal?
- Does the presence of oxidants affect treatment performance or hydraulic characteristics?
- What is the microbial community structure based on molecular microbial analysis, and does it change with changing water quality or operating conditions?
- What footprint would the full-scale system require?
- What are conceptual-level costs for a full-scale system?

- What type of unintended consequences of biotreatment might be expected? Examples are manganese release, production of unwanted degradation by-products, increased disinfectant demand, and odors.
- What operator attention and training is a full-scale system likely to require?

Demonstration-scale testing can be performed when only minor adjustments to the existing full-scale plant are needed to convert a portion of the plant to biological treatment. An example would be eliminating chlorine feed to an isolated filter train to test biofiltration. The existing configuration and media are used, treatment flow is considerably higher, and control over variables is limited. Therefore, demonstration-scale testing is typically used only to confirm performance trends observed during pilot-scale testing, although it also can be used in lieu of pilot testing for basic test plans. Baseline performance for the existing full-scale treatment process should be well established and a rigorous monitoring strategy developed to ensure that performance, cause, effect, and outcomes are well characterized and documented.

8.5.2 Budgeting

Testing scale, duration, and monitoring requirements will determine a biotreatment testing budget. Bench-scale testing can range from a few thousand dollars to tens of thousands of dollars, whereas pilot-scale testing budgets are typically an order of magnitude higher. Full-scale testing costs typically fall between those for bench- and pilot-scale testing.

8.5.3 Experimental design

Test design begins with identifying and understanding the testing goals and key questions the study must answer. If the associated full-scale process/plant will be a conversion from an existing one and not greenfield construction, then the design of the existing process/plant should be evaluated to identify operational boundary conditions for testing, such as EBCT and backwash protocol. The goals of testing are to assess the effectiveness of biotreatment in achieving water quality and hydraulic targets and to identify process limitations and negative unintended consequences. The testing plan should consider the key factors discussed in the sections below.

8.5.3.1 Water source

Raw water quality drives treatment goals and individual process performance requirements. Relevant data and information will be generated only if testing is conducted with the actual source water. Its quality should be fully characterized before testing to better understand the range of test conditions to consider.

If source water is not available, synthetic water prepared by adding chemical constituents to simulate a specific source water can be tested at the bench-scale level to evaluate proof-of-concept. However, uncharacterized physical-chemical parameters in the synthetic matrix such as the presence or absence of trace nutrients may affect biological activity and therefore impact system performance. Thus, bench-scale results should be verified through pilot- and/or full-scale testing, using the water that will be treated at full-scale.

8.5.3.2 Process configuration and biogrowth support medium

If a particular biotreatment configuration has not already been specified, the evaluation phase may include parallel testing of multiple configurations, for example a fluidized-bed system in parallel with a fixed-bed system. Parallel testing is more common for groundwater biotreatment applications since considerably fewer configurations are available for surface water biotreatment applications.

For most biotreatment applications, different types and sizes of biogrowth support media can be considered and tested. Even though smaller media can increase the concentration of biomass in a system, which may decrease required contact times, they may exert greater hydraulic demands while the system is in service. Relative to inert media, adsorptive media such as GAC may increase biomass concentration and decrease sensitivity to process upsets, but it tends to be more expensive and, if the biomass control strategy is overly aggressive, can break down to smaller particles over time, increasing hydraulic demands. Media depth also can be tested to characterize the impact of contact time and/or filtration characteristics such as biofiltration application. Multiple contact times can be tested with a single system using multiple sampling ports across a bioreactor.

8.5.3.3 Biological acclimation

Biological acclimation is the time required to establish a consistent, mature, and active biomass in a bioreactor, when the bioreactor demonstrates consistent removal of the target pollutant with relatively stable hydraulic performance. Biological activity can be observed within a week or two, although full bioacclimation may take months.

The test plan should give time for full bioacclimation to ensure pseudo-steady-state conditions. Special attention should be given to challenges that may occur during bioacclimation, including biological fouling of in-line instrumentation and sampling lines, decreased hydraulic performance, shortened filter runtimes, additional turbidity breakthrough, and/or increased disinfectant demand. For surface water applications, seasonal temperature effects on the duration of bioacclimation should be considered. For example, the time required for warmer month bioacclimation will typically be shorter than that for colder months.

8.5.3.4 Pre-oxidation

Oxidation processes are often used upstream of surface water biofiltration. Pre-oxidation can break down organic constituents into lower molecular weight products such as carboxylic acids, aldehydes, and ketones, which are more readily biodegradable. Therefore, pre-oxidation may enhance biological activity in a surface water biofilter.

Conversely, the presence of pre-oxidants in any bioreactor feed water may inhibit microbial activity, thereby dampening system performance. This is especially important during biological acclimation. Once a mature microbial community is established, the negative impacts of a low-level oxidant may not be measurable. However, even after a microbial community is fully established, the consistent presence of an oxidant in bioreactor feed water can influence biotreatment performance. This effect depends on the type of pre-oxidant used (LeChevallier *et al.* 1992; Weinberg *et al.* 1993).

If pre-oxidation is anticipated for the full-scale biotreatment application, the testing plan should evaluate not only the impact under steady conditions, but also the impact of excess oxidant carryover to the bioreactor/biofilter.

8.5.3.5 Biomass control strategy

A critical component of any biotreatment process is the biomass control strategy. Bioreactor overgrowth can result in reactor clogging or short-circuiting, limitations in pollutant diffusion into the biofilm, and poor overall system performance. On the other hand, an overly aggressive biomass control strategy can result in insufficient bioactivity, leading to pollutant breakthrough. Each bioreactor configuration employs a standard mechanically-based biomass control strategy such as, backwashing biofilters or other fixed-bed bioreactor systems. Oxidants such as chlorine or hydrogen peroxide also can be used to enhance a biomass control process and should be tested ahead of full-scale implementation.

While pilot testing is important for characterizing the general effectiveness and impact of various biomass control strategies, it typically is not used to develop rigorous design criteria for a full-scale biomass control system. Wall effects and other hydraulic limitations of smaller-scale reactors can limit the effectiveness of the biomass control system compared with that observed for a comparable full-scale biomass control system.

8.5.3.6 Influent water quality and seasonal variability

Temperature, pH, DO concentration, BOM concentration, solids loading, and nutrient concentration can vary seasonally and impact bioacclimation as well as long-term bioreactor performance. Thus, seasonal variability of these constituents should be well characterized, and testing should quantify the impact of this variability to the extent possible. Pollutant loading also can vary widely with season and from year to year, for example, MIB and geosmin. Consider these variations when designing a biotreatment testing plan. Chemical feed systems often are used to spike pollutants during bench- and pilot-scale testing.

Table 8.6 Monitoring considerations for biotreatment testing.

Category	Parameter	Comments
Water quality	Temperature	Can be measured in-line line or by grab sample.
	pH	Can be measured in-line line or by grab sample.
	Alkalinity	Measured by grab sample and used as an indicator of buffering capacity.
	Dissolved Oxygen	Can be measured in-line or by grab samples. Measurement upstream and downstream of a filter can provide an indication of oxygen consumption/biological activity.
	Turbidity	Continuous monitoring required.
	Total Organic Carbon (TOC)	TOC measurement can be compared to DOC as an indicator of biomass sloughing from a biofilter.
	Dissolved Organic Carbon (DOC)	DOC removal across biofilter(s) may replace need for AOC or biodegradable dissolved organic carbon (BDOC).
	Assimilable Organic Carbon (AOC)	Can be used to check whether the biofilter(s) is realizing its full potential in terms of carbon removal.
	Biodegradable Dissolved Organic Carbon (BDOC)	Can be used to check whether the biofilter(s) is realizing its full potential in terms of carbon removal.
	Ozonation By-Products (e.g. aldehydes, carboxylic acids)	Ozonation by-products tend to be readily biodegradable and therefore provide one measure of biofilter effluent regrowth potential.
	DBP Formation Potential	Can be used to monitor DBP precursor removal.
	Volatile organic compounds (VOCs)	Site-specific target pollutant.
	Chlorine/Chloramines Stability	Can be used to determine disinfectant demand in biofilter effluent.
	Iron/Manganese	Site-specific target pollutant.
	Geosmin/MIB	Site-specific target pollutant.
	Algal Toxins	Site-specific target pollutant.
	EDCs/PPCPs	Consider if raw water source has known wastewater or industrial influences.

	Ammonia (Free)	Important macronutrient. Can be measured in-line or by grab sample.
	Ortho-Phosphate (PO_4^{3-})	Important macronutrient.
	Nitrate/Nitrite	Site-specific target pollutant; can be measured in-line and/or by grab samples.
	Perchlorate (ClO_4^-)	Site-specific target pollutant.
	Total/Hexavalent Chromium (Cr(VI))	Site-specific target pollutant.
	Pre-Filter Oxidant Residual	Pre-filter oxidants can exert stress on the biological community in a filter.
Microbial	Adenosine Triphosphate (ATP)	Can be used to track trends in biological activity. This includes both aqueous and media-based samples.
	Phospholipids	Alternate method for tracking trends in biological activity.
	Extracellular Polymeric Substances (EPS)	Can indicate stressed biological activity.
	Heterotrophic Plate Counts	Can indicate stressed biological activity. Can also be used to measure the efficacy of a given disinfection protocol.
	Microbial Community Structure	Requires the use of molecular microbial methods; typically performed at specialty labs and universities.
Hydraulic	Flow/Loading Rate	Correlates with EBCT, which typically drives the effectiveness of pollutant biodegradation.
	Run Time	Important parameter with respect to turbidity breakthrough, headloss, and biomass control.
	Clean Bed Headloss or Pressure Drop Across the System	Can be used to detect the onset of biofouling within the biological reactor.
	Unit Filter Run Volume	Can be used to detect the onset of biofouling within the biological reactor.
	Intra- and Inter-run Headloss (or pressure drop across the system) patterns	Can be used to detect the onset of biofouling within the biological reactor.
	Biomass Control Strategy	Both under-backwashing and over-backwashing can be detrimental to biofilter operation and efficiency.
	Backwash Pressure	Can be used to detect the onset of biofouling within the media bed.

Source: Modified from Brown *et al.* (2016).

8.5.3.7 Performance enhancement strategies

Various strategies can enhance biotreatment performance. For example, macronutrients such as nitrogen or phosphorus may be limiting and can be dosed to the feed water to improve both hydraulic and water treatment outcomes (Hoyland, 2012; Lauderdale *et al.* 2012; Granger, 2013; Lauderdale *et al.* 2014). Other examples of enhancement strategies include dosing trace nutrients, dosing low levels of easily degradable substrate, pre-oxidation, and pH adjustment. Essential to any biotreatment enhancement testing are (1) a baseline characterization of system performance prior to enhancement, and (2) testing of an unenhanced parallel bioreactor as a control so that the impact of each enhancement strategy can be rigorously quantified.

8.5.4 Monitoring

As shown in Table 8.6, a wide variety of water quality, microbial, and hydraulic parameters can be considered for biological treatment testing. The breadth and frequency of sampling will be determined by testing scale, performance targets, the number of parallel systems being tested, budget, availability of on-site analytical capabilities, and staffing. Data should be plotted and evaluated as frequently as possible to detect performance variations that could lead to operational changes. Ideally, key performance plots should be updated and evaluated daily. Automated data management tools are available to minimize the time required for this effort (Nyfennegger *et al.* 2016).

8.6 REFERENCES

Alexander M. (1999). Biodegradation and Bioremediation. Academic Press, San Diego, CA.

Andersson A., Laurent P., Kihn A., Prévost M. and Servais P. (2001). Impact of temperature on nitrification in biological activated carbon (BAC) filters used for drinking water treatment. *Water Research*, **35**(12), 2923–2935.

Basu O. D., Dhawan S. and Black K. (2016). Applications of biofiltration in drinking water treatment – a review. *Journal of Chemical Technology & Biotechnology*, **91**(3), 585–595.

Beech I. B. and Sunner J. (2004). Biocorrosion: towards understanding interactions between biofilms and metals. *Current Opinion in Biotechnology*, **15**(3), 181–186.

Berbenni P., Pollice A., Canziani R., Stabile L. and Nobili F. (2000). Removal of iron and manganese from hydrocarbon-contaminated groundwaters. *Bioresource Technology*, **74**(2), 109–114.

Boles A. R., Conneely T., McKeever R., Nixon P., Nüsslein K. R. and Ergas S. J. (2001). Performance of a pilot-scale packed bed reactor for perchlorate reduction using a sulfur oxidizing bacterial consortium. *Biotechnology and Bioengineering*, **109**(3), 637–646.

Bourgine F. P., Gennery M., Chapman I. I., Kerai H., Green J. G., Ellis S. and Gaumard C. (1994). Biological Processes at Saints Hill water-treatment plant, Kent. *Journal of the Institution of Water Environment Management*, **8**, 379–392.

Bouwer E. J. and Wright J. P. (1988). Transformations of trace halogenated aliphatics in anoxic biofilm columns. *Journal of Pollutant Hydrology*, **2**(2), 155–169.

Brown J. (2012). Overcoming Conventional Wisdom in the Drinking Water Industry. Keynote address given at the 18th Annual University of Illinois Environmental Engineering & Science Spring Symposium, April 6, Urbana, IL.

Brown J., Murphy B. and Hart V. (2013). Biofilter Acclimation & Operation at 10°C for the Removal of Ammonia, Iron, and Manganese. Paper presented at the American Water Works Association Annual Conference and Exhibition, June 9–13, Denver, CO.

Brown J., Riley J., Cleveland C., Carlo P., Upadhyaya G., Silva B. and Mount D. (2015). Two-Stage Fixed-Bed Biological Treatment: Unlocking Contaminated Groundwater Sources. Proceedings of the American Water Works Association Annual Conference and Exposition, June 9, Anaheim, CA.

Brown J., Upadhyaya G., Carter J., Brown T. and Lauderdale C. (2016). North American Biofiltration Knowledge Base. Water Research Foundation, Colorado.

Brown J. C., Snoeyink V. L., Raskin L. and Lin R. (2003). The sensitivity of fixed-bed biological perchlorate removal to changes in operating conditions and water quality characteristics. *Water Research*, **37**(1), 206–214.

Brown J. C., Safely J. and Carlson P. (2009). Efficient Biological Removal of Nitrate from Reverse Osmosis By-Pass Water at the Arlington Desalter Water Treatment Plant. Proceedings from the American Water Works Association Annual Conference and Exhibition, June 14–18, San Diego, CA.

Bruins J. H., Petrusevski B., Slokar Y. M., Kruithof J. C. and Kennedy M. D. (2015a). Manganese removal from groundwater: characterization of filter media coating. *Desalination and Water Treatment*, **55**(7), 1853–1863.

Bruins J. H., Petrusevski B., Slokar Y. M., Huysman K., Joris K., Kruithof J. C. and Kennedy M. D. (2015b). Biological and physico-chemical formation of Birnessite during the ripening of manganese removal filters. *Water Research*, **69**, 154–161.

Cai Y., Li D., Liang Y., Luo Y., Zeng H. and Zhang J. (2015). Effective start-up biofiltration method for Fe, Mn, and ammonia removal and bacterial community analysis. *Bioresource Technology*, **176**, 149–155.

Cai Y., Li D., Liang Y., Zeng H. and Zhang J. (2016). Operational parameters required for the start-up process of a biofilter to remove Fe, Mn, and NH3-N from low-temperature groundwater. *Desalination and Water Treatment*, **57**(8), 3588–3596.

Cardenas-Gonzalez M., Osorio-Yanez C., Gaspar-Ramirez O., Pavkovic M., Ochoa-Martinez A., Lopez-Ventura D., Medeiros M., Barbier O. C., Perez-Maldonado I. N., Sabbisetti V. S., Bonventre J. V. and Vaidya V. S. (2016). Environmental exposure to arsenic and chromium in children is associated with kidney injury molecule-1. *Environmental Research*, **150**, 653–662.

Carlson K. H. and Amy G. L. (1998). BOM removal during biofiltration. *Journal of American Water Works Association*, **90**(12), 42–52.

Carter S. R. and Jewell W. J. (1993). Biotransformation of tetrachloroethylene by anaerobic attached-films at low temperatures. *Water Research*, **27**(4), 607–615.

Chen H. M. and Wu M. T. (2017). Residential exposure to chlorinated hydrocarbons from groundwater contamination and the impairment of renal function-An ecological study. *Scientific Reports*, **7**, 40283.

Clement B. G., Luther G. W. and Tebo B. M. (2009). Rapid, oxygen-dependent microbial Mn(II) oxidation kinetics at sub-micromolar oxygen concentrations in the Black Sea suboxic zone. *Geochimica et Cosmochimica Acta*, **73**(7), 1878–1889.

Coppola E. N. (2000). ARA perchlorate treatment experience. In: Proceedings of the 5th Annual Joint Services Pollution Prevention & Hazardous Waste Management Conference & Exhibition, sponsored by the Air Force Research Laboratory, August 23–24, 2000, San Antonio, TX, Cook L. (ed.), Air Force Research Laboratory, Tyndall AFB, FL.

de Bruin W. P., Kotterman M. J. J., Posthumus M. A., Schraa G. and Zehnder A. J. B. (1992). Complete biological reductive transformation of tetrachloroethene to ethane. *Applied and Environmental Microbiology*, **58**(6), 1996–2000.

de Vet W. W. J. M., Rietveld L. C. and Van Loosdrecht M. C. M. (2009). Influence of iron on nitrification in full-scale drinking water trickling filters. *Journal of Water Supply Research and Technology-Aqua*, **58**(4), 247–256.

de Vet W. W. J. M., Dinkla I. J. T., Rietveld L. C. and van Loosdrecht M. C. M. (2011). Biological iron oxidation by *Gallionella* spp. in drinking water production under fully aerated conditions. *Water Research*, **45**(17), 5389–5398.

de Vet W. W. J. M., van Loosdrecht M. C. M. and Rietveld L. C. (2012). Phosphorus limitation in nitrifying groundwater filters. *Water Research*, **46**(4), 1061–1069.

Dong W. Y., Wang H. J., Li W. G., Ying W. C., Gan G. H. and Yang Y. (2009). Effect of DO on simultaneous removal of carbon and nitrogen by a membrane aeration/filtration combined bioreactor. *Journal of Membrane Science*, **344**(1–2), 219–224.

Dowdell K. (2012). Trace Organic Pollutant Removal in Drinking Water Biofilters under Carbonaceous and Nitrogen-Supplemented Conditions and Evaluating Biomass with ATP and Phospholipid Methods. Master's thesis, University of Colorado-Boulder.

Du X., Liu G., Qu F., Li K., Shao S., Li G. and Liang H. (2016). Removal of iron, manganese and ammonia from groundwater using a PAC-MBR system: the anti-pollution ability, microbial population and membrane fouling. *Desalination*, **403**, 97–106.

Dugan N. R., Williams D. J., Meyer M., Schneider R. R., Speth T. F. and Metz D. H. (2009). The impact of temperature on the performance of anaerobic biological treatment of perchlorate in drinking water. *Water Research*, **43**(7), 1867–1878.

Ergas S. J. and Rheinheimer D. E. (2004). Drinking water denitrification using a membrane bioreactor. *Water Research*, **38**(14–15), 3225–3232.

Evans P. (2010). Nature Works. *Opflow*, **36**(7), 12–15.

Fathepure B. Z. and Vogel T. M. (1991). Complete degradation of polychlorinated hydrocarbons by a two-stage biofilm reactor. *Applied and Environmental Microbiology*, **57**(12), 3418–3422.

Frischherz H., Zibuschka H., Jung H. and Zerobin W. (1985). Biological elimination of iron and manganese. *Water Supply*, **3**, 125–136.

Fu J. X., Shang J. and Zhao Y. H. (2011). Simultaneous Removal of Iron, Manganese and Ammonia from Groundwater in Single Biofilter Layer Using BAF. *Advanced Materials Research*, **183–185**, 442–446.

Gouzinis A., Kosmidis N., Vayenas D. V. and Lyberatos G. (1998). Removal of Mn and simultaneous removal of NH3, Fe and Mn from potable water using a trickling filter. *Water Research*, **32**(8), 2442–2450.

Granger H. (2013). Manganese removal from surface water using bench-scale biofiltration. Master Thesis submitted to Dalhousie University, Halifax, Nova Scotia, Canada.

Granger H. C., Stoddart A. K. and Gagnon G. A. (2014). Direct biofiltration for manganese removal from surface water. *Journal of Environmental Engineering*, **140**(4), 1–8.

Gu J. D., Ford T. E. and Mitchell R. (2011). Microbiological corrosion of metallic materials. In: Uhlig's Corrosion Handbook, Winston Revie R. (ed.), 3rd edn, John Wiley & Sons, Inc., Hoboken, New Jersey.

Guillette L. J. and Edwards T. M. (2005). Is nitrate an ecologically relevant endocrine disruptor in vertebrates? Integrative and comparative biology. *Integrative and Comparative Biology*, **45**(1), 19–27.

Halle C., Huck P. and Peldszus S. (2015). Emerging pollutant removal by biofiltration: temperature, concentration, and EBCT impacts. *Journal American Water Works Association*, **107**(7), E364–E379.

Hanert H. H. (1981). The genus *Gallionella*. In: The Prokaryotes. A Handbook on Habitats, Isolation and Identification of Bacteria, Starr M. P., Triiper H. G., Balows A. & Schlegel H. G. (eds), Springer, Berlin, Vol. 1, pp. 509–515.

Hanert H. H. (1992). The genus Siderocapsa (and other iron- and manganese-oxidizing eubacteria). In: The Prokaryotes. Triiper H. G., Balows A., Dworkin M., Harder W. and Schleifer K. H. (eds), Springer-Verlag, New York, NY, USA, Vol. 4.

Hanert H. H. (2006). The genus *Gallionella*. In: The Prokaryotes, A Handbook on the Biology of Bacteria, Dworkin M., Falkow S., Rosenburg E., Scleifer K.-H. and Stackebrandt E. (eds), Springer-Verlag, New York, NY, pp. 990–995.

Hasan H. A., Abdullah S. R. S., Kamarudin S. K. and Kofli N. T. (2011). Response surface methodology for optimization of simultaneous COD, NH_4^+–N and Mn^{2+} removal from drinking water by biological aerated filter. *Desalination*, **275**(1–3), 50–61.

Hasan H. A., Abdullah S. R. S., Kofli N. T. and Kamarudin S. K. (2012). Effective microbes for simultaneous bio-oxidation of ammonia and manganese in biological aerated filter system. *Bioresource Technology*, **124**, 355–363.

Hasan H. A., Abdullah S. R. S., Kamarudin S. K. and Kofli N. T. (2013). On-off control of aeration time in the simultaneous removal of ammonia and manganese using a biological aerated filter system. *Process Safety and Environmental Protection*, **91**(5), 415–422.

Haston Z. C. and McCarty P. L. (1999). Chlorinated Ethene Half-Velocity Coefficients (KS) for Reductive Dehalogenation. *Environmental Science & Technology*, **33**, 223–226.

Hatzinger P. B., Greene M. R., Frisch S. and Togna A. P. (2000). Biological Treatment of Perchlorate-Contaminated Groundwater Using Fluidized Bed Reactors. Conference proceedings: 2nd International Conference on Remediation of Chlorinated and Recalcitrant Compounds, May 22–25, 2000, Monterey, California.

Hauck R. and Hegemann W. (2000). Anaerobic degradation of 1,2-dichloropropane in batch and continuous culture. *Water Science & Technology*, **41**, 7–14.

Hoyland V. W. (2012). Evaluating the Use of Manganese-Oxidizing Bacteria in Surface Water Treatment Plants. Master's thesis. Virginia Polytechnic Institute and State University, Blacksburg, Virginia.

Hozalski R. M., Bouwer E. J. and Goel S. (1999). Removal of Natural Organic Matter (NOM) from Drinking Water Supplies by Ozone-Biofiltration. *Water Science & Technology*, **40**(9), 157–163.

Hu C. Y., Zhang J., Xu B., Lin Y. L., Zhang T. Y. and Tian F. X. (2016). Effect of pipe corrosion product–goethite–on the formation of disinfection by-products during chlorination. *Desalination and Water Treatment*, **57**(2), 553–561.

Hwu C. S. and Lu C. J. (2008). Continuous dechlorination of tetrachloroethene in an upflow anaerobic sludge blanket reactor. *Biotechnology Letters*, **30**(9), 1589–1593.

ITRC (Interstate Technology & Regulatory Council) (2007). Remediation Technologies for Perchlorate Contamination in Water and Soil. PERC-2. Interstate Technology & Regulatory Council, Perchlorate Team, Washington, D.C.

Jannat M., Fatimah R. and Kishida M. (2014). Nitrate (NO3-) and nitrite (NO2-) are endocrine disruptors to downregulate expression of tyrosine hydroxylase and motor behavior through conversion to nitric oxide in early development of zebrafish. *Biochemical and Biophysical Research Communications*, **452**(4), 608–613.

Katsoyiannis I. A. and Zouboulis A. I. (2004). Biological treatment of Mn(II) and Fe(II) containing groundwater: kinetic considerations and product characterization. *Water Research*, **38**(7), 1922–1932.

Koenig J. C., Lee M. J. and Manefield M. (2012). Successful microcosm demonstration of a strategy for biodegradation of a mixture of carbon tetrachloride and perchloroethene harnessing sulphate reducing and dehalorespiring bacteria. *Journal of Hazardous Materials*, **219–220**, 169–175.

Kohl P. M. and Dixon D. (2012). Occurrence, Impacts, and Removal of Manganese in Biofiltration Processes. Water Research Foundation, Denver, CO.

Lauderdale C., Chadik P., Kirisits M. J. and Brown J. (2012). Engineered biofiltration: enhanced biofilter performance through nutrient and peroxide addition. *Journal American Water Works Association*, **104**(5), 298–309.

Lauderdale C., Scheitlin K., Nyfennegger J., Upadhyaya G. and Brown J. C. (2014). Optimizing Engineered Biofiltration. WRF, Denver, CO.

LeChevallier M. W., Becker W. C., Shorr P. and Lee R. G. (1992). Evaluating the performance of biologically active rapid filters. *Journal American Water Works Association*, **84**(4), 136–146.

LeChevallier M. W., Schneider O. D., Weinrich L. A., Jjemba P. K., Evans P. J., Hooper J. L. and Chappell R. W. (2015). An Operational Definition of Biostability in Drinking Water. WRF, Denver, CO.

Leung A. M., Pearce E. N. and Braverman L. E. (2014). Environmental perchlorate exposure: potential adverse thyroid effects. *Current Opinion in Endocrinology & Diabetes and Obesity*, **21**(5), 372–376.

Li D., Zhang J., Wang H., Yang H. and Wang B. (2005). Operational performance of biological treatment plant for iron and manganese removal. *Journal of Water Supply: Research and Technology – AQUA*, **54**(1), 15–24.

Li X., Upadhyaya G., Yuen W., Brown J., Morgenroth E. and Raskin L. (2010). Changes in the structure and function of microbial communities in drinking water treatment bioreactors upon addition of phosphorus. *Applied and Environmental Microbiology*, **76**(22), 7473–7481.

Li X. K., Chu Z. R., Liu Y. J., Zhu M. T., Yang L. and Zhang J. (2013). Molecular characterization of microbial populations in full-scale biofilters treating iron, manganese, and ammonia containing groundwater in Harbin, China. *Bioresource Technology*, **147**, 234–239.

Liu J. R., Liu C. T., Edwards E. A. and Liss S. N. (2006). Effect of phosphorus limitation on microbial floc structure and gene expression in activated sludge. *Water Science & Technology*, **54**(1), 247–255.

Liu Y. and Liu Q. S. (2006). Causes and control of filamentous growth in aerobic granular sludge sequencing batch reactors. *Biotechnology Advances*, **24**(1), 115–127.

Logan B. E. and LaPoint D. (2002). Treatment of perchlorate- and nitrate-contaminated groundwater in an autotrophic, gas phase, packed-bed bioreactor. *Water Research*, **36**(14), 3647–3653.

Logan B. E., Min B., Kim K., Miller J., Lapoint D., Batista J., Liu J., Evans P. J., Chu A. and Price S. (2004). Bioreactor Systems for Treating Perchlorate-Contaminated Water: Bench- and Pilot-Scale Investigations. Awwa Research Foundation, Denver, CO.

Losi M. E., Giblin T., Hosangadi V. and Frankenberger W. T. Jr. (2002). Bioremediation of perchlorate-contaminated groundwater using a packed-bed biological reactor. *Bioremediation Journal*, **6**(2), 97–103.

Madoni P., Davoli D., Fontani N., Cucchi A. and Rossi F. (2001). Spatial distribution of microorganisms and measurements of oxygen uptake rate and ammonia uptake rate activity in a drinking water biofilter. *Environmental Technology*, **22**(4), 455–462.

Mauclaire L., Schurmann A., Thullner M., Gammeter S. and Zeyer J. (2004). Sand filtration in a water treatment plant: biological parameters responsible for clogging. *Journal of Water Supply: Research and Technology – AQUA*, **53**(2), 93–108.

McCarty P. L. and Meyer T. E. (2005). Numerical model for biological fluidized-bed reactor treatment of groundwater. *Environmental Science & Technology*, **39**(3), 850–858.

Melin E. and Ødegaard H. (2000). The effect of biofilter loading rate on the removal of organic ozonation by-products. *Water Research*, **34**(18), 4464–4476.

Metz D. H., Pohlman R. C., Vogt J. and Summers R. S. (2006). Removal of MIB and Geosmin by full-scale biological sand filters. In: Recent Progress in Slow Sand and Alternative Biofiltration Processes, Gimbel R., Graham N. and Collins M. R. (eds), International Water Association, London, UK.

Miller J. P. and Logan B. E. (2000). Sustained Perchlorate Degradation in an Autotrophic, Gas-Phase, Packed-Bed Bioreactor. *Environmental Science & Technology*, **34**(14), 3018–3022.

Min B., Evans P. J., Chu A. K. and Logan B. E. (2004). Perchlorate removal in sand and plastic media bioreactors. *Water Research*, **38**(1), 47–70.

Mitra S. and Gupta S. K. (2014). Pilot-scale treatment of a trichloethylene rich synthetic wastewater in anaerobic hybrid reactor, with morphological study of the sludge granules. *Clean Technologies and Environmental Policy*, **16**(5), 947–956.

Mouchet P. (1992). From conventional to biological removal of iron and manganese in France. *Journal of American Water Works Association*, **84**, 158–166.

Najm I., Brown N. P., Gallagher B., Seo E. and Gramith K. (2014). Minimizing Waste Backwash Water from a Biological Denitrification Treatment System. Water Research Foundation, Denver, CO.

Nozawa-inoue M., Weaver D. E. and O'Connell Joseph E. (2010). Perchlorate Reduction Using Fine Media Fluidized Bed Bioreactor with Oxidation-Reduction Potential-Based Feed Control. Conference Proceedings: Annual International Conference on Soils, Sediments, Water, and Energy, Vol. 15, pp. 7–21.

Nyfennegger J., Burns S., Ross N., Upadhyaya G., Lauderdale C., Kirisits M. J., Keithley S., Bae S. W. and Brown J. (2016). Full-Scale Engineered Biofiltration Evaluation and Development of a Performance Tracking Tool. WRF, Denver, Colo.

Pacini V. A., Ingallinella A. M. and Sanguinetti G. (2005). Removal of iron and manganese using biological roughing up flow filtration technology. *Water Research*, **39**(18), 4463–4475.

Polk J. and Murray C. (2001). Case Study of Ex-situ Biological Treatment of Perchlorate-Contaminated Groundwater. Conference Proceedings: 4th Tri-Services Environmental Technology Symposium, June 18–20, San Diego, CA.

Prest E. I., Hammes F., van Loosdrecht M. C M. and Vrouwenvelder J. S. (2016). Biological stability of drinking water: controlling factors, methods, and challenges. *Frontiers in Microbiology*, **7**(Feb), 1–24.

Ravnjak M., Vrtovšek J. and Pintar A. (2013). Denitrification of drinking water in a two-stage membrane bioreactor by using immobilized biomass. *Bioresource Technology*, **128**, 804–806.

Richard B. (1988). Operating Experiences of Full-scale Biological and Ion-Exchange Denitrification Plants in France. *Journal of institution of Water & Environmental Management*, **3**, 154–167.

Richter D., Massmann G. and Duennbier U. (2008). Behaviour and Biodegradation of Sulfonamides (p-TSA, o-TSA, BSA) during Drinking Water Treatment. *Chemosphere*, **71**(8), 1574–1581.

Rittman B. and McCarty P. (2001). Environmental Biotechnology: Principles and Applications. McGraw-Hill, Boston, MA.

Rogalla F., Ravarini P., De Larminat G. and Couttelle J. (1990). Large-scale biological nitrate and ammonia removal. *Journal IWEM*, **4**(8), 319–320.

Scheutz C., Durant N. D., Hansen M. H. and Bjerg P. L. (2011). Natural and enhanced anaerobic degradation of 1,1,1-trichloroethane and its degradation products in the subsurface-a critical review. *Water Research*, **45**(9), 2701–2723.

Schollhorn A., Savary C., Stucki G. and Hanselmann K. W. (1997). Comparison of different substrates for the fast reductive dechlorination of trichloroethene under groundwater conditions. *Water Research*, **31**(6), 1275–1282.

Seredyńska-Sobecka B., Tomaszewska M., Janus M. and Morawski A. W. (2006). Biological activation of carbon filters. *Water Research*, **40**(2), 355–363.

Servais P., Prevost M., Laurent P., Joret J. C., Summers R. S., Hamsch B. and Ventresque C. (2005). Biodegradable organic matter in drinking water treatment. In: Biodegradable Organic Matter in Drinking Water Treatment and Distribution, Prevost M., Laurent P., Servais P. and Joret J. C. (eds), Amer. Water Works Assoc., Denver, CO.

Sparrow L. A. and Uren N. C. (1987). Oxidation and reduction of Mn in acidic soils: effect of temperature and soil pH. *Soil Biology and Biochemistry*, **19**(2), 143–148.

Steinmaus C. M. (2016). Perchlorate in water supplies: sources, exposures, and health effects. *Current Environmental Health Reports*, **3**(2), 136–143.

Štembal T., Markić M., Briški F. and Sipos L. (2004). Rapid start-up of biofilters for removal of ammonium, iron and manganese from groundwater. *Journal of Water Supply: Research and Technology – AQUA*, **53**(7), 509–518.

Štembal T., Markić M., Ribičić N., Briški F. and Sipos L. (2005). Removal of ammonia, iron and manganese from groundwaters of northern Croatia – pilot plant studies. *Process Biochemistry*, **40**(1), 327–335.

Stucki G. and Thuer M. (1995). Experiences of a Large-Scale Application of 1,2-Dichloroethane Degrading Microorganisms for Groundwater Treatment. *Environmental Science & Technology*, **29**(9), 2339–2345.

Summers R. S., Shimabuku K. and Zearley T. (2014). A review of biologically-based drinking water treatment processes for organic micropollutant removal. In: Progress in Slow Sand and Alternative Biofiltration Processes – Further Developments and Applications, Collins M. and Graham N. (eds), IWA, London, pp. 17–24.

Tekerlekopoulou A. G. and Vayenas D. V. (2007). Ammonia, iron, and manganese removal from potable water using trickling filters. *Desalination*, **210**(1–3), 225–235.

Tekerlekopoulou A. G. and Vayenas D. V. (2008). Simultaneous biological removal of ammonia, iron, and manganese from potable water using a trickling filter. *Biochemical Engineering Journal*, **39**(1), 215–220.

Tekerlekopoulou A. G., Papazafiris P. G. D. and Vayenas D. V. (2010). A full-scale trickling filter for the simultaneous removal of ammonium, iron, and manganese from potable water. *Journal of Chemical Technology and Biotechnology*, **84**(7), 1023–1026.

Tekerlekopoulou A. G., Pavlou S. and Vayenas D. V. (2013). Removal of ammonium, iron and manganese from potable water in biofiltration units: a review. *Journal of Chemical Technology & Biotechnology*, **88**(5), 751–773.

Terry, L. and Summers, R. S. (2017). Biodegradable Organic Matter and Biofilter Performance: A Review. Paper Accepted to Water Research.

Upadhyaya G. and Brown J. (2016). Biologically Mediated Simultaneous Removal of VOCs and Inorganic Pollutants from Drinking Water Sources. Proceedings of AWWA Annual Conference and Exposition, June 19–22, 2016, Chicago, IL.

Upadhyaya G., Kotlarz N., Hayes K. F., Raskin L. and Brown J. (2013). Efficient Removal of Multiple Pollutants Using Two-Stage Biologically Active Carbon Bioreactors. AWWA Water Quality Technology Conference, November 3–7, 2013. Long Beach, CA.

Vandenabeele J., Vande W. M., Houwen F., Germonpré R., Vandesande D. and Verstraete W. (1995). Role of autotrophic nitrifiers in biological manganese removal from groundwater containing manganese and ammonium. *Microbial Ecology*, **29**(1), 83–98.

Vinceti M., Wei E. T., Malagoli C., Bergomi M. and Vivoli G. (2001). Adverse health effects of selenium in humans. *Reviews on Environmental Health*, **16**(4), 233–252.

Vinceti M., Bonvicini F., Rothman K. J., Vescovi L. and Wang F. (2010). The relation between amyotrophic lateral sclerosis and inorganic selenium in drinking water: a population-based case-control study. *Environmental Health*, **9**(1), 77.

Viswanathan M. N. and Boettcher B. (1991). Biological removal of iron from groundwater. *Water Science and Technology*, **23**(7–9), 1437–1446.

Webster T. S. and Togna P. (2009). Demonstration of a Full-Scale Fluidized Bed Bioreactor for the Treatment of Perchlorate at Low Concentrations in Groundwater. Environmental Security Technology Certification Program, SERDP/ESTCP Support Office, Reston, VA.

Weinberg H. S., Glaze W. H., Krasner S. W. and Sclimenti M. J. (1993). Formation and removal of aldehydes in plants that use ozonation. *Journal of American Water Works Association*, **85**(5), 72–85.

Wert E. C., Neemann J. J., Rexing D. J. and Zegers R. E. (2008). Biofiltration for removal of BOM and residual ammonia following control of bromate formation. *Water Research*, **42**, 372–378.

Wu H. and Xie Y. (2005). Effect of EBCT and water temperature on HAA removal using BAC. *Journal American Water Works Association*, **97**(11), 94–101.

Xie Y. and Zhou H. (2002). Use of BAC for HAA removal Part 2, column study. *Journal American Water Works Association*, **94**(5), 126–134.

Yulan T., Juanjuan H., Xingguan M., Kun Y., Rognxin Z., Weibing W., Jinxiang F. and Haipeng Z. (2010). Simultaneous Biological Removal of Iron, Manganese, and Ammonium Nitrogen in Simulated Groundwater Using Biological Aerated Filter. Conference Proceedings: International Conference on Bioinformation and Biomedical Engineering (ICBBE), June 18–20, 2010, Chengdu, China.

Zearley T. and Summers R. S. (2012). Removal of trace organic micropollutants by drinking water biological filters Environ. *Science and Technology*, **46**(17), 9412–9.

Chapter 9

Future aspects of hazardous pollutants and their biological removal

Ulaş Tezel

Boğaziçi University, Institute of Environmental Sciences, 34342 Bebek, Istanbul, Turkey

Corresponding author: ulas.tezel@boun.edu.tr.

9.1 NEW CHEMICALS, NEW REGULATIONS

Contamination of water resources with thousands of chemical compounds is one of the challenging environmental problems facing humanity. Although most of these chemicals are present at very low concentrations, most of them are threatening both human and environmental health, particularly when present as components of complex mixtures (Schwarzenbach *et al.* 2006). The problem with chemical pollution will aggravate in the future, as the impact of global climate change becomes more drastic. As evaporation increases, so does the concentration of those chemicals in water bodies (Schwarzenbach *et al.* 2010).

Today, about 60 million organic and inorganic chemical substances are registered in the Chemical Abstract System (CAS). Trends show that about 4000 chemicals are registered daily and the total number is growing exponentially (Binetti *et al.* 2008). Of course, not all of those chemicals end up in the environment, neither are they all pollutants. Whether or not a particular compound is a pollutant is decided based on exposure to that compound and its impact on organisms. Assessment of exposure and impact is challenging due

 Hazardous Pollutants in Biological Treatment Systems: Fundamentals and a Guide to Experimental Research
Ferhan Çeçen, Ulaş Tezel
doi: 10.2166/9781780407708_359

to the number of chemicals introduced into the market as well as duration of experimentation (see **Chapter 2**).

Given the high rate of production of new chemicals and slow process of their environmental/human impact assessment, as well as the presence of old chemicals that have not yet been investigated, the number of potential water pollutants is enormous. Among those chemicals, the ones that enter freshwaters due to their presence in wastewaters are in the scope of this book. Those chemicals undergo biological treatment and pass through wastewater treatment plants. Identification of chemicals that blend into wastewater is a challenge and needs prediction strategies. Software such as The Chemical Screening Tool for Exposures and Environmental Releases (ChemSTEER), can be used to estimate workplace exposures and environmental releases for chemicals manufactured and used in industrial/commercial settings.

Classification of chemicals in wastewater is another challenge. Generally, the ones that are degraded completely during wastewater treatment are not regarded as pollutants. However, the ones that are not degradable under conventional treatment conditions and are released into water bodies are generally classified as pollutants if they threaten living organisms. Water pollutants are robustly classified as priority pollutants (PPs) or chemicals with emerging concern (CECs). PPs are both recalcitrant or slowly degradable and present harm to living organisms. For instance, the European Union (EU) has classified 45 chemicals as priority pollutants in its Water Framework Directive and set environmental quality standards (EQS) for priority pollutants in inland and other surface waters. On the other hand, CECs are elected based on their occurrences in surface waters and potential for risk to the environment and humans (Altenburger *et al.* 2015; Brack *et al.* 2017). Strategies that need to be followed for the prioritization of chemicals with respect to environmental impact have been discussed in detail by Brack and co-workers (Brack *et al.* 2017). Currently, the NORMAN network, which is a consortium to monitor chemical substances in water and develop strategies to prioritize chemicals, has identified around 1000 CECs which are surfactants, flame retardants, pharmaceuticals and personal care products (PPCPs), gasoline additives and their degradation products, biocides, polar pesticides and their degradation products, and various proven or suspected endocrine disrupting compounds (EDCs). This information has to be disseminated efficiently.

CAS is the most comprehensive database of chemical substances. CHEMLIST®, which is a part of CAS, collects the record of regulated substances all around the world. Today, CHEMLIST® contains about 350,000 chemicals that are listed as regulated chemicals in the national chemical inventories such as the Toxic Substances Control Act (TSCA) of the U.S., European Inventory of Existing Commercial Chemical Substances (EINECS), Existing and New Chemical Substances (ENCS), and Inventory of Existing Chemical Substances in China (IECSC). Based on the rate of entries to the CHEMLIST, about 50 chemicals are added to this list weekly. In the future, CHEMLIST® may also integrate data from the NORMAN Network for classification of chemicals as CEC.

9.2 BARRIER OF "LIMIT OF QUANTIFICATION" AND STANDARDIZATION OF INSTRUMENTAL METHODS

Instrumental methods are important for classification and prioritization of pollutants as well as assessment of biodegradation potential and pathways of pollutants. Hyphened methods such as gas chromatography-mass spectrometry (GC-MS) and liquid chromatography-mass spectrometry (LC-MS) are used for identification and quantification of pollutants at low concentrations (see **Chapter 3**). GC-MS is more pronounced for the measurement of apolar compounds whereas LC-MS is suitable for analysis of a wider range of both apolar and polar compounds. Given the wastewater matrix is aqueous, LC-MS is more suitable and straight-forward instrumental method for measurement of pollutants in wastewater since sample extraction which is a must pretreatment for aqueous samples in GC-MS analysis is not required most of the time in LC-MS.

Hazardous pollutants appear in wastewater at concentrations ranging between ng and μg/L. In addition, biotransformation of those compounds produces by-products most of which are unknown and need to be identified and quantified in wastewater and during the biodegradation tests in order to predict their fate and effect in the environment. A range of different LC-MS technologies are available to meet the challenges posed when analyzing a mixture of many known and unknown compounds at low concentrations in complex matrices. In particular, the combination of LC with tandem high-resolution mass spectrometry (LC-HRMS^n) has enhanced the identification capability as well as the sensitivity of pollutant analysis. In addition, use of multiple ionization techniques such as electrospray (ESI), atmospheric pressure chemical ionization (APCI) and atmospheric-pressure photoionization (APPI) has broadened the spectrum of pollutants that can be analyzed by MS. Now almost any organic chemical can be analyzed by LC-MS including the apolar chemicals that used to be measured only by GC-MS, previously (Krauss *et al.* 2010).

On the other hand, limit of quantification (LOQ) is still a challenge. Many hazardous pollutants have very low predicted non-effect concentrations (PNECs), so they need to be measured at below ng/L in wastewater treatment plant effluents or during biodegradation tests. For instance, EQS for dichlorvos and cypermethrin are 0.7 and 0.6 ng/L in surface waters which are lower than the LOQs reported to date. In such cases, passive sampling is offered to concentrate pollutants for accurate measurements (Brack *et al.* 2017).

After the introduction of ultra high performance liquid chromatography (UHPLC), sample analysis run time dropped extensively, however, the analysis time needs to be reduced even more. In the future, LC-HRMS systems may be automatized and become high throughput. A high throughput analysis system may include a syringe pump which infuses samples taken from microplates directly to HRMS. With a faster HRMS, chromatographic separation may not be required for identification and accurate quantification of pollutants.

In addition, although a standardized extensive MS database is available for GC-MS identification of organic compounds, a comprehensive database is missing for LC-MS identification. The NORMAN network compiled a mass database for pollutants detected in surface waters under the name "EU Mass Databank", however, an interface is needed to integrate that database to the data acquisition and evaluation software of common LC-MS^n equipment. Therefore, future efforts should be on compilation of high resolution exact mass and mass fragmentation pattern data for chemicals and developing a standardized database for LC-MS identification of chemicals in unknown samples.

9.3 BIODEGRADATION TESTING NEEDS STANDARDIZATION AND HIGH THROUGHPUT

The chemicals that end up in wastewater finally enter wastewater treatment plants. To mitigate the adverse effect of those chemicals, they have to be removed from the wastewater during treatment. Unfortunately, current wastewater treatment systems are not designed to degrade most of the hazardous pollutants. But this does not mean that those chemicals are not degradable. Biodegradation is the processes that can transform potentially toxic chemicals into less harmful products and can lead to their complete mineralization into carbon dioxide, water, and inorganic nutrients and ions. Testing biodegradability of chemicals is not only crucial for classification of chemicals as pollutants but also to predict the fate of the chemicals in wastewater treatment plants and environment. The Organisation for Economic Co-operation and Development (OECD) 301 Ready Biodegradability Tests (RBTs, see **Chapter 5**) are used to assess aerobic biodegradability of chemicals in aqueous media. They are one of the standards of international regulatory screening of manufactured chemicals. Major bottlenecks of OECD RBTs are their low-throughput as well as long test durations, variable inoculum source, low inoculum size and the high pollutant concentration used in the tests. Development of a standardized high throughput biodegradation testing method that verifies biodegradability and gives the rate of biodegradation is necessary.

Such a method should employ microplates and reactions should be set at microlitre volume. Multiple chemicals can be tested simultaneously (Martin *et al.* 2017). A standardized inoculum should be used in this test method. This standard inoculum should be developed by an international effort based on the microbial community analyses of multiple biological treatment systems located at geographically different places (Paixao *et al.* 2003; Paixao *et al.* 2006). The test should be set at very low pollutant concentration and pollutants should be monitored with High Throughput-High Resolution Mass Spectrometry

(HT-HRMS) during the time course. HT-HRMS measurements during the time course would not only give the rate of biodegradation but also products produced or accumulated during biodegradation of the parent pollutant.

9.4 ENGINEERS CANNOT UNDERESTIMATE "OMICS" NOW

Wastewater treatment plants are the hotspots of evolution of genes that are involved in biotransformation of hazardous chemicals. Therefore, monitoring of both microbial community structure and function in biological treatment systems is necessary (see **Chapter 7**). Next-generation sequencing methods such as amplicon sequencing, metagenomics and transcriptomics are tools to assess diversity and function of the microbial community in a biological treatment system. Metagenomics allows us to sequence the complete community genome as a single unit and it provides an opportunity for linking microbial composition with concurrent function. On the other hand, metatranscriptomics is used to profile all genes actively expressed in a community which helps identify active metabolic processes in that community at the specified time. However, metagenomics and metatranscriptomic datasets are complicated and need advanced bioinformatics in order to interpret them (Kowalczyk *et al.* 2015).

On the contrary, monitoring should be high throughput and reflect the performance of a treatment plant in removing hazardous pollutants. The genes related to degradation of many hazardous chemicals are known and available in many databases (see **Chapter 5**). A microarray chip similar to PhyloChip but targeting genes involved in pollutant degradation can be designed and utilized to elucidate the capacity of a plant in removing hazardous pollutants as well as the impact of hazardous pollutants on the microbial community.

Another important point is identification of emerging degraders in a microbial community. As many new chemicals end up in wastewater, microorganisms that can degrade those chemicals would also evolve. Identification of new degraders has biotechnology applications which was discussed in Section 9.5. Metagenomics coupled to single cell isolation by labeling the microorganisms present in the community and high throughput biodegradation screening (see Section 9.3) will be a strategy to identify and culture the degraders of new pollutants in biological treatment systems.

9.5 EMERGING TECHNOLOGIES FOR ADVANCED TREATMENT OF HAZARDOUS POLLUTANTS

Upgrading or optimization of existing treatment technologies or development of new technologies are two options for achieving better removal of a large number of chemically very different micropollutants in a broad range of water matrices. Many hazardous pollutants are either not or partially removed in conventional biological treatment systems. This may be due to the recalcitrance of the molecule, but also the short contact time or absence of the degraders in the reactor. Clara and co-workers showed that solid retention time is the most effective parameter that controls the removal of hazardous pollutants (Clara *et al.* 2005). As a result, membrane bioreactors (MBRs) are good alternatives for enhanced removal of hazardous pollutants (Siegrist & Joss, 2012). However, the degrader that is responsible for degradation of a certain pollutant may not be present in the microbial community. In that case, adding a degrader into the system may facilitate biodegradation. This process is called bioaugmentation. Nguyen and co-workers reported that the performance of an MBR system in removing micropollutants significantly increased when white-rot fungi were amended into the reactor. They suggested that the laccase enzyme was responsible for degradation (Nguyen *et al.* 2013).

As special degraders and catabolic enzymes that degrade pollutants are discovered, new treatment technologies will emerge that utilize them. As an advanced treatment technology, degraders that can utilize a hazardous pollutant or a group of pollutants can be immobilized and used in a reactor coupled to conventional biological treatment systems.

In addition, oxidase enzymes have great potential to transform hazardous pollutants. Two oxidases, namely Cytochromes P450 type oxygenase and laccase, efficiently transform many antibiotics, biocides and PPCPs etc. Now, the use of enzymes for the removal of hazardous pollutants from domestic and industrial wastewater is recognized as an emerging waste treatment technology (Demarche *et al.* 2012). As an advanced treatment technology, also the application of reactors packed with immobilized enzymes for removal of hazardous pollutants will dominate in the future.

In conclusion, biological removal of hazardous pollutants from wastewater has become a very recent topic in wastewater treatment. A framework responsible for identification of hazardous pollutants to development of new technologies for removal of them from wastewater has to be developed (Figure 9.1). Based on current gaps in the information and technology, this framework should be composed of five emerging areas on biological removal of hazardous pollutants:

(1) Synchronization of chemical registry systems;
(2) Prediction of chemicals and their flux into wastewater and listing of potential pollutants;
(3) Standardization and development of high throughput biodegradation testing systems;
(4) Development of a comprehensive Mass Databank and high throughput $HRMS^n$ system ($HT\text{-}HRMS^n$) for monitoring of hazardous pollutants;
(5) Design of new treatment technologies.

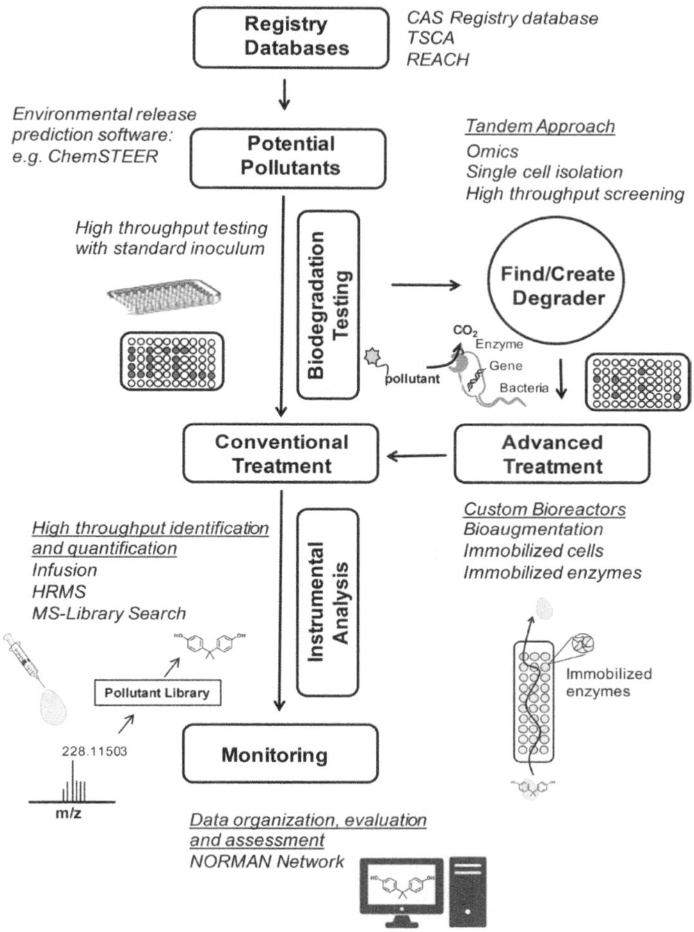

Figure 9.1 Framework proposed for advancement of biological removal of hazardous pollutants from wastewater.

9.6 REFERENCES

Altenburger R., Ait-Aissa S., Antczak P., Backhaus T., Barcelo D., Seiler T. B., Brion F., Busch W., Chipman K., Lopez de Alda M., Umbuzeiro Gde A., Escher B. I., Falciani F., Faust M., Focks A., Hilscherova K., Hollender J., Hollert H., Jager F., Jahnke A., Kortenkamp A., Krauss M., Lemkine G. F., Munthe J., Neumann S., Schymanski E. L., Scrimshaw M., Segner H., Slobodnik J., Smedes F., Kughathas S., Teodorovic I., Tindall A. J., Tollefsen K. E., Walz K. H., Williams T. D., Van den Brink P. J., van Gils J., Vrana B., Zhang X. and Brack W. (2015). Future water quality monitoring – adapting tools to deal with mixtures of pollutants in water resource management. *Science of the Total Environment*, **512–513**, 540–551.

Binetti R., Costamagna F. M. and Marcello I. (2008). Exponential growth of new chemicals and evolution of information relevant to risk control. *Annali Dell Istituto Superiore Di Sanita*, **44**(1), 13–15.

Brack W., Dulio V., Agerstrand M., Allan I., Altenburger R., Brinkmann M., Bunke D., Burgess R. M., Cousins I., Escher B. I., Hernandez F. J., Hewitt L. M., Hilscherova K., Hollender J., Hollert H., Kase R., Klauer B., Lindim C., Herraez D. L., Miege C., Munthe J., O'Toole S., Posthuma L., Rudel H., Schafer R. B., Sengl M., Smedes F., van de Meent D., van den Brink P. J., van Gils J., van Wezel A. P., Vethaak A. D., Vermeirssen E., von der Ohe P. C. and Vrana B. (2017). Towards the review of the European Union Water Framework Directive: recommendations for more efficient assessment and management of chemical contamination in European surface water resources. *Science of the Total Environment*, **576**, 720–737.

Clara M., Kreuzinger N., Strenn B., Gans O. and Kroiss H. (2005). The solids retention time – a suitable design parameter to evaluate the capacity of wastewater treatment plants to remove micropollutants. *Water Research*, **39**(1), 97–106.

Demarche P., Junghanns C., Nair R. R. and Agathos S. N. (2012). Harnessing the power of enzymes for environmental stewardship. *Biotechnology Advances*, **30**(5), 933–953.

Kowalczyk A., Martin T. J., Price O. R., Snape J. R., van Egmond R. A., Finnegan C. J., Schafer H., Davenport R. J. and Bending G. D. (2015). Refinement of biodegradation tests methodologies and the proposed utility of new microbial ecology techniques. *Ecotoxicology and Environmental Safety*, **111**, 9–22.

Krauss M., Singer H. and Hollender J. (2010). LC-high resolution MS in environmental analysis: from target screening to the identification of unknowns. *Analytical and Bioanalytical Chemistry*, **397**(3), 943–951.

Martin T. J., Goodhead A. K., Acharya K., Head I. M., Snape J. R. and Davenport R. J. (2017). High Throughput Biodegradation-Screening Test to Prioritize and Evaluate Chemical Biodegradability. *Environmental Science and Technology*, **51**(12), 7236–7244.

Nguyen L. N., Hai F. I., Yang S. F., Kang J. G., Leusch F. D. L., Roddick F., Price W. E. and Nghiem L. D. (2013). Removal of trace organic contaminants by an MBR comprising a mixed culture of bacteria and white-rot fungi. *Bioresource Technology*, **148**, 234–241.

Paixao S. M., Santos P., Baeta-Hall L., Tenreiro R. and Anselmo A. M. (2003). Alternative inocula as activated sludge surrogate culture for a toxicity test. *Environmental Toxicology*, **18**(1), 37–44.

Paixao S. M., Saagua M. C., Tenreiro R. and Anselmo A. M. (2006). Biodegradability testing using standardized microbial communities as inoculum. *Environmental Toxicology*, **21**(2), 131–140.

Schwarzenbach R. P., Escher B. I., Fenner K., Hofstetter T. B., Johnson C. A., von Gunten U. and Wehrli B. (2006). The challenge of micropollutants in aquatic systems. *Science*, **313**(5790), 1072–1077.

Schwarzenbach R. P., Egli T., Hofstetter T. B., von Gunten U. and Wehrli B. (2010). Global Water Pollution and Human Health. *Annual Review of Environment and Resources*, **35**, 109–136.

Siegrist H. and Joss A. (2012). Review on the fate of organic micropollutants in wastewater treatment and water reuse with membranes. *Water Science and Technology*, **66**(6), 1369–1376.

Index

IWA Publishing's authorised EU representative for General Product Safety Regulations is Diane D'Arras, 15 rue Duret, 75116 Paris, France, e-mail: safety@iwap.co.uk.

Printed and bound by CPI Group (UK) Ltd, Croydon, CR0 4YY
12/05/2026
02108863-0003